Two Manuals Combined into One Book

Marine Operations on Urbanized Terrain

And

FM 7-85 Ranger Unit Operations

Department of Defense

Pentagon
Publishing

Military Operations on Urbanized Terrain

Table of Contents

Chapter 1

Modern Urban Warfare

"... the likelihood is high that in the future, the National Command Authorities will again commit Marines to missions in urban areas."

A Concept for Future Military Operations on Urbanized Terrain

1001. Importance of Urban Areas. Throughout history, military planners have viewed cities as centers of gravity. As such, in war, cities are something to be either protected or taken away, depending upon one's perspective (MCDP 1, *Warfighting*). Cities house the population centers, transportation hubs, seats of government, sources of wealth, centers for industry, information networks, and key nodes of communication within a nation. Recent forecasts based on population statistics and the worldwide migration trend from agrarian to industrialized societies predict that 85 percent of the world's population will reside in urbanized areas by the year 2025. As the world trend toward urbanization increases, the military significance of cities is likely to increase proportionally.

Urbanized areas, themselves, may be significant sources of future conflict. Cities historically are where radical ideas ferment, dissenters find allies, mixtures of people cause ethnic friction, and discontented groups receive media attention. Adversaries may focus on the capture of radio and television stations in an attempt to influence public opinion and attain their political goals. Our political leaders may take advantage to neutralize or stabilize some extremely volatile political situations, or to provide assistance to allies in need of support, by deploying U.S. forces into urban environments,

1002. The Marine Corps Role in Urban Warfare. As the Nation's force in readiness, forward deployed with expeditionary forces, Marines must be prepared to fight on urbanized terrain. In the past two decades, MAGTFs ranging in size from MEFs (Saudi Arabia, Desert Shield/Desert Storm; Somalia, Restore Hope) through Marine expeditionary units (MEUs) (Beirut, Lebanon; Grenada, Urgent Fury; Somalia, Eastern Exit and Restore Hope) have participated in MOUT. The task-organization and combined-arms aspect of the MAGTF makes it well suited for combat on urbanized terrain.

The results of geographical studies show that 60 percent of politically significant urban areas outside allied or former Warsaw Pact territory are located along or within 25 miles of a coastline; 75 percent are within 150 miles; 87 percent are within 300 miles; 95 percent are within 600 miles; and all are within 800 miles. U.S. embassies and diplomatic facilities are primarily located in cities where the host country's political and economic leadership is concentrated. The Marine Corps will continue to play a prominent role in future evacuations of U.S. citizens, as well as the conduct of peace, counterinsurgency, and contingency operations centered on urbanized areas.

Today's Marine air-ground task forces (MAGTFs) are deployed as part of naval expeditionary forces (NEFs) that maintain a *global forward presence for rapid crisis response*. These integrated combined-arms forces are part of the Nation's proven contingency and naval power projection force. Therefore, Marines may find themselves rapidly deployed and employed in actions across the spectrum of military operations. Many of these trouble spots will likely be located in or around large urban centers.

In the years since World War II, the United States has employed military force more than 200 times. Of these, four out of five involved naval forces, and the majority of the naval efforts included Marines embarked in amphibious ships. The reasons are straightforward: availability and adaptability. Availability derives from the loiter time of forward deployed forces embarked on amphibious shipping. Adaptability comes from the Marine Corps' MAGTF organization, doctrine, training, and equipment, which prepare us for expeditionary missions from the sea in support of a variety of missions, including forcible entry. Enhancing our adaptability are the maritime prepositioning forces (MPFs). MPFs provide a rapid buildup of combat and logistics equipment that is joined with Marines on a distant shore, creating a substantial combat force. Despite our availability and adaptability, the prospect of urban warfare combined with an amphibious assault is a complex task which requires special preparation.

At the outset of a developing situation, forward-deployed expeditionary forces can move quickly within range of a crisis that threatens the political stability of a country. Urban intervention operations must often be planned and executed in a matter of hours or days (rather than weeks or months) to take advantage of the internal turmoil surrounding a developing crisis. Navy and Marine forces should anticipate deployment to urbanized areas on a day-to-day basis.

1003. Distinguishing Features of Urbanized Terrain. Urbanized terrain is a complex and challenging environment. It possesses all of the characteristics of the natural landscape, coupled with manmade construction, resulting in an incredibly complicated and fluid environment that influences the conduct of military operations in unique ways.

Military operations on urbanized terrain (MOUT) is defined as *all military actions planned and conducted on a topographical complex and its adjacent terrain where manmade construction is the dominant feature. It includes combat in cities, which is that portion of MOUT involving house-to-house and street-by-street fighting in towns and cities* (Marine Corps Reference Publication (MCRP) 5-12A, *Operational Terms and Graphics*). MOUT effects the tactical options available to a commander. A built-up area is *a concentration of structures, facilities, and populations, such as villages, cities, and towns,* that form the economic and cultural focus for the surrounding area. (MCRP 5-12A)

a. Cities. Cities are centers of finance, politics, transportation, communication, industry, and culture. They generally have large population concentrations ranging from tens of thousands to millions of people. Because of their psychological, political, or logistical value, control of cities have often been the scenes of pitched battles.

(1) Operations in built-up areas are normally conducted to capitalize on the operational or tactical significance of a particular city. In developing nations, control of only a few cities is often the key to the control of national resources. The side that controls a major city usually has a psychological advantage, which can be enough to significantly affect the outcome of a countrywide conflict.

(2) The abundance of guerrilla and terrorist operations conducted in built-up areas (e.g., Santo Domingo, Caracas, Belfast, Managua, and Beirut) demonstrates the importance many insurgent groups place on urban warfare.

(3) In the past 40 years, many cities have expanded dramatically, losing their well-defined boundaries as they extended into the countryside. New transportation systems (highways, canals, and railroads) have been built to connect population centers. Industries have grown along those connectors, creating "strip areas." Rural areas, although retaining much of their farmlike character, are connected to the towns by a network of secondary roads.

b. Multiple Avenues of Approach. Urbanized terrain is a unique battlespace that provides both attacker and defender with numerous and varied avenues of approach and fields of fire. The urban battlespace is divided into four basic levels: *building*, *street*, *subterranean*, and *air*. Operations can be conducted from above ground, on ground level, inside buildings, or below the ground. Most operations will include fighting on all levels simultaneously.

(1) Building Level. Buildings provide cover and concealment; limit or increase fields of observation and fire; and canalize, restrict, or block movement of forces, especially mechanized forces. They provide optimum perches for snipers and antiair weapons. Buildings also provide antitank weapons optimum positioning to allow engagement from above, exploiting an inherent weakness found in most armored vehicles.

(2) Street Level. While streets provide the means for rapid advance or withdawal, forces moving along streets are often canalized by buildings and have little space for off-road maneuver. Because they are more difficult to bypass, obstacles on streets in urbanized areas are usually more effective than those on roads in open terrain.

(3) Subterranean Level. Subterranean systems are easily overlooked but can be important to the outcome of operations. These areas may be substantial and include subways, sewers, cellars, and utility systems (Figure 1-1 on page 1-4). The city of Los Angeles alone has more than 200 miles of storm sewers located under the city streets. Both attacker and defender can use subterranean avenues to maneuver to the rear or the flanks of an enemy. These avenues also facilitate the conduct of ambushes, counterattacks, and infiltrations. (See Appendix E)

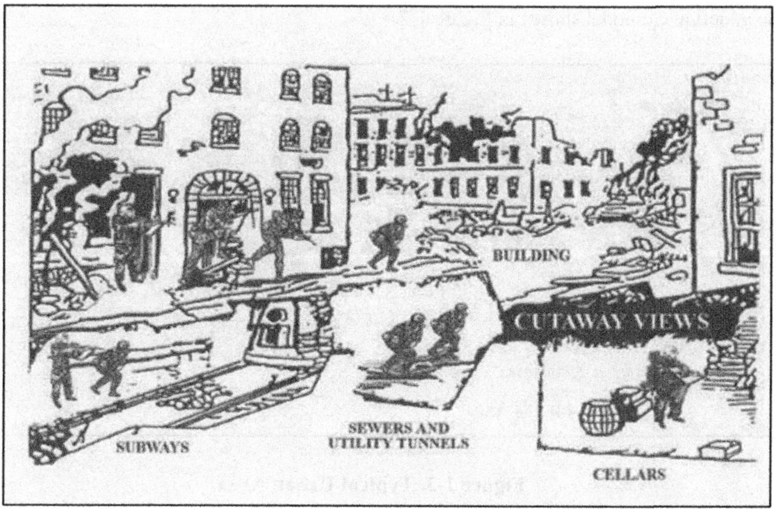

Figure 1-1. Subterranean Systems

(4) Air Level. The air provides another avenue of approach in urbanized areas. Aviation assets can be used for high speed insertion or extraction of troops, supplies, and equipment. While aviation assets are not affected by obstacles on the streets, they are affected by light towers, signs, power lines, and other aerial obstructions. They are also vulnerable to the man-portable surface-to-air missile threat, crew served weapons, and small arms fire.

c. Categories of Built-Up Areas. Built-up areas are generally classified as:

- Villages (populations of 3,000 or less)
- Strip areas (industrialized zones built along roads connecting towns or cities)
- Towns or small cities (populations of up to 100,000 and not part of a major urban complex)
- Large cities with associated urban sprawl (populations in the millions, covering hundreds of square kilometers).

d. Characteristics of Urbanized Areas. A typical urban area consists of combinations of the *city core*, *commercial ribbon*, *core periphery*, *residential sprawl*, *outlying industrial areas*, and *outlying high-rise areas*. Each of the urban area's regions has distinctive characteristics that may weigh heavily in planning for MOUT. Most urbanized areas resemble the generalized model shown in Figure 1-2.

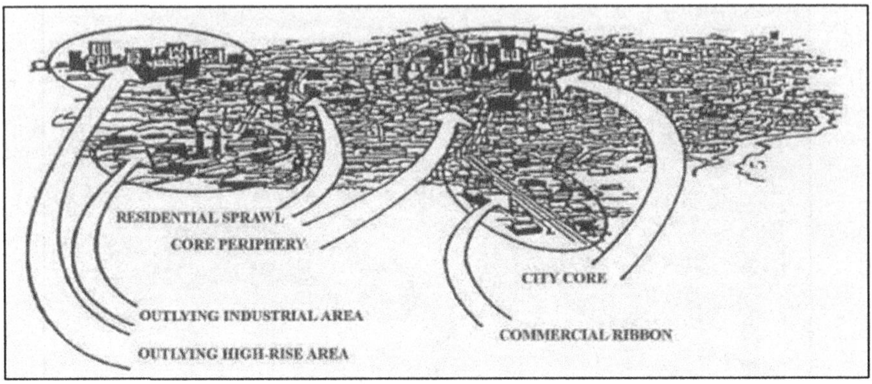

Figure 1-2. Typical Urban Area

(1) City Core. In most cities, the city core has undergone more recent development than the core periphery. As a result, the two regions are often quite different. Typical city cores are made up of high-rise buildings which vary in height. Modern urban planning for built-up areas allows for more open spaces between buildings than in old city cores or in core peripheries (Figure 1-3).

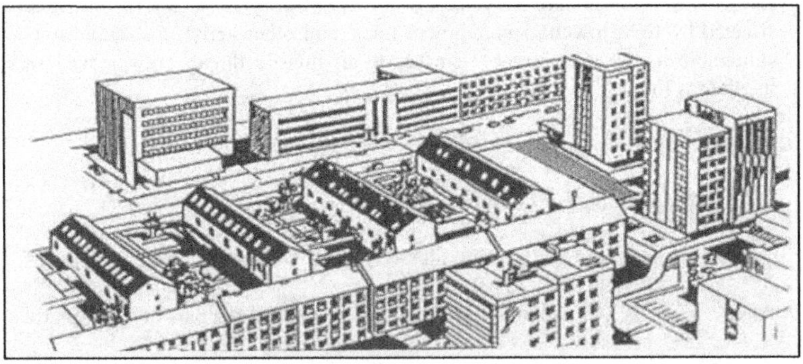

Figure 1-3. City Core

(2) Commercial Ribbon. Commercial ribbons are composed of rows of stores, shops, and restaurants that are built along both sides of major streets through built-up areas.

Typically, such streets are 25 meters wide or more. The buildings in the outer areas are uniformly two to three stories tall—about one story taller than the dwellings on the streets behind them (Figure 1-4).

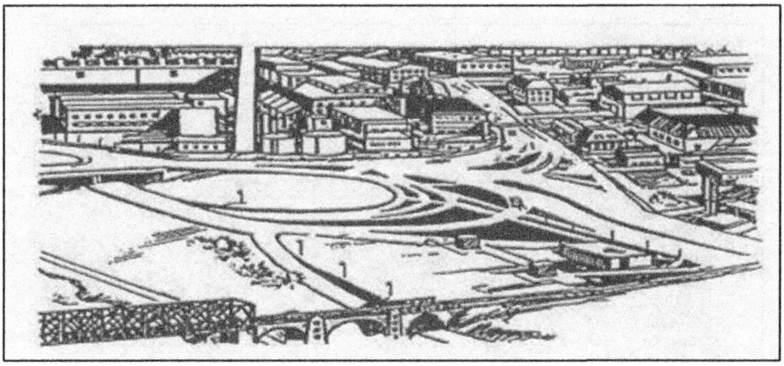

Figure 1-4. Commercial Ribbons

(3) Core Periphery. The core periphery generally consists of streets 12 - 20 meters wide with continuous fronts of brick or concrete buildings. The building heights are fairly uniform—2 or 3 stories in small towns, 5 to 10 or more stories in large cities (Figure 1-5).

Figure 1-5. Core Periphery

(4) Residential Sprawl. Residential sprawl areas consist mainly of low houses or apartments that are one to three stories tall. The area is primarily composed of detached dwellings that are usually arranged in irregular patterns along streets, with many smaller open areas between structures (Figure 1-6).

Figure 1-6. Residential Sprawl

(5) Outlying Industrial Areas. These areas generally consist of clusters of industrial buildings varying from one to five stories in height. Buildings generally vary dramatically in size and composition to match the needs of the particular businesses they house. Industrial parks are good examples of this category (Figure 1-7).

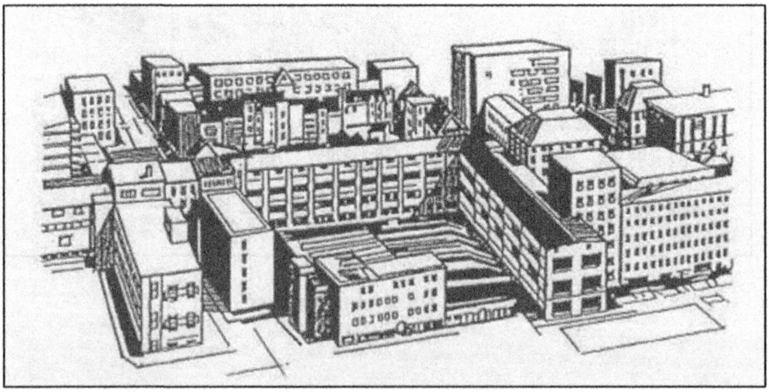

Figure 1-7. Outlying Industrial Areas

(6) Outlying High-Rise Areas. These areas are similar in composition to city core areas, but may be composed of clusters of more modern multistory high-rise buildings in outlying parts of the city. Building height and size may vary dramatically (Figure 1-8). Generally, there is more open space between buildings located in the outlying high-rise areas than is found within the city core area.

Figure 1-8. Outlying High-Rise Area

1004. Modern Battles for Urbanized Terrain. Urban warfare is as old as war itself. Since man began building villages, he has fought battles in and around them. Geography, politics, and economics dictate that cities will continue to be an objective of armies in warfare. From the armies that invaded and liberated Europe twice during the 20th century, to the forces that fought in Korea and Vietnam, to our most recent urban battles in the Middle East and Southwest Asia, *the basic principles of combat in built-up areas have essentially remained unchanged in this century.* While the principles remain the same, the introduction of helicopters, fixed-wing aircraft, armor, and precision-guided munitions (PGMs) has altered some of the techniques associated with urban combat.

Twenty two modern urban battles are discussed to illustrate the trends, dominant factors, and principles of combat in urbanized areas. (The majority of this material is extracted from the 1987 study, *Modern Experience in City Combat,* produced by Abbott Associates.)

• Aachen	• Manila
• Arnhem	• Ortona
• Ashrafiyeh	• Quang Tri City I
• Ban Me Thuot	• Quang Tri City II
• Beirut Port/Hotel (I)	• Seoul
• Beirut 1982 (II)	• Sidon
• Berlin	• Stalingrad
• Cherbourg	• Suez City
• Hue	• Tel Zaatar
• Jerusalem	• Tyre
• Khorramshahr	• Zahle

A brief description of each battle is provided to illustrate that battle's significance.

a. Stalingrad (1942 - 1943). The tenacious Soviet defense of Stalingrad cost the attacking Germans dearly in every way and set up conditions for a decisive counteroffensive. This classic urban battle involved large forces and resulted in innovative urban combat techniques and the creation of the highly successful storm groups (task-organized assault units). (Length of battle: greater than 30 days) (Casualties: 1,630,000+)

b. Ortona (1943). In this Italian town, determined resistance by a battalion of the elite German 3rd Parachute Regiment against Canadian Army attackers demonstrated the difficulty of overcoming a well-prepared defense. The Canadians were unfamiliar with urban combat and had to develop urban fighting techniques during the battle. After the town was largely destroyed and the defender had extracted a high cost in time and casualties from Canadian forces, the German parachute battalion withdrew. (Length of battle: 6 - 13 days) (Casualties: estimated in the hundreds)

c. Aachen (1944). The battle for Aachen, Germany, in the fall of 1944 developed during the U.S. First Army's offensive to breach the Westwall fortifications and the vaunted Siegfried Line. Aachen, the ancient capital of Charlemagne, had symbolic political and psychological significance to both the Germans and Americans. Furthermore, it was the first city on German soil to face an assault by the Allies. This first major battle on German soil foreshadowed bitter resistance against the American attackers in subsequent battles. The German defenders surrendered only after the city was destroyed. Although the U.S. Army had achieved a clear tactical victory, the German defense of Aachen cost the U.S. First Army valuable time and delayed the planned attack to the Rhine river. (Length of battle: 14 - 30 days) (Casualties: 8,000+)

d. Arnhem (1944). On September 17, 1944, Operation Market-Garden, the largest airborne operation in history, was launched in the Netherlands. The plan was to land three airborne divisions to seize key bridges along a 100-kilometer-long corridor through which allied mechanized forces would pass as the first step in the final offensive into Germany. The British 1st Airborne Division made a surprise landing near the Dutch city of Arnhem in order to seize a bridge over the Rhine river for advancing British forces. An unexpected German armor force counterattacked and eliminated all footholds, virtually destroying the British division before a linkup could occur. (Length of battle: 6 - 13 days) (Casualties: estimated in the thousands)

e. Cherbourg (1944). By June 17, 1944, U.S. forces advancing toward Cherbourg from the Normandy beachhead succeeded in cutting off defending German forces in the Cotentin Peninsula. Four German divisions withdrew to a perimeter surrounding Cherbourg. After much fighting, particularly in strongpoints outside the city, the German garrison surrendered to the Americans. Unfortunately for the Americans, the port facilities were destroyed which denied their early use by Allied forces. (Length of battle: 6 - 13 days) (Casualties: estimated in the thousands)

f. Berlin (1945). The long, bloody Soviet offensive to seize the German capital city effectively concluded the last battle of World War II in Europe. Bitter fighting occurred, but the defense was never well coordinated due in part to poor preparation by the Germans. (Length of battle: 14 - 30 days) (Casualties: estimated in the thousands)

g. Manila (1945). Japanese Army troops evacuated Manila under pressure from advancing American forces, but the local Japanese naval commander independently decided to hold the city at all costs. Despite defending Manila with poorly trained and equipped personnel, the determined resistance resulted in a high number of casualties to attacking U.S. forces as well as the destruction of the city and much of its population. (Length of battle: 14 - 30 days) (Casualties: 22,000+)

h. Seoul (1950). Following the Inchon landing, U.S. and Republic of Korea (ROK) forces recaptured the South Korean capital from the North Koreans. The fighting was unusual in that combat was largely centered on seizure of street barricades rather than buildings. (Length of battle: 6 - 13 days) (Casualties: Marines, 2,383; others, estimated in the thousands)

i. Jerusalem (1967). Israeli forces seized Jerusalem in a well prepared and well executed operation. Despite an uncoordinated Jordanian defense, Israeli casualties in this battle were the highest of those encountered during the Six Day War. Regular Jordanian forces withdrew during the latter stages of the battle, effectively ending organized resistance. (Length of battle: 48 hours to 5 days) (Casualties: Israeli forces, 400+; Jordanian forces, estimated in the hundreds)

j. Hue (1968). On January 31, 1968, the North Vietnamese Army (NVA) and Viet Cong (VC) forces launched their Tet Offensive at targets throughout South Vietnam. As part of this operation, two NVA/VC regiments and two sapper battalions conducted a surprise attack and seized part of the walled city (Citadel) of Hue. The NVA/VC held this part of the walled city for about three weeks against determined U.S. and South Vietnamese forces before finally succumbing. The battle for Hue is considered one of the most intense and savage battles of the Vietnam War. (Length of battle: 14 - 30 days) (Casualties: Marines, 433; others, 5000+)

k. Quang Tri City I and II (1972). An objective of the North Vietnamese 1972 winter-spring offensive was the capture of Quang Tri, the northernmost major city in South Vietnam. The NVA overwhelmed the Army, Republic of Vietnam (ARVN) defenders (I). Later, the city was recaptured (II) by a smaller ARVN force using extensive artillery and air support. The large conventional forces involved on both sides made Quang Tri I and II the major urban battles of the Vietnam War. (Length of battle: Quang Tri I, 6 - 13 days; Quang Tri II, 30 days or greater) (Casualties: battles combined, 30,000+)

l. Suez City (1973). Israeli Defense Forces (IDF) attempted to seize this Egyptian city before the anticipated United Nations (U.N.) cease-fire to end the Yom Kippur War. IDF armored shock tactics led to disaster against a well-prepared Egyptian defense. High casualties forced the IDF to withdraw. (Length of battle: less than 24 hours) (Casualties: Israeli forces, estimated 100 - 500; others, unknown)

m. Ban Me Thuot (1975). This South Vietnamese highlands town was the first strategic city to fall in the final North Vietnamese general offensive in 1975 South Vietnamese forces were surprised and overwhelmed. The fall of Ban Me Thuot resulted in a rout that the North Vietnamese exploited to achieve total victory in Vietnam. (Length of battle: 24 - 48 hours) (Casualties: estimated in the hundreds)

n. Beirut I (1976). When the Lebanese civil war broke out in the spring of 1975, combat in the capital city of Beirut assumed a central role. The battle for Beirut I was a series of small, local operations between largely irregular Christian and Muslim forces fighting over control of the hotel and port districts. Combat was not decisive, but led to changes in the boundary, called the "Green Line." This separated the antagonists and lead to the stagnation of the Lebanese conflict. (Length of battle: greater than 30 days) (Casualties: estimated in the hundreds)

o. Tel Zaatar (1976). Lebanese Christian attackers encircled and methodically besieged this Palestinian camp before overcoming its defenders with a final assault. (Length of battle: greater than 30 days) (Casualties: estimated in the hundreds)

p. Ashrafiyeh (1978). The Syrian forces occupying portions of Lebanon faced a complex political situation in which the power of the Christian militia was seen as a clear threat to stability. In an attempt to weaken the Christian militia by an attack on their center of power, the Syrians laid siege to the Christian militia stronghold of East Beirut (Ashrafiyeh). This urban battle was essentially an artillery bombardment without air attacks. Syria failed to break the will of the defenders and final positions remained unchanged. (Length of battle: greater than 30 days) (Casualties: estimated in the hundreds)

q. Khorramshahr (1980). Iranian regular forces initially evacuated this port city in the face of an Iraqi offensive. Irregular Iranian forces, however, continued to fight. They offered prolonged resistance and inflicted heavy casualties. Iraq eventually won this battle, but at a high cost in time and resources that ultimately served to halt the entire offensive against Iran. The intensity of fighting during the battle for Khorramshahr earned the city the nickname, "City of Blood." (Length of battle: 14 - 30 days) (Casualties: Iraqi, 3,000 - 9,000; Iranian, estimated in the thousands)

r. Zahle (1981). Syria laid siege to the Lebanese regular forces and militia in this key crossroads town. Fighting was inconclusive and ended in a negotiated settlement whereby the Lebanese defenders evacuated the town. (Length of battle: greater than 30 days) (Casualties: estimated in the hundreds)

s. Beirut II (1982). The siege of Beirut culminated the Israeli campaign to evict the Palestine Liberation Organization (PLO) from Lebanon. Fighting under domestic and world political pressures, the IDF besieged the PLO, selectively applying heavy ground and air firepower in conjunction with psychological warfare and limited-objective ground operations. The fighting resulted in a negotiated PLO evacuation from the city. (Length of battle: greater than 30 days) (Casualties: 2,300+)

t. Sidon (1982). Israeli forces easily seized this PLO southern headquarters during the invasion of Lebanon. The IDF was fully prepared for major urban combat using lessons learned from earlier battles, but resistance was unexpectedly light as PLO forces had largely withdrawn from the city. (Length of battle: 48 hours to 5 days) (Casualties: unknown)

u. Tyre (1982). The Israeli attack on the PLO in this Lebanese coastal city was well planned, with excellent intelligence on the target. All branches of the IDF participated in an operation that included naval fire support and amphibious landings. PLO resistance was uncoordinated and easily overcome. (Length of battle: 48 hours to 5 days) (Casualties: Israeli forces, 120+; others unknown)

1005. Modern Urban Battle Analysis and Observations. The 20th-century urban warfare experience has caused us to reevaluate old factors and consider new developments that affect the way we fight in this environment. The following discussion, based on the historical analysis of these select urban battles, illustrates the importance of maneuver warfare and combined arms philosophies in the urban environment. Factors that have had an impact on the manner in which urban warfare has been conducted are:

Intelligence is imperative to success in urban warfare. "Maneuver warfare requires a firm focus on the enemy" (MCDP 2, *Intelligence*). Few subsequent tactical changes can overcome the far-reaching impact of a major intelligence error.

a. Intelligence. The historical review of modern urban battles discloses that the attacker will usually win. Failures to win generally reflect classic military errors not characteristically unique to cities. *However, of the battles studied, battles lost were attributed to errors in* *initial intelligence*. The battles of **Arnhem** and **Suez City** probably would never have occurred had the attacker known the strength and locations of the defender's forces. At **Stalingrad**, the attacking Germans were aware of the defending forces facing them in the Sixth Army's zone, but the buildup of Soviet forces in other areas of the front was not anticipated and resulted in tactical surprise at those points, diluting the German offensive push to seize the city.

Surprise is a combat multiplier and can substantially reduce the cost of urban warfare.

b. Surprise. Surprise is a combat multiplier and can substantially reduce the cost of urban warfare. It can be achieved through deception, stealth, and ambiguity (MCDP 1-3, *Tactics*). Surprise was achieved by the attacker at **Aachen** and **Ban Me Thuot** and by the defender at **Suez City**. Surprise can be an important asset to increase leverage, but, as the failure at **Arnhem** shows, not necessarily a decisive one. When surprise is employed as a means to overcome other disadvantages, it is important to maintain accurate intelligence. In urban areas, tactical surprise by the attacker can be used to preempt effective defensive preparation of a city.

c. Combined Arms. The MAGTF must capitalize on one of the key means for gaining advantage in maneuver warfare — the use of combined arms (MCDP 1-3, *Tactics*). The use of combined arms places the enemy in a dilemma. Any action the enemy takes to avoid one combat arm makes him more vulnerable to another. An analysis of categories of weapons systems found in a MAGTF helps to illustrate it is a warfighting organization well suited for MOUT:

Combat in urban areas is primarily a small unit, infantry intensive operation

(1) Infantry. Combat in urban areas is primarily a small-unit, infantry intensive operation. Restrictions on maneuver, particularly for mechanized units, increases oportunities for infiltration. Urban combat requires small-unit leadership, initiative, and skill. Decentralized actions and difficulties in command , control, and communications are typically encountered. Built-up areas, like close terrain found in other operational environments, are generally considered to be most suited for operations conducted by infantry. Infantry units can be organized, trained and equipped to negotiate urbanized terrain that restricts observation, fields of fire, and mechanized movement.

(2) Armor. The role of armor in urban warfare can be significant. Of the 22 battles studied, armor participated in 21. In three-fourths of these battles, organic tank support was a central element when special assault teams were employed. *Overall, special assault units supported by tanks were more successful than any other task organization.*

The use of tanks to the attacker inside a city has been effective only when they were protected by infantry. Tanks in support of infantry act as an "assault gun" that delivers concentrated, sustained fires to reduce held strongpoints.

The U.S. experience in **Hue** demonstrated the key role that armor can play on a combined-arms team fighting inside the city. The Marines' most effective weapons during the battle were the M48A1 Patton tank and the M-50 Ontos. Both were protected by infantry. The M48A1, with its 90-mm main gun, was used extensively to reduce fortified positions. The Ontos, an armor-protected tracked vehicle mounting six 106-mm recoilless rifles, was highly effective against concrete and steel structures. The munitions of these armored systems provided breaches that the infantry could exploit.

Armor providing direct-fire support inside a built-up area requires protection by infantry.

In contrast, during the **Suez City** battle, Israeli armor forces attacked on "armor thrust avenues" into the city, outpacing their armored personnel carrier (APC) mounted paratroop/infantry support. The Egyptian defenders lacked organic artillery (except limited antiaircraft artillery (AAA) and mortars) and had no air support and virtually no armor support. The Egyptians prepared "kill zones" on the principal avenues down which the IDF armored forces attacked. As the lead IDF armor battalion entered the second of the three road intersection objectives, the Egyptians engaged with Sagger missiles, RPGs, ZSU-23 antiaircraft guns, antitank grenades thrown from balconies, and small arms. All of the tank commanders in the lead battalion were killed or wounded. Disabled vehicles blocked the road. Vehicles veering into the narrow side streets became trapped and were destroyed. *The lack of infantry to protect the armor proved disastrous to the attacking armor battalion.*

(3) Artillery. Artillery has played an important role in most major urban conflicts. At **Aachen,** U.S. forces combined infantry with antitank teams and artillery (in a direct-fire role) down to the squad and fire team level. Also, artillery firing shells with "delay" fuzes in an indirect-fire role were used to penetrate one or more floors before exploding, thus driving the enemy to the ground where infantry and armor could attack. Artillery was also positioned to fire *perpendicular* to the direction of movement of assaulting forces. Thus, fratricide from artillery range errors was alleviated.

Artillery has two distinct roles: outside the built-up area to isolate or prevent isolation with indirect-fire; and within the built-up area to provide direct-fire support.

Artillery proved most useful for interdicting enemy supplies, enemy evacuation, and the movement of reinforcements in the enemy's rear; for physically and psychologically harassing the enemy; and in direct-fire roles within a city.

Artillery employed in the indirect-fire role has been effective in disrupting defenders in half of the studied battles. However, some indirect-fire roles have proved more effective than others in urban combat. Artillery was most effective in the interdiction of supplies, enemy evacuation, movement of reinforcements in the enemy's rear (outside the city), and for indiscriminate physical and psychological harassment of the enemy. Artillery used as an indirect-fire siege weapon, as was done at **Ashrafiyeh** and **Zahle**, proved ineffective. Artillery can also cause problems for the attacker. The rubble resulting from indirect artillery fires can create considerable obstacles for the attacker while providing the defender with obstacles, materials, cover, and concealment.

Artillery employed in the direct-fire role proved useful in the reduction of strongpoints. Self-propelled artillery was used effectively inside **Aachen** and **Stalingrad** and more recently by the IDF at **Beirut II**.

(4) Mortars. The mortar is the most used indirect-fire weapon in urban combat. The mortar's high angle of fire allows the round to reach the street level accurately without being masked by surrounding structures. During the battle for **Hue** city, the most effective fire support provided to Marines was the indirect fire from 60-mm, 81-mm, and 4.2-inch mortars. Approximately 20,000 rounds of high explosive (HE) 60-mm and 81-mm mortar ammunition were expended during the battle. The 4.2-inch mortar was used primarily to deliver riot control agent munitions. It was discovered that 4.2-inch riot control agent munitions could be fired with great effectiveness through the tops of buildings to drive the enemy out. When the enemy emerged from their concealed positions, HE rounds were fired to complete the attack.

(5) Antiaircraft Artillery. Antiaircraft artillery (AAA) was extremely useful in a ground fire role in some urban battles. AAA was used only rarely in World War II, and then generally against assaulting personnel rather than against structures. AAA has been used more frequently in more recent events, but against buildings rather than people. *The high rates of fire of modern AAA make it excellent in terms of shock and destructive potential.* However, ammunition supply can be a problem because the volume of fire. During **Hue**, the Duster, an AAA vehicle mounting twin 40-mm guns, was provided to the Marines by the U.S. Army. These guns were extremely effective in the suppression of enemy positions due to the lethality of the 40-mm round and the quantities in which it could be delivered. During the **Suez City** battle, the Egyptians used the ZSU-23 antiaircraft gun in the direct-fire role against armored columns. This gun proved to be frightening and effective. Its high volume of explosive power created shock among IDF armor personnel akin to the shock that the IDF armor thrust tactics were designed to create. During **Sidon,** the IDF used 20-mm antiaircraft Vulcans in the direct-fire role against ground targets with great success.

(6) Aviation. Historically, aviation assets have played an important role in helping to isolate the objective and to interdict the flow of the defender's supplies and reinforcements. However, aviation has been relatively ineffective when not used in conjunction with ground forces. In past wars, bombing operations have been used in attempts to reduce the defender's will to resist and destroy their physical capabilities. In the majority of urban battles, aerial bombing, by itself, did not erode the defender's will to resist, nor did it significantly degrade the defender's military capabilities.

Aviation plays an important role in interdicting movement of the defender's supplies and reinforcement into the built-up area.

The Marine air-ground task force is employed as a combined arms team. As such, the aviation combat element will not usually be called upon to conduct independent operations in MOUT. It will function as part of a MAGTF and, regardless of operating on or over urbanized terrain, will still execute the six functions of Marine aviation (i.e., offensive air support, assault support, antiair warfare, electronic warfare, air reconnaissance, and control of aircraft and missiles) in support of the MAGTF.

In future urban warfare, aviation will be even more effective due to advances in fixed- and rotary-wing aircraft, unmanned aerial vehicles, precision guided munitions, improved munitions, communications, sensors, and targeting systems. Our battle study already indicates a trend toward more extensive aviation participation in MOUT. As an example, the IDF in **Beirut II** employed bombing by fixed-wing aircraft using cluster bomb units, "smart" bombs, phosphorous, and other munitions. Attack helicopters operated on the outskirts of the built up area with impunity, and medical evacuation (MEDEVAC) proceeded swiftly and efficiently using helicopter support.

d. Combat Forces. Whether attacking or defending, the size of the force relative to the enemy can be critical to success. When provided with adequate forces, the attacker can isolate and encircle the defender and prevent a breakout or linkup. The defender can use them to create a mobile defense or to create strong reserves for counterattacks.

Other factors that impact on the size of forces required are the degree of surprise achieved and the firepower utilized (aviation, armor, artillery, mortars, etc.). They should be weighed against the sophistication of the prepared defense. Ample consideration should be given to the local population, degree of external support, and utilization of existing services (communications, water, etc.).

Regardless of the size or quality of defensive forces, the defender usually extracts large costs from the attacker in time, resources, and casualties.

In the historical examples reviewed, the defender was usually outnumbered by the attacker, the quality of the defender's available forces was inferior, and defeat of the forces defending the city was usually certain. However, regardless of the size or quality of defensive forces, the defender can extract enormous costs to the attacker in time, resources, and casualties. As was seen at **Khorramshahr**, the Iranian defenders, outnumbered 4 to 1, still held the city for 26 days. Although the Iranian defenders eventually lost the city, its defense allowed the remaining Iranian forces time to organize and redeploy. Furthermore, the winter rains that followed the battle turned much of the region into a sea of mud and largely halted further Iraqi efforts. The Iraqi army's offensive thrusts into Iran lost momentum as a result of the defense of Khorramshahr.

e. Special Assault Teams. In these battle studies, "shock units" or "special assault teams" have been used by attackers (and often by defenders) with great success. These organizations were characterized by the integration of combined arms at the battalion level and below. Control was decentralized to lower echelons (down to the squad level in some cases). Assault teams typically contained infantry with various combinations of armor, artillery, and engineers. Hence, all combat arms should plan, train, and develop common tactics, techniques, and procedures for use on urbanized terrain. Before the battle of **Aachen**, U.S. ground forces conducted intensive urban training with antitank teams and artillery pieces down to the squad and fire team levels. In contrast, the defender, in most cases having to fight with whatever forces were locally available, had no opportunity to conduct combined-arms training.

The use of combined arms warfare on urbanized terrain is imperative.

f. Time. In most cases, successful conclusion of an urban battle took two to three times longer than the initial estimates. Consequently, the additional tome resulted in the expenditure of more logistics and the loss of more personnel then initially anticipated. This often had adverse affects on the overall campaign. Well-planned urban defense, even if the defender is isolated or lacking in aviation, armor, or artillery weapons, can be time consuming to the attacker. Time can allow the defender to reorganize, redeploy, or marshal resources in other areas.

Generally, urban warfare is time consuming.

Three battles where time played an unanticipated critical role in the attacker's strategic timetable were **Aachen**, **Khorramshahr**, and **Stalingrad**. In these battles, the defenders delayed the attacker longer than was estimated, resulting in the modification of operational or strategic plans.

g. Isolation. *The attacker won all urban battles where the defender was totally isolated.* Even the partial isolation of the defenders resulted in attackers enjoying a success rate of 80 percent. Conversely, attackers won only 50 percent of the battles in which defenders were not significantly isolated, and those victories came at great cost.

No single factor is more important to the attacker's success than isolation of the urban area.

In most urban battles, some form of isolation occurred as a result of the attacker's actions. While it is unrealistic to envision complete isolation of a city until enemy forces to the rear of the city are pushed far beyond its outermost boundaries, *total isolation does not appear necessary.* The key to the attacker's success is in stemming the *unimpeded* flow of manpower, supplies, and weapons to replace the defender's losses.

h. Cost. The cost of conducting urban warfare is relative to the percentage of total expended resources, the time elapsed, and the results achieved. The cost to the attacker was considered high in the majority of urban battles. A high cost does not necessarily imply that the results were not worth the price. *The attacker and defender must thoroughly evaluate the overall cost prior to committing to an urban battle.*

From the offensive point of view, several factors can be associated with cost. First, and most importantly, isolating the urban area is critical to the attacker's success. Second, overwhelming superiority is needed if all costs are to be minimized. Third, the operation should be carefully planned. Fourth, intelligence is invaluable. Knowing where and how the city has been prepared for defense is also important. Fifth, attacking forces should understand the unique nature of urban combat. Clear tactics, techniques, and procedures for urban combat is required. Every aspect, from taking a building to using destructive technology and coordinating combined arms, must be thoroughly understood. Careful consideration of these five factors can minimize the cost of urban warfare to the attacker.

Attacker cost was generally high in casualties, time, and resources in the majority of urban battles studied.

From the standpoint of the defender, the critical variable is the defensive preparation of the city. Defensive preparations should include measures to prevent isolation of the city by the attackers. The capture of a prepared city can be made to be extremely costly. Preparations can include creating kill zones, clearing fields of fire, constructing canalizing obstacles, establishing reinforcing and fall-back positions, decentralizing command and control, and organizing multiple movement routes above ground, between rooftops, and below ground in subways and sewers. Artillery and aviation support can also be valuable force multipliers in defense.

i. Rules of Engagement (ROE). The nature of the military operation may restrict our use of weapons. The majority of urban battles since 1967 (such as **Beirut II**, **Hue**, **Jerusalem**) have had one or more of the following restrictions imposed on the attacking force:

(1) Minimizing civilian casualties and/or collateral destruction in order to:

- Avoiding alienation of the local population
- Reducing the risk of adverse world or domestic opinion
- Preserving facilities for future use
- Preserving cultural facilities and grounds.

(2) Limiting the use of specific ground or air weapons.

j. Logistics. Timely combat service support, particularly in the areas of ammunition resupply and casualty treatment and evacuation, is a critical element in MOUT.

(1) Logistics Support. Historically, combat in urban environments has seen a dramatic increase in the amounts of Class IV (shoring, sandbags, concertina wire), Class V (ammunition), and Class VIII (medical material). Unique items, such as rope, grappling hooks, and ladders are required for operations on urbanized terrain. Intense close-quarter combat requires a continuous flow of ammunition, particularly small-arms, tank, antitank, mortar, and artillery ammunition, as well as mines, grenades, and demolition explosives. Medical supplies must be readily available to treat the anticipated increase in casualties. Once battle the has been initiated, combat forces will require continuous supply.

Urban warfare requires a flexible, balanced logistics system capable of sustaining the close-quarter fighting evolutions within the built-up area.

One method of providing continuous logistics support is to establish a "push system." Essentially, this system pushes supplies to fighting units without their having to request them. A "push system" should negate the inherent delays of a "pull system," which requires units to request supplies then await their arrival. Optimally however, a balanced approach should be taken to sustain engaged forces (MCDP 4.). Combat service support, in terms of its timeliness and anticipation of the true needs of the combat units, should sustain the tempo of operations. It should avoid the delivery of unwanted or unnecessary supplies. To achieve this balanced approach, logisticians must develop flexible support plans that readily adapt to the ebb and flow of urban combat.

Mobile aid stations and surgical teams should be positioned as far forward as the tactical situation allows and be provided with dedicated evacuation vehicles.

(2) Health Service Support (HSS). Responsive treatment and evacuation plans should be established to handle the expected increase in casualties. Both immediate treatment and evacuation are critical in maintaining the morale and confidence of forces engaged in urban warfare. To meet casualty treatment and evacuation needs, mobile aid stations with surgical teams should be placed as far forward as the tactical situation allows. Dedicated vehicles and aircraft for the evacuation of casualties should be provided.

In addition to casualties resulting from physical injury on the battlefield, there are often considerable psychological casualties in modern battles, particularly urban battles. This can be attributed to the intense, sustained, and close-quarter combat associated with urban warfare. In many battles, attacking troops were not properly trained for urban warfare and felt inadequate in their ability to fight on equal terms with the defender. Many felt moral or ethical dilemmas associated with collateral damage and civilian casualties. Finally, the constant threat of being killed from any quarter coupled with the extreme fatigue of fighting in an urban environment can create psychological casualties. Leaders at all levels should be attuned to the symptoms associated with psychological casualties in order to get effected individuals prompt treatment so that they can be returned to their units.

1006. Implications of Urban Warfare. The commander charged with making decisions needs to understand the operational and strategic implications of a tactical struggle in an urban area (MCDP 1-1, *Strategy*, and MCDP 1-2, *Campaigning*.). Three urban battles (**Stalingrad**, **Hue**, and **Beirut II**) illustrate the importance of seeing beyond the tactical nature of the battle.

Commanders must see beyond the immediate tactical implications of fighting an urban battle.

The battle for **Stalingrad** had major operational and strategic implications. The entanglement of German forces at Stalingrad bought time for the Soviets to mobilize and prevent a major combined effort of two German Army Groups in the south. While Stalingrad resulted operationally in the destruction of the German Sixth Army, the strategic results were even greater. Stalingrad caused a complete change of German strategy in the east. Hitler made major changes in his General Staff, and from this point on in the war, he was a man estranged from his military leadership. The loss of enough men and equipment to field one-fourth of the German Army shook the foundations of the Third Reich.

The Battle for **Hue**, although only one of over one hundred different attacks of the Tet Offensive of 1968, had a negative impact on the will of both the American people and their political leadership. Hue marked a revolution in the coverage of war by modern mass media. It was the first time Americans could and watch an ongoing battle from their living room on the evening news. Hue was a television bonanza for almost a month. When North Vietnamese leadership directed that Hue be held for at least seven days, it was clearly *not* their intent to win a tactical battle, but to strike at the *strategic center of gravity*—in this case, the will of the American people. Although the battle for Hue was a tactical victory for the U.S., the North Vietnamese had achieved their strategic goal of making the American public question the costs associated with the war.

During the battle of **Beirut II**, the IDF's objective was to drive the PLO from Lebanon. The ability of the PLO to leverage the media to gain an advantage was one of the most significant weapons in their arsenal. Despite an Israeli tactical victory, the costs in image, prestige, allies, and, most importantly, its own national will were enormous. The results of this battle eventually caused a change of political leadership at the highest levels of government.

1007. Key Insights. Analysis of modern urban battles provides insight into the major factors affecting the course and outcome of combat. Consideration of these key insights are required in order to effectively and efficiently plan and execute MOUT:

a. MOUT is infantry intensive.

b. A tactical battle may have far-reaching implications. No longer are the strategic, operational, and tactical levels overlap, creating the situation where tactical actions can have operational and possibly strategic repercussions (MCDP 1-1, *Strategy*, and MCDP 1-2, *Campaigning*).

c. Commanders at all levels must understand the impact that media representation will have on the accomplishment of operational and strategic objectives.

d. Maneuver warfare doctrine must be applied to the environment.

e. Intelligence is imperative to success in urban warfare.

f. Surprise is a combat multiplier.

g. Armor, artillery, and aviation are effective at the outer perimeter of built-up areas for causing isolation or preventing reinforcement.

h. Armor operating inside a built-up area must be protected by infantry.

i. Artillery providing direct fire inside a built-up area can be effective in the reduction of strongpoints.

j. As force ratio increases in favor of the attacker, combat duration decreases.

k. Urban warfare is time consuming.

l. Isolation of an urban defender ultimately ensures his defeat.

m. Attack of an urban area is costly to the attacker in terms of resources and casualties.

1008. Necessity for Preparation. Since MAGTFs are usually forward-deployed forces, it is inevitable that MAGTFs will be tasked to deploy to urban areas and conduct military operations. Fighting in this environment will be violent, close, and personal. Fighting in villages, towns, and cities are likely to become more frequent for military forces as the United States responds to global crises. Forces will be expected to exercise adaptability in responding to a wide variety of missions in the world's built-up areas. Marines may soon find themselves operating in the conceptual "three block battlefield" within a city grid: feeding refugees in one block, separating combatants in the adjoining block, and returning hostile fire in a third block.

MOUT is infantry intensive. The remaining chapters and appendices of this publication focus on how the GCE fights and integrates supporting elements in an urban environment.

Chapter 2

Offensive Operations

Urban combat isolates and separates units. Operations are reduced to a series of small-unit actions, placing a premium on small-unit leadership, initiative, and skill.

2001. Introduction. The Marine Corps' manuever warfare philosophy guides the conduct of offensive operations in MOUT. This warfighting philosophy serves to guide commanders through the development, planning, and execution of missions. A detailed discussion of maneuver warfare can be found in MCDP 1, *Warfighting*, MCDP 1-3, *Tactics*, MCWP 3-1, *Ground Combat Operations* (under development), and other Marine Corps doctrinal publications.

This chapter focuses only on offensive operations in an urban environment where collateral damage is of minimal concern. It describes tactics, techniques and procedures to be employed for seizing and clearing buildings and other urbanized areas. Chapter 6 addresses some considerations associated with fighting under constraints and restraints on urbanized terrain.

Section I

Planning

2101. Considerations. Identification of the adversary's centers of gravity and critical vulnerabilities will allow the commander to focus his efforts on those portions of the built-up area essential for mission accomplishment.

a. Reasons for Attacking a Built-Up Area. A commander considers the following before deciding to attack a built-up area:

(1) Tactical Advantage. Cities control key routes of commerce and provide a tactical advantage to the commander who controls them. Control of features such as bridges, railways, and road networks can have a significant impact on future operations. Urbanized areas may be used by the enemy as a base of operations from which they launch their own offensive operations. It may be advantageous to attack those bases and separate the enemy from their support infrastructure.

(2) Political Advantage. The political importance of a built-up area may justify the use of time and resources to liberate it. Capturing a city could destroy the seat of local and national government. At the very least, it could deal the enemy a decisive psychological blow.

(3) Economical Advantage. The destruction or capture of key industrial and commercial cities with the resulting denial of production and distribution of equipment and supplies strikes at the enemy's future ability to wage war. The requirement for a logistics base, especially a port or airfield, may play a pivotal role in the enemy's ability to continue the conflict. Capture of such cities may prove extremely beneficial to the attackers, who can use these resources to their advantage.

(4) Potential Threats to Operations

(a) Enemy Threat Too Great To Bypass. Though the terrain around a built-up area may facilitate its bypass, the enemy within that urbanized area may remain a threat capable of interdicting lines of communications. This may require the enemy force to be contained or destroyed.

(b) Terrain Does Not Allow Bypass. The urbanized area may sit between two natural slopes on the avenue of approach and thus require capture in order to secure the main supply route. Additionally, the urbanized area, itself, may sit on dominating terrain that threatens combat support and CSS elements.

b. Reasons for Not Attacking a Built-Up Area. The commander considers the following reasons for not attacking a built-up area:

(1) Built-Up Area Not Required To Support Future Operations. The attacker may have adequate supply support and resources established at another site from which combat forces can be supported. Therefore, during the estimate process, commanders may assess that the urbanized area is not necessary to support future operations.

(2) Enemy Not a Threat. The commander may decide to bypass upon determination that no substantial threat exists in the built-up area that could affect the unit's ability to accomplish its mission.

(3) Time or Risk Unacceptable. The commander's intent may dictate that speed of movement is essential to the mission. Because MOUT can be time-consuming, the commander may choose to bypass the urbanized area to maintain tempo. Furthermore, the potential for numerous casualties, the expenditure of critical resources, or the restrictions placed on attacking forces may result in unacceptable risks to the commander's primary mission.

(4) Declared an Open City. The area may have been declared an "open city" because it is undefended or of religious or historical significance. By international agreements, open cities are demilitarized and must be neither defended nor attacked. (See HR, 25, Annex to Hague Convention No. IV, *Embodying Regulations Respecting the Laws and Customs of War on Land*, 18 October 1907.) The attacking force must assume civil administrative control and treat the civilians as noncombatants in an occupied country. The defender must immediately evacuate and cannot arm the civilian population. A city can be declared open only before it is attacked. Other reasons for not defending could be the presence of large numbers of noncombatants, hospitals, or wounded personnel or the city's cultural, religious, or historical significance.

2102. Commander's Estimate. Once assigned the mission to attack a built-up area, a thorough mission analysis is conducted. Marines follow the same planning process in MOUT as for other operational environments by evaluating the mission, enemy, terrain and weather, troops and support available, and time available (METT-T). Based on METT-T and application of maneuver warfare concepts, the commander can decide on a plan that defeats the enemy by exploiting identified critical vulnerabilities. The commander may not have sufficient forces to fight everywhere and so must focus on the enemy and concentrate resources to decisive ends. This may involve making conscious decisions to accept risk at some place or time and to seize opportunities at others. The challenges of an urbanized environment dictate an even greater need to fully understand maneuver warfare.

This section addresses METT-T considerations as they pertain to the offense in urban warfare. METT-T is covered in detail in MCRP 3-11.5A, *Marine Troop Leader's Guide.*

a. Mission. Mission analysis is the first step of the estimate process. One of the important mission analysis considerations in urban conflict is the scope of clearance required to accomplish the mission. Commanders and planners should ask themselves the following questions:

- Do I need to clear every building?
- Should I clear only certain blocks?
- Should I only control certain areas?
- What level of protection is required for my lines of communication?

b. Enemy. Maneuver warfare focuses on the enemy. The intelligence preparation of the battlefield (IPB) process should be used to analyze the enemy and terrain in detail. (See Field Manual (FM) 34-130, *Intelligence Preparation of the Battlefield.*) The commander decides whether threat forces are conventional or unconventional.

(1) Conventional Forces. Most potential adversaries have adopted techniques of urban combat from either the United States or the former Soviet Union. Therefore, potential adversaries will build their urban defense to counter the attack of a combined arms force. Potential enemy forces will normally organize their defense in depth. Prepared strong points will usually form the perimeter of a larger defense while reserves locate in a separate position within the perimeter. You can expect ambushes to fill gaps in the perimeter while dummy strongpoints are established to deceive the attacker. Positions for securing the entrances to and exits from underground structures and routes are normally established. Security positions will normally be located forward of first-echelon defensive positions.

(2) Unconventional Forces. Urban areas have become a haven for unconventional forces. The large noncombatant population provides cover and concealment for unconventional force operations. Conventional forces operating in MOUT will normally be placed under restrictive ROE to minimize collateral damage. Unconventional forces will often use our restrictive ROE and the noncombatant population to their advantage when devising an urban defense. (See FM 34-130, FM 7-98, *Operations in a Low-Intensity Conflict*, and Chapter 6 of this manual.)

c. Terrain and Weather

(1) Terrain. Offensive operations should be tailored to the urban environment based on a detailed analysis of urbanized terrain (see Appendix I). Commanders and subordinate leaders use **KOCOA** (**k**ey terrain, **o**bservation and fields of fire, **c**over and concealment, **o**bstacles, and **a**venues of approach) to identify important terrain factors:

(a) K: Key Terrain. Key terrain surrounding an urban area can facilitate entry or deny escape. Within the city, airports or airfields, stadiums, parks, sports fields, school playgrounds, public buildings, road junctions, bridges, or industrial facilities may be key terrain.

Critical public buildings are identified during the terrain-analysis phase of an IPB. Hospitals, clinics, and surgical facilities are important because the laws of war prohibit their attack when not being used for military purposes other than medical support. The locations of civil defense, air raid shelters, and food supplies are critical in dealing with civilian affairs. Additionally, population size, locations, and density; density of the built-up area; firefighting capabilities; the location of hazardous materials; police and security capabilities; civil evacuation plans; and key public buildings should be identified.

(b) O: Observation and Fields of Fire. Urbanized terrain is characterized by restrictive observation and fields of fire. Weapon ranges can be greatly reduced because of buildings and other manmade structures. On the other hand, high ground or tall buildings can provide perches which enhance line of sight (LOS) for observation and communications as well as for individual and crew-served weapons. This includes man-portable surface-to-air missiles.

(c) C: Cover and Concealment. Buildings, sewers, and subways can provide excellent cover and concealment for enemy and friendly forces. They also provide covered and/or concealed maneuver routes within the built-up area. The civilian population can also offer cover and concealment to enemy forces.

(d) O: Obstacles. Natural or manmade obstacles restrict or deny maneuver within the urban area. Bridges, walls/fences, canals, streams, rivers, as well as rubble created by the effects of weapons should be thoroughly analyzed. Construction sites and commercial operations such as lumberyards, brickyards, steelyards, and railroad maintenance yards are primary sources of obstacle and barrier construction materials.

These sites can also supply engineers with materials to strengthen existing obstacles or to set up antitank hedgehogs or crib-type roadblocks.

(e) A: Avenues of Approach. Avenues to the city should support maneuver and be concealed either by terrain, darkness, smoke, or a combination of the three. Avenues that canalize or choke maneuver, due to the density of built-up area or natural terrain, should be identified and avoided. Roads, rivers, streams, and bridges provide high-speed avenues for movement. **Generally, military maps do not provide enough detail for urbanized terrain analysis.** They usually do not show the underground sewer system, subways, underground water system, mass transit routes, and power plants. Local maps of intracity road networks and subway systems and city hall/department of public works blueprints of all city buildings (if available), coupled with aerial photos, should be used.

(2) Weather. As in any military operation, weather affects equipment, terrain, and visibility, but its greatest impact is on the individual Marine. Snow, ice, dust, wind, rain, humidity, and temperature extremes reduce human efficiency. Weather extremes coupled with stress and the physical strain of urban combat can be minimized with effective small-unit leadership. Weather factors include:

(a) Precipitation. Rain or melting snow may flood basements and subway systems. This is especially true when automatic pumping facilities that normally handle rising water are deprived of power. Flooding makes storm and other sewer systems hazardous or impassable. In an NBC environment, chemical agents can be washed into underground systems by precipitation. As a result, these systems may contain chemical agent concentrations that are much higher than surface areas and thus become contaminated "hot spots." Hot spot effects become more pronounced as agents are absorbed by brick or unsealed concrete sewer walls.

(b) Fog. Many major cities experience fog, especially those located in low-lying areas and along canals or rivers. Industrial and transportation areas are the most likely to be affected by fog because of their proximity to waterways. Fog adversely affects vision and optical aids. It may also be used to help conceal friendly movement.

(c) Inversion Layers. Air inversion layers are common over cities, especially cities located in low-lying "bowls" or in river valleys. Inversion layers trap dust, smoke, chemical agents, and other pollutants, reducing visibility and often creating a greenhouse effect, which raises ground and air temperatures.

(d) Temperature. Built-up areas are often warmer than surrounding open areas during both summer and winter. This difference can be as great as 10 to 20 degrees hotter throughout the year.

(e) Wind Effects. Generally, wind chill is not as pronounced in built-up areas. However, the configuration of streets, especially in the city core and outlying high-rise areas, can cause wind tunneling. This increases the effects of the wind along streets that parallel the wind direction, while cross-streets remain relatively well protected. Wind tunneling can have a negative impact on aviation support within the urbanized area by creating dangerous wind effects down streets and between buildings.

(f) Day/Night Differences. Night and periods of reduced visibility favor surprise, which in turn, may facilitate infiltration, detailed reconnaissance, attacks across open areas, seizure of defended strongpoints, and reduction of defended obstacles. Because of the difficulties of night navigation in restrictive terrain (usually without reference points and in close proximity to the enemy) forces may consider using simple maneuver plans with easily recognizable objectives.

Rotary-wing aircraft are best utilized at night when operating on urbanized terrain. The most effective method of protecting helicopters from a successful engagement by hostile ground forces is to remain unseen. During night flights, the use of night vision goggles (NVGs) should be weighed carefully. Artificial lights can render NVGs partially ineffective and can increase the possibility of mid-air collisions by degrading the goggles to the point where nearby aircraft and aerial obstructions are not seen.

(g) Aviation Weather Considerations. Weather conditions can be a critical factor in determining the amount of aviation support the aviation combat element can provide in an urbanized area. Aviation forces should consider the following:

- Presence or absence of fog, industrial haze, low clouds, heavy rain, and other factors that limit visibility for pilots
- Illumination and moon phase/angle during night vision goggle operations
- Ice, sleet, and freezing rain that degrade aerodynamic efficiency
- Updrafts and downdrafts in the urban canyon that may present an unpredictable hazard
- High temperatures and/or high-density altitudes that degrade aircraft engine performance and lift capability
- High winds (large gust spreads) and crosswinds, to include the tunneling effect, that may create localized and unpredictable hazards to aviation
- Weather conditions that create hazards on pick-up zones and LZs, such as blowing dust, sand, or snow.

d. Troops and Support Available. The commander must be aware of all of his combat power (e.g., types of weapons systems, their numbers, and their capabilities) when performing analysis for operations in a built-up area. In the initial planning phase, force size must be evaluated in relation to urban size, enemy forces, and mission assignment. In the attack of a built-up area (population 100,000+), the GCE of a MEF would be a Marine division. However, in the initial stages of an amphibious or MPF operation, a reinforced infantry battalion or regiment may provide the initial forces. No matter what the size GCE, operations will be conducted by task-organized, reinforced battalions and their companies/platoons. Therefore, by using the infantry battalion as the basis for projecting the size of an overall force, the number of regiments or divisions required to secure a built-up area can be determined. Whatever size MAGTF is required for operations in an urbanized area, much of the fighting may be generally conducted by small teams of Marines. Consequently, our success in urban fighting may largely depend upon small unit and individual tactical skills. Specifically, leaders at all levels must analyze the following factors:

- Number and type of available units
- Task organization for urban combat (See Appendix A.)
- Availability of critical weapons systems
- State of training and discipline (training for urban warfare is imperative)
- Strength in terms of men and materiel in relation to enemy and size of built-up area
- Aviation support available
- State of maintenance and supplies
- Available combat support (based on availability and mission requirements)
- Available CSS (critical logistical and maintenance items)
- Host nation support available. (If the city's occupants are friendly to the attacker, then support in intelligence, deception, and diversion may be possible.)

e. Time Available. The following issues should be considered when analyzing the time available for an attack on urbanized terrain:

- A significant amount of time is required for clearing buildings, blocks, or axes of advance.
- Marines tire more quickly when clearing buildings because of stress and additional physical exertion.

- Adequate time should be allocated for thorough reconnaissance, planning, and rehearsals. Ultimately, this can result in higher tempo in the execution of the plan.

Once an estimate of the situation (using METT-T) has been conducted, planning for offensive operations can proceed. It is important to remember that the initial estimate of the situation must be continually updated throughout the planning process and the conduct of operations to account for changes in METT-T. (See MCWP 5-1 *Marine Corps Planning* [under development])

2103. Phases of the Attack. Attacks are categorized as either hasty or deliberate. Both hasty and deliberate attacks should take advantage of as much planning, reconnaissance, and coordination as time and the situation permit. Regardless of the size of the attacking force or of the objective to be secured, the phases of the attack (i.e. reconnoiter, isolate, secure foothold, and control the objective) remain constant.

a. Deliberate Attack. A deliberate attack *is a type of offensive action characterized by preplanned coordinated employment of firepower and maneuver to close with and destroy or capture the enemy* (Jt Pub 1-02). It is a fully coordinated operation that employs the MAGTF's combined arms team against the enemy's defense. It is used when enemy positions are well prepared, when the built-up area is large or severely congested, or when the element of surprise is lost. Given the nature of urbanized terrain, the deliberate attack of a built-up area is similar to the assault of a strongpoint (see MCWP 3-1, *Ground Combat Operations* [under development]). The deliberate attack of a built-up area is conducted in the following phases:

(a) Phase I: Reconnoiter the Objective. Intelligence gathering and reconnaissance/ surveillance are critical to the planning process and success of the operation. All available intelligence resources should be used to gain vital information on the enemy. Whenever possible, a leader makes a personal reconnaissance of the objective area to collect first-hand information regarding the area to be attacked. A thorough reconnaissance yields a wealth of information to use in the continued development of the plan. Avenues of approach, observation posts (OPs), supply routes, and the emplacement positions of direct- and indirect-fire weapons systems are all examples of information that may be amassed during the reconnaissance of the objective area. Composition and structure of buildings and roadbeds, cover and concealment opportunities, and other information not apparent in a map study may have a significant impact on the plan.

(b) Phase II: Isolate the Objective. The objective can be isolated by seizing natural and man-made features that dominate the area. Isolation may also be accomplished by coordinated use of supporting arms to seal off enemy lines of communication. This phase may be conducted simultaneously with Phase III (securing a foothold). Figure 2-1 shows a task-organized, reinforced infantry battalion isolating the objective.

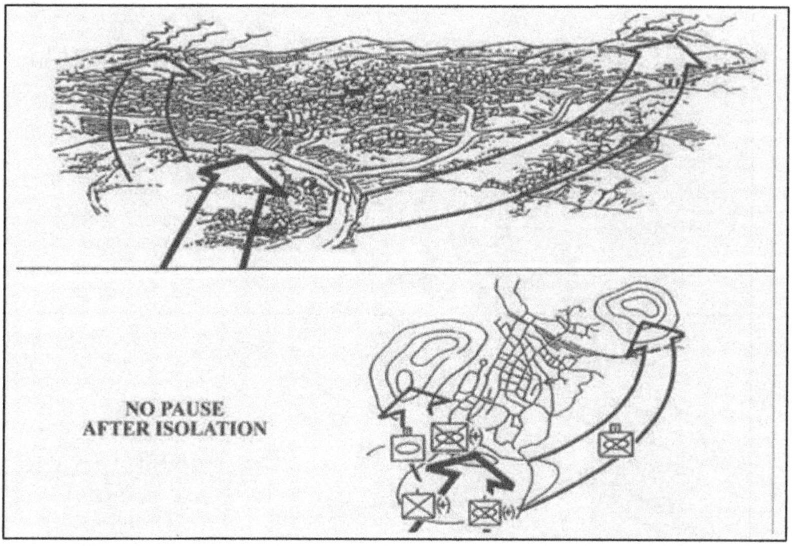

(c) Phase III: Secure a Foothold. Once the objective is isolated, a foothold should be secured as soon as possible in order to maintain tempo. The seizure of a foothold provides the attacking force with a position from which they can continue the assault through the objective area. The assault is supported by direct and indirect suppressive fires and smoke. Figure 2-2 (pg. 2-10) shows the battalion objective being isolated. One company is providing suppressive fires while another company seizes an initial foothold. Fire support assets are providing suppressive fires on the objective.

Figure 2-1. Isolation of the Objective Area

(d) Phase IV: Seizing the Objective. Once a foothold is seized and consolidated, supporting forces move to the built-up area to support the seizing of the objective area. To maintain tempo, the transition between the phases should be seamless. Once the foothold has been established, forward units continue the attack through the objective area. Supporting units assist as required. The momentum of the assault is continued until the objective area is cleared or controlled.

The assault force should establish limited objectives to ensure that the attacking forces do not get strung out along the axis of advance. Gaps may give the enemy the opportunity to infiltrate along the line of advance or make isolated friendly forces vulnerable to attack.

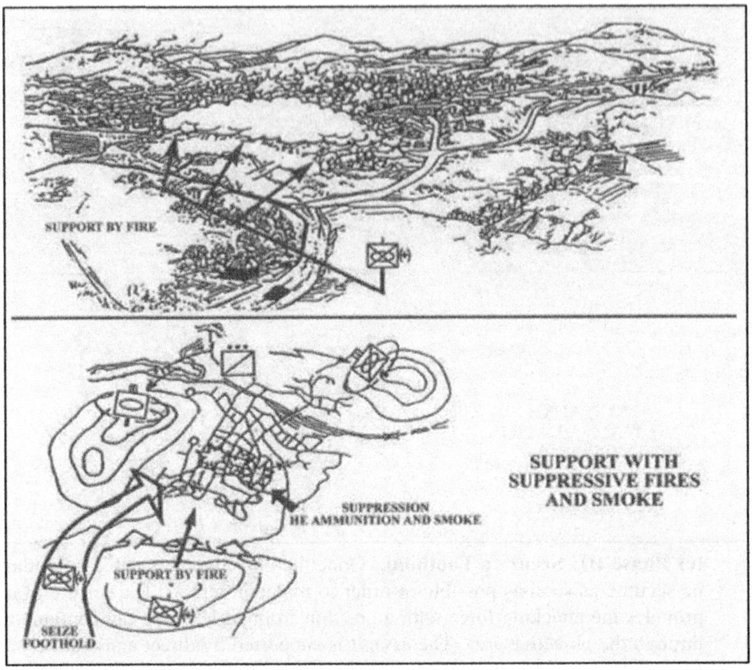

This phase also includes consolidation of the objective area and reorganization for future taskings. By setting limited objectives, the attacking forces have an opportunity to reorganize and defend against counterattacks while maintaining momentum. Once the limited objectives have been secured, the urbanized area may be sectored for detailed clearing. Clearing by sectors may help in the allocation of forces and in the prevention of fratricide. In addition to defensive preparation against counterattack, preparation should be made for follow-on missions (e.g., restoration of civilian facilities, marking and clearing minefields, etc.).

Figure 2-2. Seizing a Foothold

Seizing may consist of a systematic house-by-house, block-by-block advance through the entire zone of action, or it may be a rapid advance through a lightly defended area to seize a key objective. The difference between the two techniques is made by the commander on the basis of mission requirements and METT-T analysis. Seizing an objective in urbanized terrain requires detailed planning, coordination, decentralized command and contol, and small-unit execution. It is a continuation of the assault until all mission requirements are met.

(1) Rapid Advance. Rapid advance (Figure 2-3) may be used to maintain

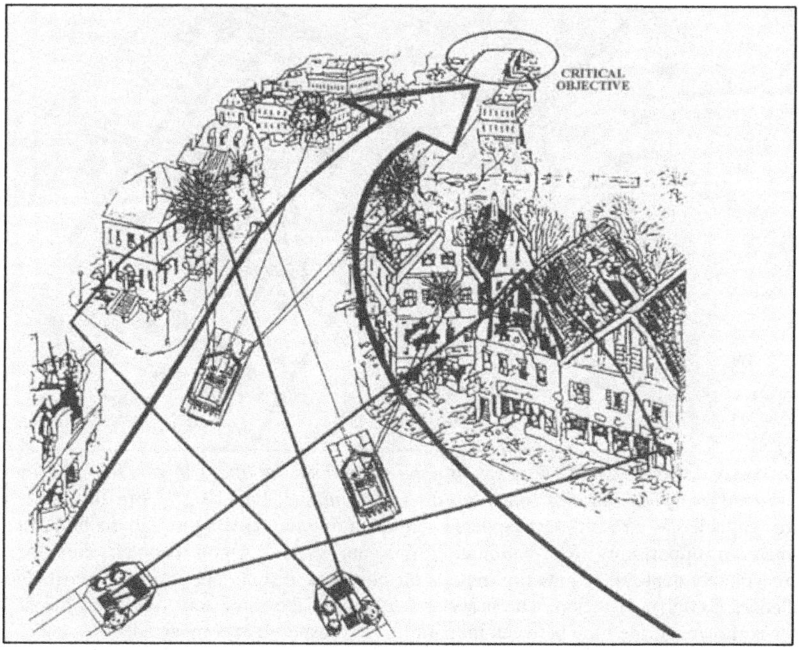

momentum through the zone of action and to seize key objectives. Some enemy positions may be isolated or bypassed in order to maintain tempo. This procedure may be used when:

- A key objective is identified (an enemy critical vulnerability or an objective that supports the friendly scheme of maneuver)
- Time is critical in reaching the objective
- Only selected buildings on the route of advance require clearing.

Figure 2-3. Rapid Advance En Route to a Key Objective

(2) Systematic Clearance. Systematic clearance (figure 2-4 on pg. 2-12) involves the deliberate reduction of all significant enemy positions throughout the objective area. It is usually used when time is not a critical factor.

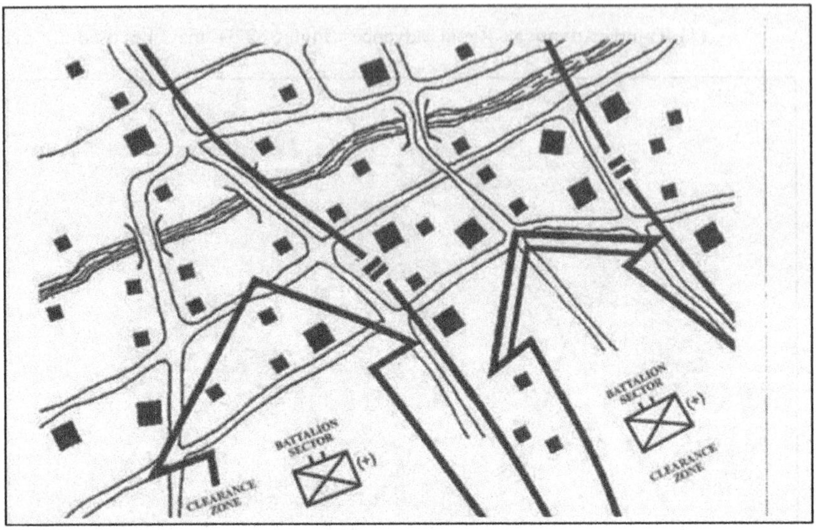

b. Hasty Attack. A hasty attack is, *in land operations, an attack in which preparation time is traded for speed in order to exploit an opportunity* (Jt Pub 1-02). Normally, hasty attacks are conducted as a result of unexpected contact during movement through the built-up area or when an opportunity for exploitation presents itself. Upon contact, the commander immediately deploys, returns fire, reports the contact and situation, develops the situation, and chooses a course of action. The hasty attack follows the same four phases as the deliberate attack; however, the time between preparation and response is compacted.

2104. Warfighting Functions. The warfighting functions (WFs) encompass all activities in the battlespace. Planners consider and integrate the warfighting functions when analyzing how best

Figure 2-4. Systematic Clearance Within Assigned Sectors

to accomplish the mission. The WFs include maneuver, intelligence, fires, logistics, command and control, and force protection.

a. Command and Control. The commander positions himself where he can best observe and influence the battle. Positioning is relative to the level of command. For example, a squad leader locates where he can observe and direct his fire teams, while the battalion commander locates where he can observe and direct his companies.

Close-quarter fighting in and around buildings makes command and control exceedingly difficult. Commanders and small-unit leaders rely on guidance and intent provided by higher

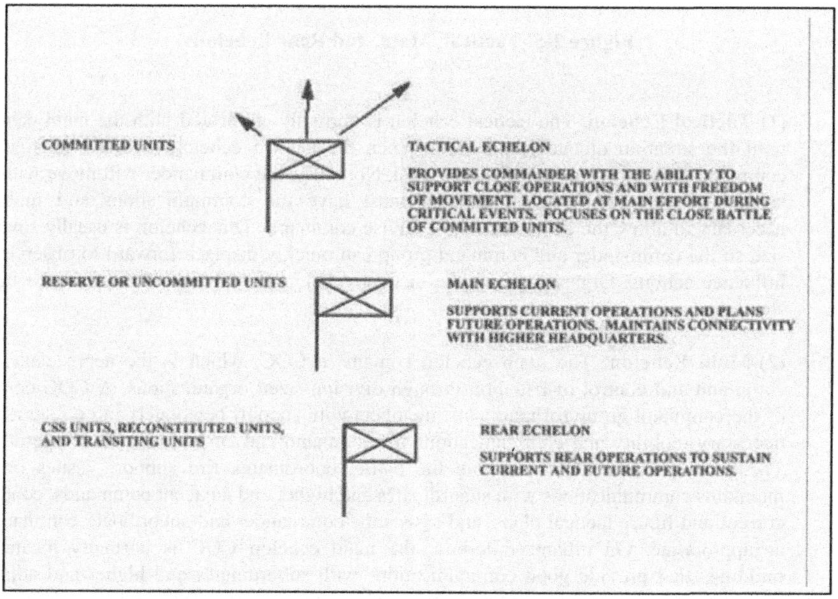

headquarters to facilitate decentralized execution while maintaining control and direction. In close-quarter fighting, commanders may have to rely upon messengers to communicate with their units and higher headquarters.

At battalion and higher levels, command and control is accomplished through three command echelons, which can be collocated or echeloned across the battlefield. The three echelons are tactical, main, and rear (see fig. 2-5). Each echelon may have a combat operations center (COC) capability. The COCs are the facilities (buildings, vehicles, or tents) used by the commander or staff at each echelon to plan, direct, control, and coordinate operations.

Figure 2-5. Tactical, Main, and Rear Echelons

(1) Tactical Echelon. The tactical echelon is normally collocated with the main echelon until the situation dictates separation. When the tactical echelon separates, designated command group members displace forward. Normally, the commander will move with the tactical echelon. The tactical echelon must have the communications and mobility necessary to allow the commander to exercise command. This echelon is usually small in size so the commander and command group can quickly displace forward to observe and influence actions. Organization of the tactical echelon should be addressed in the unit's standing operating procedures (SOPs).

(2) Main Echelon. The main echelon contains a COC, which is the nerve center for command and control of battalion through division-sized organizations. A COC consists of the command group (primary staff members with support personnel) and possesses the necessary mobility and communications for command and control of tactical operations. The main echelon COC monitors the battle, coordinates fire support, issues orders, maintains communications with subordinates and higher and adjacent commands, conducts current and future tactical plans, and assists the commander and subordinate commanders as appropriate. On urbanized terrain, the main echelon COC is normally located in buildings that provide good communications with subordinates and higher and adjacent commands. The command post (CP) is where the commander is physically located. Therefore, the CP can be located at any of the three echelons based on the commander's presence.

(3) Rear Echelon. The rear echelon's COC consists of the logistical and administrative personnel required to plan, coordinate, and execute logistics support. During urban warfare, the rear echelon COC is usually closer to combat forces than normal as it uses the city's captured buildings and facilities to provide cover and concealment for stocking supplies, conducting maintenance repair, and treating the injured.

Attack of a built-up area requires detailed planning due to restrictive terrain, close proximity to opposing forces, fire coordination requirements, and reduced communications capability. The GCE will frame a centralized plan of attack, fire support, and logistics support to conduct urban operations. However, the actual attack of a built-up area is a series of isolated, close-quarter battles carried out by small units. The execution of these attacks should be decentralized.

Communications planning procedures, as outlined in MCWP 3-40.3, *Communications and Information Systems* (under development), are applicable to the urban battle. Planners should be meticulous in their scrutiny of the environment and the effect it will have on communications. Manmade structures can create problems for single-channel radios. These structures inhibit LOS radio communications by absorbing or reflecting transmitted signals. However, the urban environment may have exploitable advantages such as the availability of electrical power, commercial telecommunications networks, and environmental control systems. Electrical power generation stations and other emergency power systems are normally found in protected structures and are probably usable. Due to their value to the enemy, these areas are more likely to be heavily defended. Enclosed areas offer excellent concealment and protection of communications and other command and control support equipment. Extensive commercial communications networks composed of miles of underground protected cable connecting central telephone exchanges are likely to be available, as well as a multitude of public service radio nets (police, fire, civil defense, taxi, etc.) complete with existing antenna systems and retransmission stations. To communicate effectively and continuously, commanders must minimize limitations imposed by the urban environment and maximize the advantage of existing civil communications.

b. Intelligence. Collection of intelligence information on urbanized terrain is difficult. Urbanized terrain provides overhead cover and concealment from observation by aircraft, reconnaissance personnel, and satellites.

(1) Information Requirements. In addition to information about the location of enemy units, nature of the terrain, trafficability of roads, etc., that all operations require, urban warfare requires detailed information on the following:

- Population density, characteristics, and culture
- Location and quality of water supply
- Width and construction of streets
- Layout of road networks
- Location, width, and load capacity of bridges
- Layout of underground networks such as subways, sewers, and utility ducts
- Types and construction of buildings
- Location of key communications and transportation facilities
- Location of canals and waterways.

Information on port facilities should include details on harbor sites, pier networks, unloading capabilities, warehouse/dockside facilities, harbor currents, and sandbars or other midstream obstructions.

(2) Collection. Accurate, detailed, and timely information is vital to the success of any urban operation. Collection operations conducted by national intelligence organizations may provide valuable information on the layout of the city, recent changes to transportation routes, political affiliations of various groups, and locations of military units. This information needs to be current and must be verified and updated as attacking forces close on the city. Reconnaissance activities such as the study of large-scale maps, aerial photographs, and background intelligence reports keep the commander's information current. The latest information on the enemy and the terrain should be continuously fed to the intelligence section using long-range reconnaissance patrols, human intelligence (HUMINT) reports, unmanned aerial vehicle (UAV) missions, and electronic intelligence. Use of reconnaissance assets should be tempered with a risk-versus-gain judgment. Reconnaissance assets should not be risked indiscriminately if the information can be collected by other means.

(3) Maps. Tactical maps should be supplemented by city street maps that contain street names and information on the location of important buildings and transportation terminals. These city street maps should be distributed to all units operating within the city. Although tactical maps are key in controlling fire support missions, simple street maps assist ground troops in maintaining their orientation within the city and in tracking buildings and areas that have been cleared.

c. Maneuver. Maneuver is the employment of forces on the urban battlespace through movement in combination with fire or fire potential to achieve a position of advantage (see MCWP 3-1). Maneuver is an essential element of combat power. Maneuver in itself cannot produce decisive results. Combined with mass, offensive action, economy of force, and surprise, maneuver provides favorable conditions for closing with the enemy. Maneuver contributes significantly to sustaining the initiative, exploiting success, preserving freedom of action, and reducing vulnerability. It is through maneuver that an inferior force can achieve decisive superiority at the necessary time and place. In many cases, maneuver is made possible only through the control of tempo and effective employment of firepower. The commander integrates supporting fires with the scheme of maneuver to create a dilemma for the enemy. Likewise, movement without fires exposes the force to effective enemy counteraction and risks losing the initiative and momentum. Maneuver that does not include violent action against the enemy will not be decisive. At all levels, successful application of this principle in the urban environment requires flexible and innovative plans and actions.

(1) Infantry/Armor Movement. Armored vehicles are vulnerable in built-up areas where streets and alleys provide ready-made fire lanes for defenders. Motorized traffic is greatly restricted, canalized, and vulnerable to ambush and close-range fire. Tanks are at a further disadvantage because their main guns cannot be depressed sufficiently to fire into basements or be elevated to fire into upper floors of buildings at close range (Figure 2-11). Whenever tanks are used in the direct-fire role, infantry must provide security against enemy ground attack. Movement of armor down narrow streets or down narrow paths through debris requires close coordination with infantry. Infantry and armor may use bounding overwatch techniques during movement (See MCWP 3-12, *Marine Tank Employment* [under development]). With this technique, infantry moves ahead of armored vehicles to clear the buildings on each side. Lead infantry units should not get so far ahead of armor that mutual support is lost. Conversely, if armor gets too far ahead of infantry, it may be quickly damaged or destroyed. Coordinated movement also contains a rear security element to protect armor from behind.

During movement down a street tanks should deploy with one tank on each side of the street focusing on the street and lower levels of buildings. *In MOUT, tanks are best deployed in pairs.* Other tanks of the tank platoon or section should move behind the infantry and fire at targets in the upper stories of buildings. When space is inadequate to deploy abreast, tanks may deploy in single file. Movement across open areas should be well coordinated to ensure mutual support. Suppression fire and smoke are used to cover force movements.

Armor can also be used in the reduction of obstacles. For example, grappling hooks and ropes attached to a tank can be used to pull down wire obstacles.

(2) Cover and Concealment. Walls and buildings can provide cover, concealment, and protection from enemy ATGM or heavy machine gun fire. Before armor enters a building, infantry should clear the building and check the ground floors to ensure that there is no basement into which an armored vehicle could fall and become trapped.

Tank and AAV generated smoke may be useful in concealing the location and movement of assaulting forces. Additionally, the thermal sights of the tank are very effective as they can use heat differentials to detect the movement or presence of concealed enemy soldiers, weapons, and vehicles.

(3) Tactical Considerations for Helicopterborne Movement. Helicopterborne operations are planned and conducted using the same combat fundamentals that apply to ground operations. However, there are considerations that apply only to helicopterborne operations. See MCWP 3-24, *Assault Support* (under development) for a detailed discussion on helicopterborne operations. These considerations include:

(a) Helicopterborne forces, once they have landed, lack vehicular mobility and heavy weapons. For this reason, it is important that the forces land on or near the objective. Such a force may be isolated for a time pending linkup with ground forces.

(b) Helicopters may be used for tactical deception. Helicopters may be used to make demonstration landings in several different zones to deceive the enemy as to the true objective of an operation.

(c) Helicopterborne assaults are normally conducted against undefended or lightly defended objectives. When attacking a well-defended objective, an LZ should be selected which provides cover and concealment for the assault force and is close to the objective.

(d) Helicopterborne forces are vulnerable to attack helicopters, fixed-wing aircraft, and surface-to-air missiles. To counter this vulnerability, friendly attack helicopters

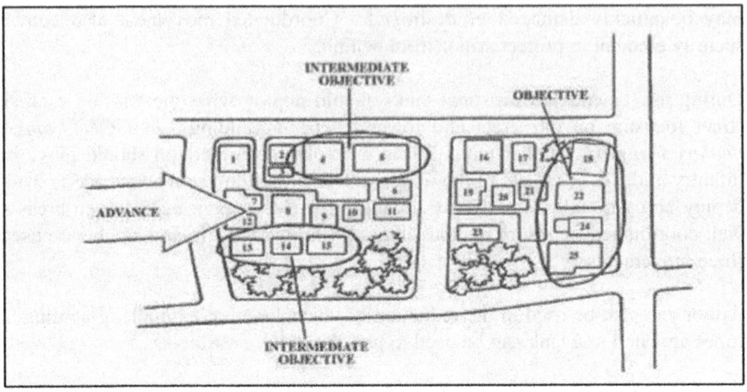

escort the helicopterborne forces during air movement, and indirect fires are used to suppress or neutralize enemy air defenses.

(4) Control Measures. The following control measures assist in the manuever of forces in MOUT:

(a) Objectives. When attacking to seize a foothold, the infantry unit may assign subordinate units the first block of buildings as their first objective. When an objective extends to a street, only the near side of the street is included. The final objective may be buildings or key terrain at the far edge of the built-up area. Key buildings, or groups

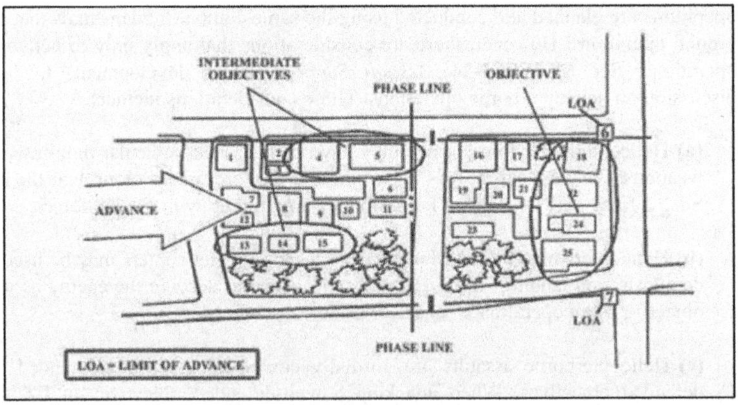

of buildings, may also be assigned as intermediate objectives. Buildings along the route of attack should be identified by numbers for clarification (Figure 2-6 on page 2-18).

During the seizing phase, bypassing buildings may increase the risk of attack from the rear or flank. It may be necessary for the unit to enter, search, and clear each building in its zone of action. A single building may be an objective for a rifle squad or, if the building is large, for a rifle platoon or company. When the commander's concept of operations is based on speed (rapid advance method) or when the force is conducting a hasty attack, a unit may be directed to bypass certain positions within its zone.

Figure 2-6. Objectives and Numbering System

(b) Phase Lines. Phase lines are control measures used to report progress or to control the advance of attacking units. Phase lines should be oriented on readily identifiable terrain features such as principal streets, rivers, and railroad lines. They should be placed on the near side of a street or open area. In an attack, a unit may have the mission to seize positions in its zone of action up to a particular phase line. (Figure 2-7).

Figure 2-7. Phase Lines

(c) Boundaries. Unit boundaries are used to define zones of action and are usually set within blocks so that a street is included in the zone. Both sides of a street should be included within the same unit's boundries.

(d) Checkpoints and Contact Points. Checkpoints aid in reporting locations and controlling movement. Contact points are used to designate specific points where units make physical contact. Checkpoints and contact points are designated by each unit as appropriate for command and control at street corners, buildings, railway crossings, bridges, or any other easily identifiable feature.

(e) Attack Position and Line of Departure (LD). A designated attack position may be occupied by forward units for last-minute preparation and coordination. The attack position is often behind or inside the last large building before crossing the LD. The LD should be located on the near side of an open area running perpendicular to the direction of attack, such as a street or rail line.

(5) Time of Attack. Ideally, the time of attack should be conducted during hours of darkness or limited visibility. Marines can exploit the poor visibility to cross open areas, to gain access to rooftops, to infiltrate enemy areas, and to gain a foothold. When attacking in unrestricted visibility, units should use smoke to conceal movement.

(6) Reserve. Reserves should be mobile and usually follow in trace of forward units. Battalion reserves normally follow one to two blocks to the rear of the lead company. If a company reserve is available, it follows within the same block so that it can immediately influence the attack. A unit with a reserve mission can be called upon to:

- Exploit an enemy weakness or friendly success
- Attack from another direction
- Attrit and/or clear bypassed enemy positions (normally requires designation of a new reserve)
- Secure the rear or a flank
- Maintain contact with adjacent units
- Counterattack

(7) Security. Each unit should ensure security of its flanks and rear (Figure 2-8 on page 2-20). Fighting in a built-up area is unique in that attacks can come from all three dimensions. Units assigned a mission to isolate can also provide security. Small units conduct security patrols and man OPs to supplement higher level reconnaissance and security operations.

Figure 2-8. Enemy Firing From Flank

(8) Engineers. Leading units should have engineers attached for immediate support. Tasks commonly assigned to engineers include:

- Preparing and using explosives to breach walls and obstacles
- Finding and exploding mines in place or helping to remove them

- Cratering roads and performing other countermobility measures
- Clearing rubble and obstacles

Requirements for engineer support in MOUT normally exceed capabilities. To compensate, all Marines should receive basic demolition training prior to conducting urban operations.

d. Fires. The attack of a built-up area may require extensive air and artillery preparation prior to the ground attack. Supporting fires suppress the defender's fire, restricts his movement, and possibly destroys his position. Consideration should be given to the rubbling effect produced by aerial and artillery bombardment. The assault should closely follow air and artillery fire to exploit its immediate effect on the defender. Maneuver units move near the final coordination line while the enemy is engaged by supporting fires. As the attacking force assaults, supporting fires lift and/or shift to block enemy withdrawal or to prevent the enemy from reinforcing their position. (See Chapter 4 for detailed discussion on air and artillery support during urban warfare.) Fire can be categorized into indirect and direct fires.

(1) Indirect Fire

(a) Artillery. Indirect artillery fire is planned to isolate objectives, to prevent reinforcement and resupply, to neutralize known and suspected CPs and OPs, and to suppress enemy defenders. Urbanized terrain requires that most indirect artillery fire use a high-angle trajectory.

(b) Mortars. Mortars are the most responsive indirect-fire weapon for the urban environment. They can be employed to hit targets of opportunity at the close ranges that are typical of combat in built-up areas.

(c) Other. Urban conflict is conducive to innovation. Many direct fire weapons can be employed in the indirect fire mode. Indirect fire weapons may also be effective at clearing the tops of buildings using variable-timed fuses.

(2) Direct Fire.

(a) Armor. Tanks are normally task organized with mechanized units for protection and security operations. The commander can employ tanks to take advantage of their long-range lethality, high-speed mobility, and survivability. They can be used outside the built-up area to cover high-speed armor avenues of approach. Tanks are normally employed in this role during the isolation phase. Tanks also support by fire the assault to seize a foothold and may support attacks to seize objectives within the built-up area. In MOUT, tanks can also be organized into special assault teams. The tank cannon's direct fire is very effective against structures. (Figure 2-9 on page 2-22 and Appendix B)

(b) Howitzers. Artillery rounds delivered by direct fire can be very effective for destroying targets in buildings. When used in this manner, the mobility, flexibility, responsiveness, and survivability of gun crews may be reduced. Artillery used in the direct fire role normally use HE shells.

(c) Infantry. Direct-fire systems, mainly machine guns, antitank guided missiles (ATGMs), shoulder- launched multipurpose assault weapons (SMAWs), and AT4s—are initially employed to support the seizure of a foothold. The positioning of antitank weapons in buildings must allow enough space for the effects of backblasts. Antitank weapons have limited as effectiveness when neutralizing targets behind walls. Designed primarily as armor penetrators, they neutralize a target only if that

Figure 2-9. Tank Direct-Fire Support Within the Built-Up Area

target is located directly behind the point of impact. Consideration must be given to weapons with minimum arming distances. Many of these factors can combine to limit the effects of weapons in the urban environment.

(d) Aviation. The vertical nature of large urban areas may block the pilot's line of sight with the target unless careful consideration is given to inbound headings and altitudes. Plan attack headings to take advantage of large open areas and street axis. Friendly ground forces must be clearly distinguishable from targets. The use of smoke, visible or infrared strobes, or other visual devices must be effectively employed by ground forces and clearly understood by friendly aircraft.

e. Logistics. Logistics support in urban operations should be timely, flexible, and responsive. Logistics planners, both in the supported units and the CSSE, should be fully integrated in the planning process. The final logistics plan must be thoroughly coordinated with the scheme of maneuver. Units may need special equipment such as ropes, grappling hooks, ladders, sandbags, and wire; medical supplies must be readied in anticipation to support increased casualties; and additional supplies must be prestocked. (See Chapter 5 for a detailed discussion on logistics and urban warfare, and MCWP 4-1, *Tactical Logistics*, for additional planning considerations.)

f. Force Protection. Force protection encompasses all those measures that preserve the fighting potential of the command. It is integrated throughout all the warfighting functions. It does not entail security operations alone. It involves the continuous implementation of active and passive measures by all units during all phases of an operation.

Some particular active and passive measures in MOUT include:

-Manmade structures in the urban environment provide cover and concealment for both the attacker and defender.

-Lines of communication can be more easily interrupted due to multiple avenues of approach.

-All units must be prepared to implement force protection in a 3-dimensional environment.

-Proximity of civilian population makes identification of non-combatants more difficult.

-ROE may be more restrictive

-Snipers and man-portable shoulder lauched air defense missiles are usually more difficult to detect.

Section II

Infantry Battalion Operations

2201. Overview. This section addresses some of the operations that a task-organized and/or reinforced infantry battalion could conduct in MOUT. These operations are seldom conducted independently and are usually part of a larger joint operation. In the example given below, the battalion has seized a foothold and is operating in its sector of a town of 100,000 people. Concurrent operations are on-going through other parts of the town. The threat in the battalion sector is a company (-) with a capability to reinforce.

2202. Seizure of a Key Objective. The tactical situation may require the immediate seizure of a building, industrial complex, or transportation site that is deemed decisive terrain. Examples of such objectives may include key bridges, intersections, railheads, or facilities that constitue the center of gravity for the enemy. As an example, a battalion may be assigned the mission of seizing the only bridge that is capable of supporting armored vehicles and resupply by heavy transports along a river bisecting a town. The intent is to isolate enemy forces and secure the bridge for follow-on forces. Intelligence indicates that the bridge is lightly defended at this time. In this case, speed of attack and surprise is critical to capture the objective. Some key intelligence information is required prior to the attack. This would include the number of enemy defending the bridge, their weapons, disposition, their ability to reinforce their defense, location of LZs near the bridge, whether the bridge has been set for demolition, and the location of enemy indirect fire assets that can range the objective.

The commander decides to conduct a penetration using a combination of helicopterborne attack to rapidly seize the bridge followed by a mechanized ground attack to link-up with the helicopterborne forces (Figure 2-10). The helicopterborne force can achieve surprise and speed of movement required while the mechanized force can thrust quickly along high-speed avenues to minimize the time the helicopterborne force must hold the objective against enemy counterattack.

For the ground attack, the battalion commander makes a determination that speed of movement outweighs the risks of rapidly extending his lines of communication. He directs, on the basis of his knowledge of the enemy situation and his mission, that his maneuver forces avoid decisive engagement with the enemy along the route to minimize the time it takes to link up with the helicopterborne force. The commander provides guidance on those enemy forces that should be bypassed and those that should be isolated and reduced.

The commander task organizes his battalion for rapid movement on two axes to allow flexibility in reacting to unexpected enemy contact. Lead and rear security is conducted throughout the operation. As lead units locate enemy positions, they fix them by fire, and quickly reduce or bypass them. They also report the contact so that follow-on forces can then deal with the threat.

As the ground attack force reaches the edge of the built-up area, Marines remain mounted so as not to slow the advance. All vehicle machine guns fire to flanks and into doors, windows, and

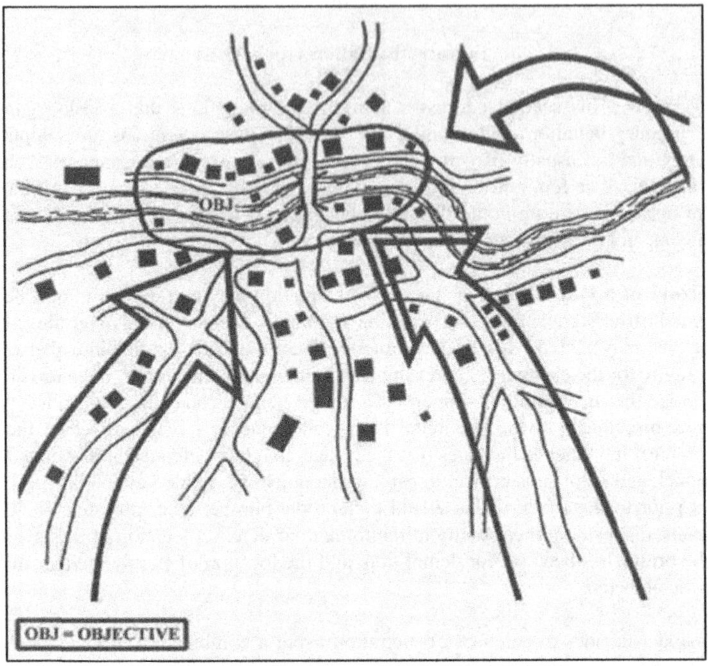

Figure 2-10. Seizure of a Key Objective

alleys to provide suppressing fires (Figure 2-11 on page 2-26). If heavy resistance is met and the column is halted, Marines dismount and the enemy and/or obstacles are cleared as quickly as possible. Once resistance and/or obstacles are cleared, the attack is immediately resumed. If necessary, units are dropped off to assume blocking positions and await follow-on clearing forces.

Once the objective is seized or linkup is accomplished, the battalion establishes a perimeter defense. The companies seize buildings around the objective and expand the size of the perimeter until it is large enough to secure the bridge. Attached engineers can then examine the bridge and clear it of any explosives.

2203. Infiltration. Infiltration is *"the movement through or into an area or territory occupied by either friendly or enemy troops or organizations. The movement is made, either by small groups or by individuals, at extended or irregular intervals. When the term is used in connection with the enemy, it infers that contact is avoided"* (Joint Pub 1-02).

Figure 2-11. Rapid Advance Movement

.3

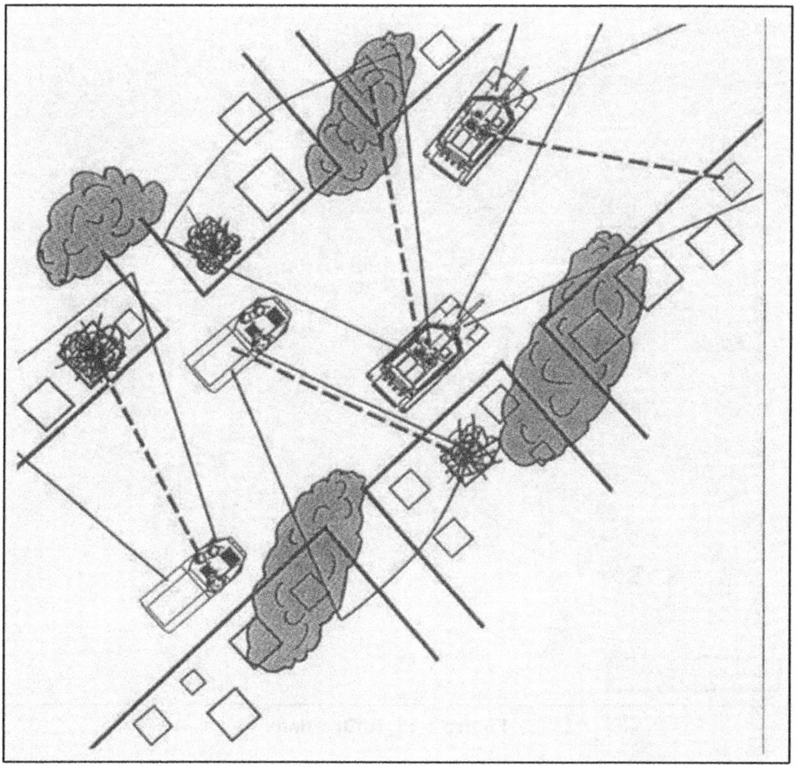

The outskirts of a town may not be strongly defended. The town's defenders may have only a series of antiarmor positions, security elements on the principal approach, or positions blocking the approaches to key features in the town. The enemy strongpoints and reserves may be positioned deeper in the city. In such a situation, an infantry battalion may be able to seize a part of the town by infiltrating companies and platoons between the enemy positions on the outskirts (Figure 2-12).

Moving by stealth on secondary streets, using the cover and concealment of alleys and buildings, the battalion may be able to seize key street junctions or terrain features, isolate enemy positions, and help following units move into the built-up area. Such an infiltration should be performed when visibility is poor.

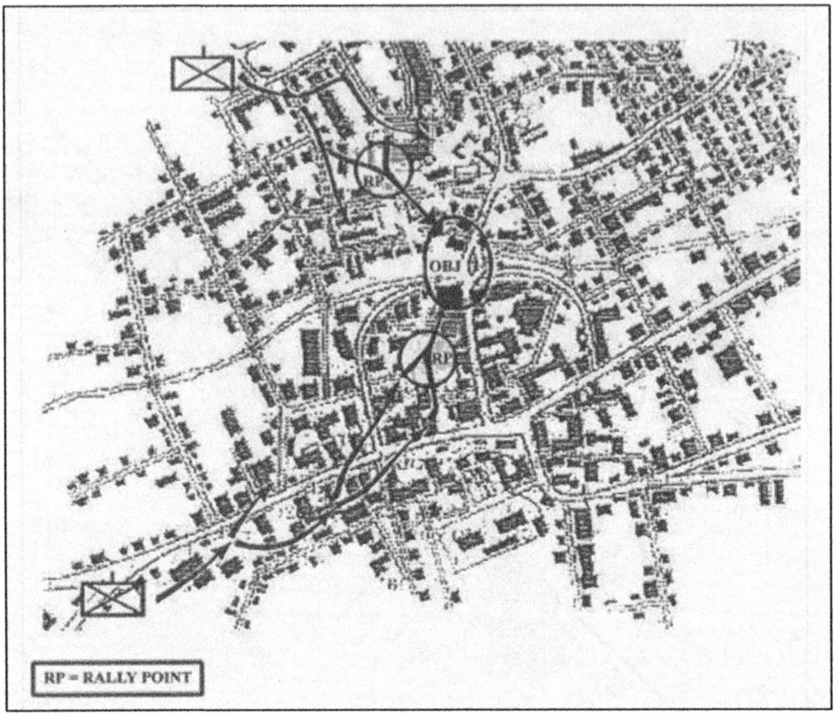

Figure 2-12. Infiltration

The infantry battalion is organized into infiltration companies with engineers attached and a designated reserve. Each company is assigned an infiltration lane that allows for infiltration by company-sized or smaller units. Depending on the construction of the built-up areas and streets, the infiltration lane may be 500 to 1,500 meters wide.

Mortar and artillery fire can be used to divert the enemy's attention and cover the sound of infiltrating forces. Armored vehicles and antitank weapons should be positioned to cover likely avenues of approach for enemy armored vehicles. Designated unit(s) provide isolation and security for infiltrating rifle companies. As the companies move into the built-up area, they secure their own flanks. Security elements may be dropped off along the route to warn of a flank attack. Engineers assist in breaching or bypassing minefields or other obstacles. Enemy positions are avoided but reported. The infiltrating companies proceed until they reach their rally points. At that time, they reorganize and make final preparations for attack of the objective.

2204. Route Security. In MOUT, a battalion may be tasked to provide route security. The process of securing the route may consist of a series of small-unit actions to clear buildings along the route. Fires should be planned to isolate the route as well. It may also require the reduction of obstacles and the emplacement of flank security along the entire route. In an outlying area, the forward units may proceed by bounds from road junction to road junction. Other units could augment flank security by moving down parallel streets and by probing the flanks.

The speed of movement required and the enemy situation determine whether the unit should be mounted or dismounted. For mechanized forces, narrow streets should be avoided if possible to avoid being trapped. Designated armored vehicles overwatch the squad to its front, keeping watch on the opposite side of the street. While armored vehicle movements and positioning provide mutual support, armored vehicles should be protected by dismounted infantry.

When contact is made with the enemy, armored vehicles seek covered positions and provide direct-fire support. Supporting fires fix and isolate enemy positions. Infantry dismount then maneuver to attack. Phase lines are used to coordinate movement along the route. For example, at each phase line, the forward companies might reestablish contact, reorganize, and continue clearing (Figure 2-13).

2205. Battalion Considerations. Some of the concerns for the commander to consider include:

a. Command and Control. CP location can be critical due to line of sight restrictions in the urban environment. Consideration should be given to maintaining a highly mobile CP with links to airborne retrans stations.

b. Intelligence. Organic collection capability resides primarily from units in contact with the enemy. The Scout Sniper Platoon also provides information but they may have obstructed visibility in the urban environment. Commanders may chose to employ them more often in a sniper role. Links to higher command intelligence sources may be key to maintaining current information on the enemy. The same line of sight restrictions encountered with command and contol will also restrict intelligence dissemination. Commanders may have to rely on messengers.

c. Manuever. Maneuver of mechanized forces in an urban environment may be restricted to the size, width, and load capabilities of the streets. Consideration must be given to plan for alternate routes for any attack. Significant manmade obstacles can be encountered throughout the battlespace. Additionally, the reserve is normally kept closer to the assault force. Movement of forces by aviation is less restrictive, but may also be vulnerable to surface to air missiles and small arms threat. Availability of LZs may also be a limiting factor.

d. Fires. A key consideration on the use of fires will be the ROE and the collateral damage that can result by their use. Rubbling of buildings can complicate an already restricted maneuver capability. Weapons effects can be more pronounced and can create a hazard for friendly forces. IPB determines potential firing positions for indirect fire assets that accounts

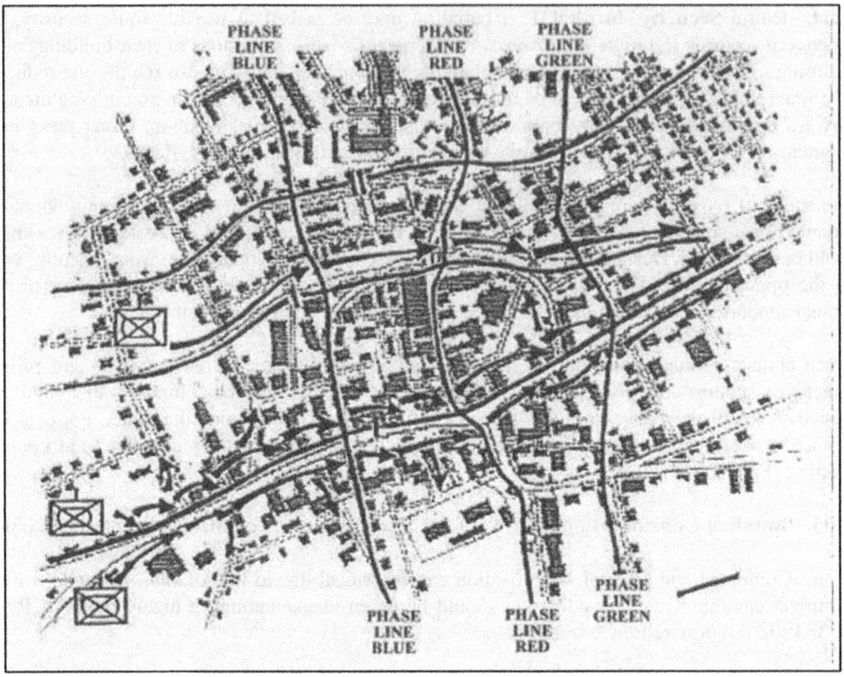

Figure 2-13. Clearing Along a Route

for the overhead masking by buildings and the ability to seat baseplates. Traditional indirect fire assets, such as artillery, may be used in the direct fire role.

e. Logistics. Historically in MOUT, the utilization of class I-V and VIII supplies is much greater. More vehicular assets should be allocated to the small unit level to transport ammunition and support the increased casualty evacuation requirements.

f. Force Protection. Lines of communication require extensive flank security in the urban environment. The commander must determine where his critical security requirements are and concentrate his force protection efforts in those areas while economizing in other areas. Manpower may further be depleted by security requirements for refugees and EPWs.

.3

Section III

Rifle Company Operations

2301. Overview. This section addresses some of the operations the infantry battalion could assign to the rifle company in MOUT. For our focus, the rifle company is a task-organized and/or reinforced unit. Company operations will normally be conducted as part of a larger battalion operation.

2302. Attacking within a Built-up Area. Attacking within a built-up area is a task the rifle company may be assigned (Figure 2-14). This task involves the isolation, attack, and clearing of a specified block of structures. It is usually conducted in conjunction with adjacent friendly attacks or security operations. (See Appendix A.)

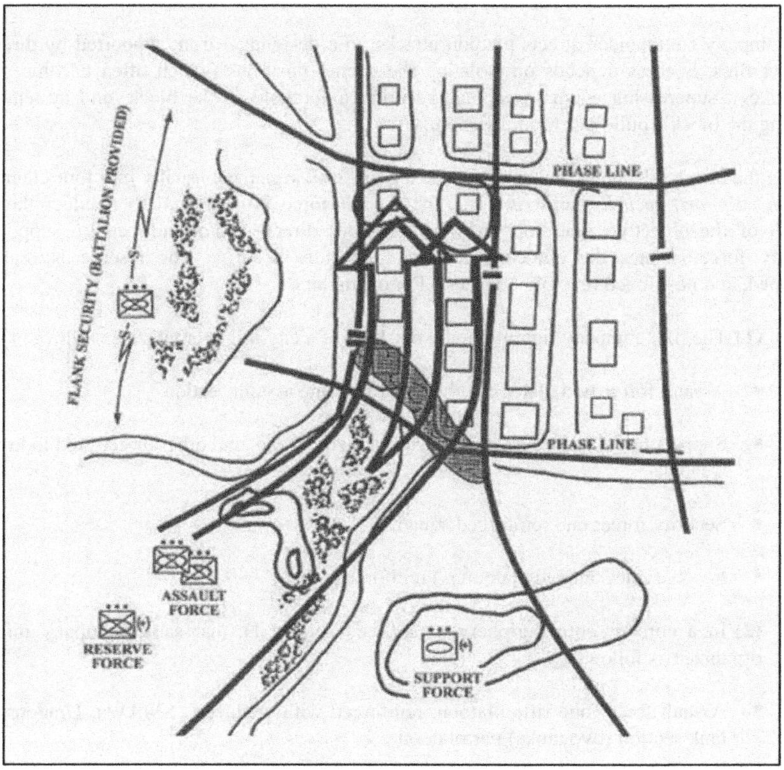

Figure 2-14. Company Attack of a Block

Before an assault, the company commander employs smoke to conceal the movements of assaulting platoons. He provides suppression fires to their flanks with direct-fire weapons. The battalion may provide additional security forces for flank protection.

(1) Concealed by smoke and supported by direct-fire weapons, an assaulting platoon attacks the first isolated building. The platoon must close on the building quickly while the enemy is still stunned by supporting fire. The company commander must closely coordinate the assault with supporting fire so that the fire is shifted at the last possible moment.

(2) The platoons continue to seize each building until the block is cleared. (See Appendix C for a detailed discussion on clearing of buildings.) After seizing the block, the company consolidates and reorganizes to repel a counterattack or to continue the attack.

The company commander directs platoon attacks on a designated front supported by direct and indirect fires. Success depends on isolating the enemy positions (which often become platoon objectives), suppressing enemy weapons, securing a foothold in the block, and systematically clearing the block's buildings room by room.

During the attack of a block, the company is usually task organized usually into four elements: *assault*, *support*, *security,* and *reserve*. The assault force is organized to conduct the actual assault of the objective. The support force provides direct- and/or indirect-fire support. The security force isolates the objective area and provides security. The reserve is organized, equipped, and positioned to exploit success. For example:

(1) The rifle company fighting in the outskirts of a city might organize as follows:

- Assault force: two platoons reinforced with the assault section
- Support force: armored vehicles, mechanized squad, and other direct- and indirect-fire weapons
- Security force: one reinforced squad.
- Reserve: mechanized platoon (-) reinforced.

(2) In a core or core periphery area (See Chapter 1), that same company might be organized as follows:

- Assault force: one rifle platoon, reinforced with engineers, SMAWs, Dragons, and a tank section (two tanks) per platoon
- Support force: One squad with direct- and indirect-fire weapons

.3

- Security force: one rifle platoon (-)
- Reservc: one rifle platoon, mechanized.

c. Armored vehicles, machine guns, and other direct-fire weapons under the company commander's control are assigned to isolate the objective and provide suppressive fires from covered positions. These weapons should not be fired for prolonged periods from one position. The gunners should use a series of positions and displace from one to another to gain better fields of fire and to avoid being targeted by the enemy. Direct-fire support tasks are assigned as follows:

- Machine guns fire along streets and into windows, doors, alleys, etc.; however, innovative use of direct fire weapons can be key in MOUT
- Tanks, ATGMs, and AT4s fire at enemy tanks and other armored vehicles.
- SMAWs can be used to make entrances in buildings and destroy identified bunker/fighting positions.
- Tanks can fire at targets protected by walls, make entrances in buildings, and engage enemy armored vehicles with their main gun and other targets with their machine guns.
- Riflemen engage targets of opportunity.

2303. Attack of an Enemy Outpost. A motorized or mechanized force may encounter enemy outposts, checkpoints, or strongpoints. In a hasty attack on an enemy outpost, the company commander makes a quick assessment based on the mission and situation, and then reacts in accordance with the commander's intent. If not able to bypass an enemy outpost, he may have to attack and reduce it.

The company occupies positions of advantage. The tanks or other armored vehicles support the dismounted infantry in the attack by assuming support-by-fire positions from which they can fire on the outpost, prevent the enemy from escaping, and destroy any reinforcements.

The infantry platoons dismount and move to engage the enemy (Figure 2-15). They should attack from a covered route to hit the outpost at a vulnerable point. As the platoons approach the outpost, smoke may be employed to screen their movement, and supporting fires are shifted. After platoons close on the outpost, they clear the buildings quickly and consolidate. The company is then ready to continue operations.

2304. Seizure of a Traffic Circle or Major Intersection. A company may have to seize a traffic circle or major road or traffic intersection (Figure 2-16 on page 2-34). This operation consists of seizing the buildings adjacent to the traffic circle or intersection while bringing direct-fire weapons into position to fire on likely avenues of approach.

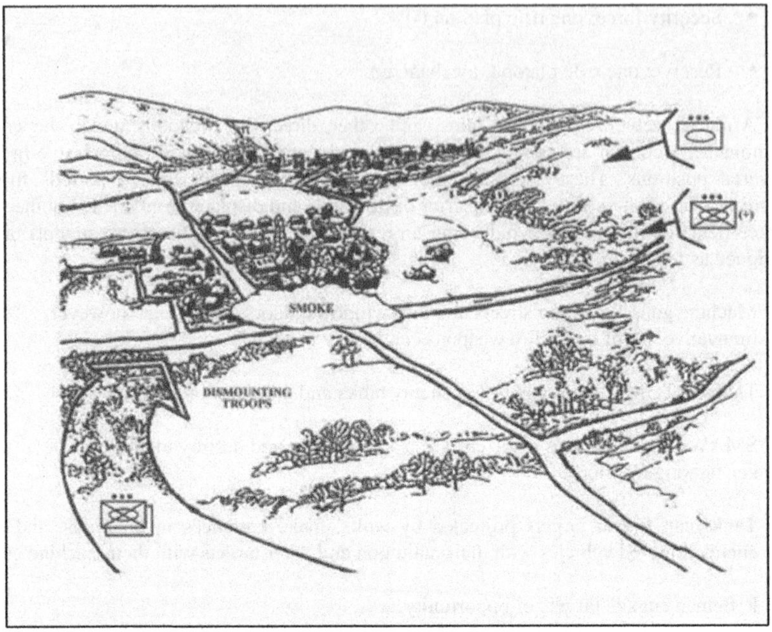

Figure 2-15. Hasty Attack of an Enemy Outpost

The company commander takes the following actions to seize a traffic circle or major intersection:

- Isolate the objective area.
- Seize and clear the buildings overlooking the traffic circle or major intersection under cover of direct-fire weapons.
- Secure the flanks against enemy attack.
- Consolidate and prepare for counterattack.

2305. Seizure of Key Terrain. The seizure of certain key terrain such as a bridge, intersection, facility, or structure may be critical to the success of an operation. Seizing key terrain for friendly use is a possible mission for a rifle company.

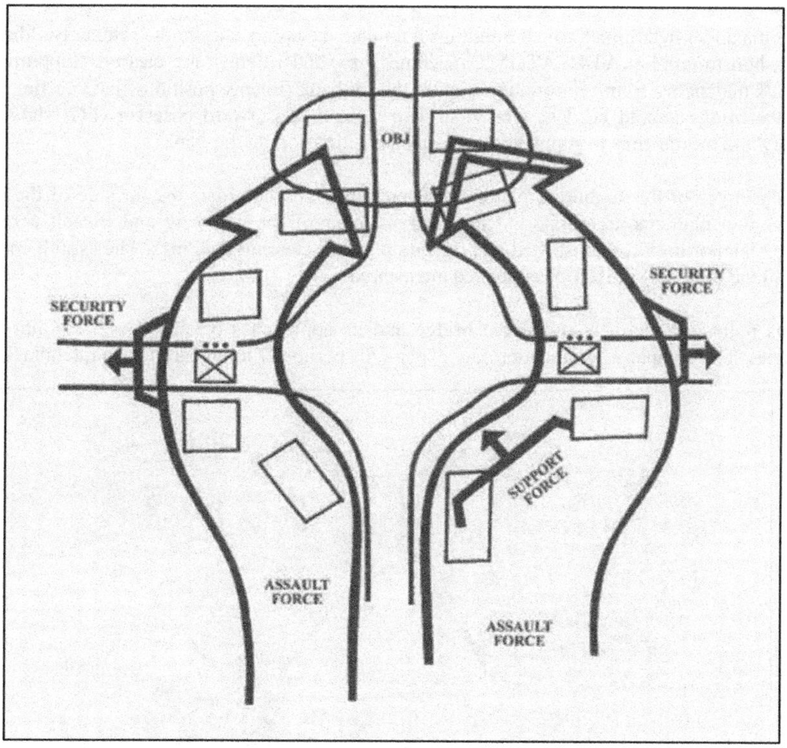

Figure 2-16. Seizure of a Traffic Circle or Major Intersection

To seize a bridge, a company should:

- Clear the buildings on the *near bank* that provide clear fields of fire for supporting weapons
- Suppress enemy weapons on the *far bank*
- Use screening smoke to obscure enemy observation
- Clear the buildings on the *far bank* that dominate the bridge
- Secure a perimeter around the bridge so that the engineers can clear obstacles or remove demolitions from the bridge.

The commander determines which buildings dominate the approaches to the bridge. Buildings that permit him to employ AT4s, ATGMs, machine guns, and riflemen are cleared. Supporting fire prevents the enemy from reinforcing units on the far bank. Enemy positions that can fire directly onto the bridge should be destroyed first. The company's forward observer (FO) should plan artillery and mortar fires to suppress the enemy's positions on the far bank.

The objectives of the assaulting platoons are buildings that dominate the far side of the bridge. One or two platoons are supported by the remainder of the company and assault across the bridge. Supporting fires are shifted and the infantry start clearing buildings. The assault continues until all the buildings in the objective area are cleared.

At this point, the engineers clear the bridge and its approaches of all mines, demolitions, and obstacles. The company commander may expand his perimeter to prepare for counterattack. Once

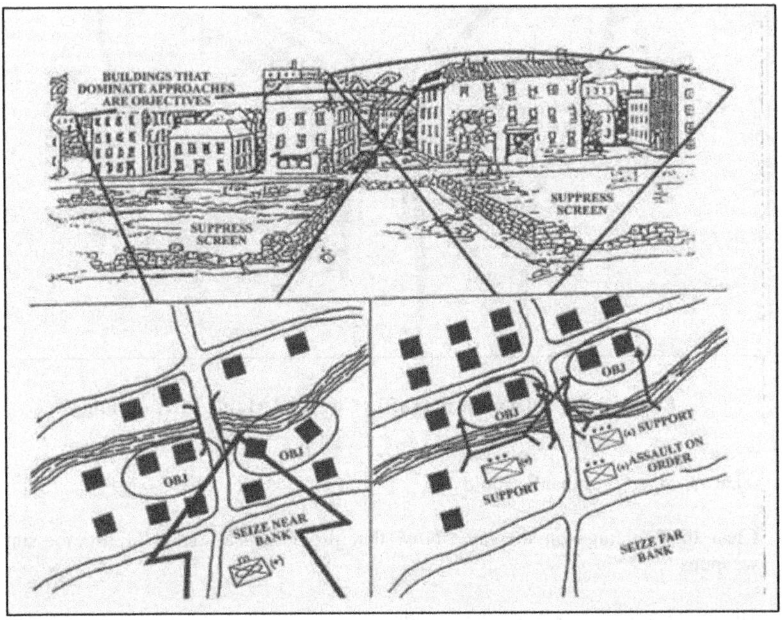

the bridge is cleared, the armored vehicles are brought across to the far bank (Figure 2-17).

Figure 2-17. Seizure of a Bridge

.3

2306. Movement to Contact. A rifle company may conduct a movement to contact along a road. Similarly, the company may have to reconnoiter such a route to prepare for a battalion attack. This type of mission can best be accomplished by a mechanized company with appropriate attachments. (Figure 2-18 on page 2-36).

Figure 2-18. Movement to Contact Along a Highway (Commercial Ribbon)

The speed of movement depends on the mission, terrain, and enemy situation. In open areas where rapid movement is possible, tanks should lead. In close terrain, the infantry may lead while being overwatched by the tanks. Artillery fire should be planned along the route. Engineers may accompany the lead platoon on the main route to help clear obstacles and mines as they may be encountered. Between key terrain, when contact is not likely, the company remains on their vehicles. At key terrain or when enemy contact is likely, the company moves on foot to clear enemy positions or to secure the key terrain. Armored vehicles support the infantry.

Section IV

Rifle Platoon Operations

2401. Overview. This section discusses some of the various tactical missions that might be assigned to a rifle platoon.

2402. Attack of a Building. One offensive mission assigned to a rifle platoon in MOUT is the attack of a building. The attack requires isolating the building and entering the building at the least-defended point or through a breached hole (Figure 2-19 on page 2-38). When possible, clear the building from top to bottom. Close coordination between the assault and support forces is required.

When attacking a building, a platoon should be organized into an assault force, a support force, and a security force. The assault force provides the forces that actually conduct the assault. A support force provides supporting fires and efforts that assist the assault force. A security force covers the platoon's flanks and rear. (See Appendix A.) The platoon may be augmented by armored vehicles and from the company's weapons platoon or the battalions weapons company. If a reserve is required it should be provided by the company.

The assault consists of three steps:

(1) Isolate the building and deny the enemy the ability to maneuver or reinforce.

(2) Enter the building to secure a foothold for further action.

(3) Clear the building systematically.

Clearing is performed by individual squads and teams, which leapfrog as rooms and floors are secured. Platoons tasked to clear buildings may be reinforced with combat engineers.

2403. Movement Down a Street. When moving in urban terrain, a rifle platoon normally follows the same principles of movement as in other environments. However, some movement techniques must be modified to adjust to the unique dangers encountered in an urban environment. (See Appendix A, Section VII.) This section describes the movement down the street of the lead platoon of a company.

The speed of movement depends on the enemy situation. In outlying or lightly defended areas, a mechanized platoon may proceed along the street mounted but send dismounted infantry forward to reconnoiter danger areas. In the center of a built-up area the platoon dismounts to move on foot with two squads leading—one on each side of the road, using all available cover. They move through the buildings, if feasible, to avoid exposure on the streets. The squads coordinate to provide each other with all-around observation and mutual support.

Figure 2-19. Attack of a Building

The platoon should organize and practice to move as two forces: a *maneuver* force and an overwatch force. The maneuver force (normally of one squad on narrow streets or two squads on wide streets) moves forward, scouts danger areas, and closes with the enemy. An *overwatch* force (the remainder of the platoon and its attached or supporting crew-served weapons), moves behind the maneuver force, secures the platoon's flanks and rear, and provides direct-fire support when needed. These two forces, can exchange roles from time to time to avoid overfatigue of the maneuver force (Figure 2-20).

2404. Utilization of the Reserve. In the offense, a rifle platoon may be assigned as a reserve. The reserve can be used to reinforce and attack or exploit success. Reserves are normally held at the company level.

.3

Figure 2-20. Movement Down a Street

In outlying areas, where the terrain is relatively open, a tank platoon reinforced with mechanized assets (tank-heavy task force) may be tasked as the reserve. However, if inside the urban area, a mechanized platoon reinforced with tanks (mech-heavy task force) would most probably be directed to provide the reserve. Regardless of what unit is designated as the reserve, the following items must be considered for their employment in an urban environment:

- The reserve should plan to use a different axis of advance
- Route selection for the reserve may be more constructive since it may have to use building, side streets, sewers, etc.
- Reserves are kept closer to assault forces in MOUT than on open terrain
- Engineers should normally be attached
- Reserves should maintain a high degree of mobility

Chapter 3

Defensive Operations

The essence of defensive tactics is to place the enemy into a position that permits his destruction through the intelligent use of terrain and firepower.

3001. Introduction. In urban combat, the defender possesses key advantages over the attacker. The defender can shape the battlespace to his advantage by maximizing the natural restrictions and obstacles found in the urban environment. ROE for the attacker can add to the defender's advantage by placing restrictions on the application of force by the attacker. Knowledge of the terrain and time available for preparing defensive positions are advantages which may enable the defender to successfully resist a numerically superior force. A spirited and stubborn defense may persuade an attacker to abandon the attack. In some battles, urban defenders can be credited with repelling or decisively delaying an attacker which ultimately influenced the course of the war in their favor.

Marines may be called upon to defend a port city, an embassy located on or near a coastline, a transportation link, or the capital of a foreign government. They may also be tasked to defend from an urban area in an economy-of-force role in order to support operations elsewhere. In combat, offensive and defensive operations are inseparable. Marines need to be capable of conducting both offensive and defensive operations in an urban environment.

The fundamentals of defense do not change in an urban environment. The defenders of a city, however, usually have detailed knowledge of the terrain. This will allow them to establish an extensive defensive network that is designed to force an attacker to expend exorbitant amounts of time, supplies, equipment, and manpower. Commanders need to recognize both the advantages and disadvantages of defensive operations in an urban environment.

3002. Decision to Defend

a. Reasons for Defending Built-Up Areas. Historically, the following are some of the reasons for defending built-up areas:

(1) Denial of Important Strategy/Political Objectives. Capitals and cultural centers can be defended strictly for psychological or national morale purposes even if they do not offer a tactical advantage to the defender. Defending a city can cause an attacker to commit a significant amount of his forces which reduces his capability to attack elsewhere. The defense of a city can delay the overall offensive capability of the attacker.

(2) Retention of Key Economic Centers. In many countries, the entire nation's economic well-being may be tied to a few key cities. These key cities usually contain the country's primary industrial, transportation, and communications base. The capture of these key centers could result in the overthrow of the current government, or deny that

.3

government the ability to adequately support combat operations against enemy regular or insurgent forces.

(3) Control of Avenues of Approach. Most avenues of approach to large cities are straddled by small towns every few kilometers which must be controlled by defending forces. These areas can be used as battle positions or strongpoints to control the avenues of approach. For the attacker to utilize these avenues of approach he must sacrifice speed and expend resources. The defense of these cities or towns may constitute a large part of the overall defense of a city.

(4) Economy of Force. The tactical advantages provided by urbanized terrain allows the defender to engage a numerically superior force. The defender can conserve the bulk of his combat power so that it is available for use in other operations.

(5) Concealment of Forces. Reconnaissance and combat identification is more difficult in an urban environment. CPs, reserves, CSS complexes, and combat forces emplaced in built-up areas are much harder to detect.

3003. Reasons for Not Defending Built-Up Areas. The commander considers the following reasons for deciding not to defend built-up areas.

a. Unnecessary to the Defensive or Offensive Plan. If the built-up area is too far forward or back in a unit's defensive sector, is isolated, or is not astride an enemy's expected avenue of approach, the commander may choose not to defend it.

b. Bypassable. If the nature of nearby terrain allows the enemy to bypass the city, then it will not make a good blocking position. Some built-up areas, mainly smaller ones, are easily bypassed by existing main road and highway systems. A built-up area that can be easily bypassed normally will be, thereby effectively isolating that area's defenders from the remainder of their forces.

c. Inadequate Structures for Defense. Extensive areas of lightly built or flammable structures offer little protection to the defender. Built-up areas near flammable or hazardous industrial areas, such as refineries or chemical plants, also may not lend themselves to successful defense.

d. Adjacent Dominating Terrain. If the built-up area is small and dominated by close, prominent terrain, the commander may choose to defend on that terrain rather than in the built-up area.

e. Better Fields of Fire Elsewhere. The commander may choose to base all or part of his defense on the better fields of fire that exist outside built-up areas. An example would be an armor-heavy force defending in sectors from multiple, small, built-up areas surrounded by open or farm-type areas.

f. Open City. See chapter 2.

Section I

Defensive Planning

3101. Commander's Estimate. As in any other defensive operation, planning begins with the commander's estimate of the situation. The complxities involved with the defense of a built-up area require detailed and centralized planning. This section details the general procedures and principles for planning and organizing the defense of an urban area.

3102. METT-T

a. Mission. Mission analysis is the first step in conducting the estimate of the situation. When conducting mission analysis commanders and their staffs should analyze their assigned tasks and commander's intent to determine the scope of the defense. For example:

- Do I need to control every building or can I control the frontage from key buildings?

- Are certain parts of the urban area more significant? (i.e. main transportation centers, communication nodes government buildings)

b. Enemy. The commander must focus on the enemy and build his defense to concentrate his strength against enemy weakness. The IPB process is used to analyze the terrain and enemy capabilities. (See FM 34-130.) For example:

- If the attacker relies heavily on aviation for mobility, the defender may defend potential LZs and emplace manpad teams to counter the air threat

- If the enemy relies heavily on mechanized movement, the defender may employ additional countermobility assets.

c. Terrain and Weather

(1) Terrain. Analysis of all manmade and natural terrain features is critical for planning a defense in built-up terrain. A defending commander has the advantage of occupying the terrain on which the fight will occur, and therefore usually knows it better than the attacker.

The defender must make best use of the battlespace that he controls: ground level (streets and parks), above ground (buildings), and below ground (subways and sewers). Defenders must be aware that built-up areas are normally interdependent. A reservoir supplying water to a city may be many miles away, electricity can come from remote power stations, and the control points for these services and others, such as communications lines, may be vulnerable.

.3

(a) K: Key Terrain. Primary examples of key terrain in the urban environment are ports, airfields, power grids, communication nodes, bridges, government building complexes, or parks. We identify key terrain so that we can better select our defensive positions and also better determine the enemy's objectives.

1 **Villages.** Villages often dominate high-speed avenues of approach. If the buildings are constructed of stone, brick, or concrete, providing good protection against both direct and indirect fires, a formidable defense can be constructed

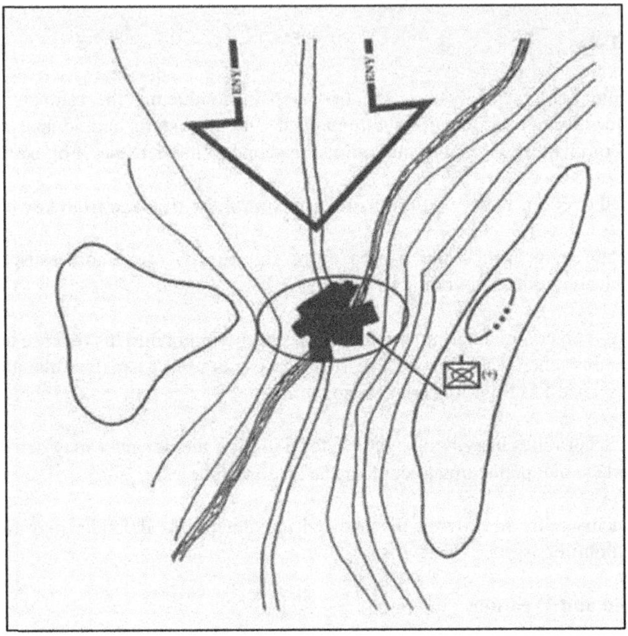

(Figure 3-1). Company-sized battle positions can be established in these small villages to block approaches into the larger urban area.

Figure 3-1. Village Defense (Chokepoint)

2 **Strip Areas.** Strip areas are built along roads or down valleys between towns and villages. They afford the defender the same advantages as villages. If effective fields of fire are available, a unit can occupy a few key positions within the strip to create an effective defense. (Figures 3-2A and 3-2B).

3 **Towns and Cities.** When a small town or city is positioned on a key enemy avenue of approach, a force can defend from the urban area while other forces

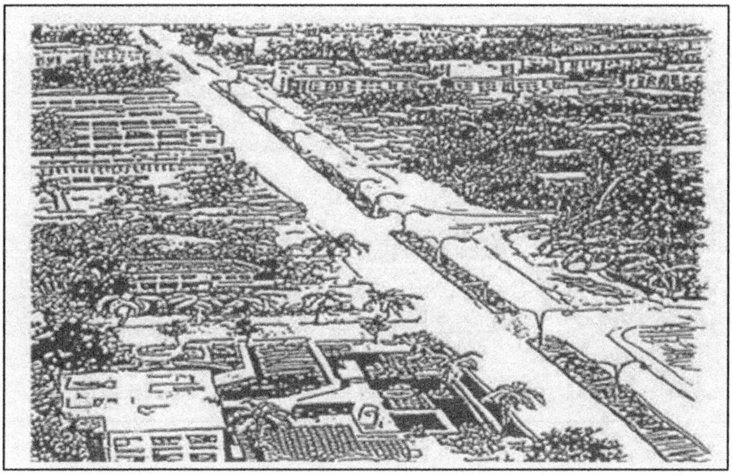

deny the enemy the ability to bypass. The town or city can be made into a strongpoint defense. (Figures 3-3A and 3-3B on page 3-6).

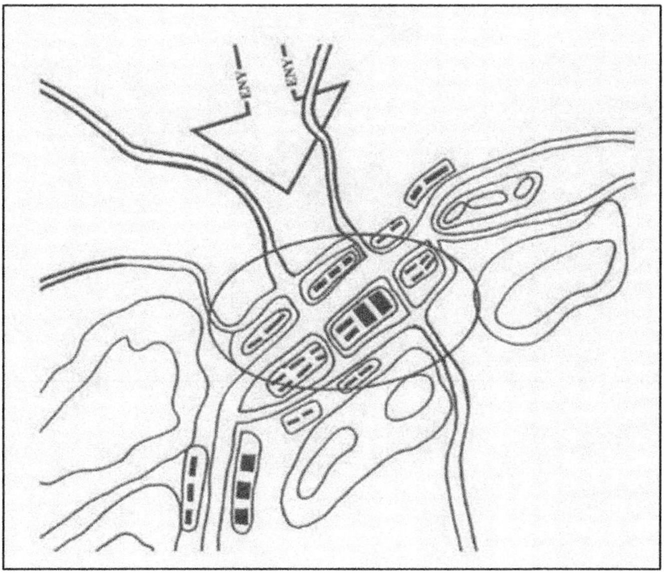

Figure 3-2A. Strip Area

.3

Figure 3-2B. Strip Area Defense

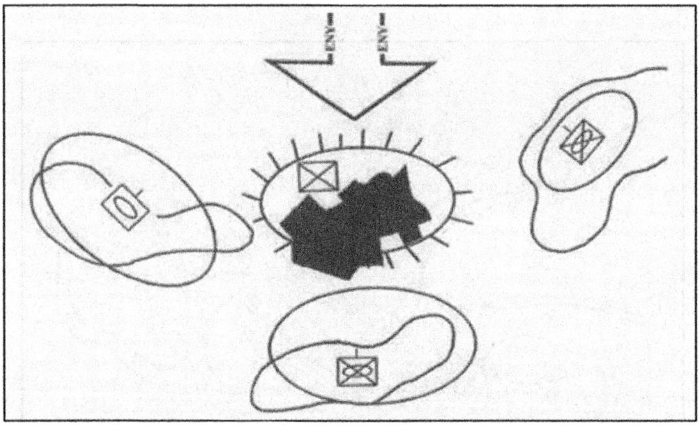

Figure 3-3A. Towns and Cities

Figure 3-3B. Defense of a Town or City

4 **Large Built-Up Areas.** Large buildings that are close together may require more forces and smaller defensive sectors than natural open terrain. The density of buildings, rubble, and street patterns will dictate the frontage of the unit.

(b) O: Observation and Fields of Fire. Attackers must generally advance by crossing streets and open areas between buildings where they are exposed to fires from concealed positions.

1 **Weapons and Range.** The defender must position weapons to obtain maximum effect and mutual support. FACs and FOs should be placed well above street level for increased observation. Fires and final protective fires should be preregistered on the most likely approaches.

2 **Limited Visibility.** The commander can expect the attacker to use periods of limited visibility to sustain or gain momentum. (See Appendix H for more detailed discussion.) The commander considers the following during periods of limited visibility:

- Unoccupied areas that can be observed and covered by fire during daylight may have to be occupied or patrolled at night.

- Remote sensors should be emplaced in dead space.

- Mines, noise-making devices, tanglefoot, and OPs should be positioned on avenues of approach to provide early warning.

- Artificial illumination to include street lamps, stadium lights, and flares should be integrated into the overall defense.

- When defenses are probed, indirect-fire weapons, grenade launchers, and hand grenades can be used to avoid disclosure of defensive position locations.

(c) C: Cover and Concealment. Although walls, floors, and ceilings can be used as protective cover, Marines should always improve these positions by using sandbags, rubble, etc. A defender can reduce his exposure by establishing routes between positions using:

- prepared breaches through buildings

.3

- reconnoitered and marked underground systems
- trenches and sewage systems
- concealment offered by smoke and darkness to cross open areas.

(d) O: Obstacles. A city is an obstacle because it canalizes and impedes an attack. Avenues of approach should be blocked by obstacles and covered by fire. (See Paragraphs 3404.d., 3404.f., 3405.e., and Appendix F.)

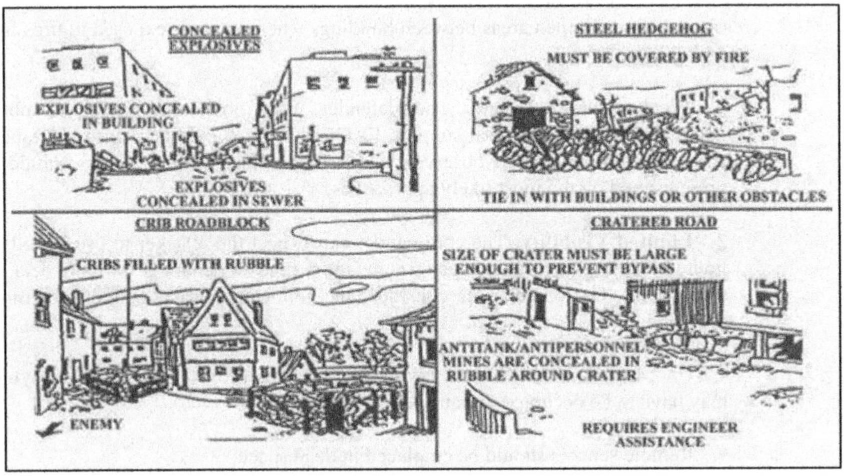

(e) A: Avenues of Approach. Avenues of approach (surface and subsurface) should be denied to the enemy in keeping with the overall defensive plan (Figure 3-4).

(2) Weather. The effects of weather as discussed in Chapter 2, Section II, pertain equally to defensive operations.

D. Troops and Support Available. The defender of a built-up area has a terrain advantage and can resist the attacker with much smaller forces. The defender stands a greater chance of success when he fights with combined arms and task organizes his forces. The discussion of troops and support available in Chapter 2, Section II, also pertains to defensive operations.

Figure 3-4. Denial of Avenues of Approach

e. Time Available. The commander establishes priorities of work, which depend on the time available. Many tasks can be accomplished at the same time, but priorities for work are accomplished according to the commander's order. An example priority-of-work sequence follows:

- Establish security and communications
- Assign sectors of responsibility, final protective fires
- Clear fields of fire
- Select and prepare initial fighting positions
- Establish and mark routes between positions (including alternate and supplementary routes)
- Emplace obstacles and mines
- Improve fighting positions.

3103. Organizing for a Defense

a. Types of Defense. The defense of a built-up area should be organized around key terrain features that preserve the integrity of the defense and provide the defender ease of movement. There are two types of defense: position and mobile. (See MCWP 3-1, *Ground Combat Operations* [under development] for details on defensive operations.) Most defenses will include a combination of position and mobile defense. The type of defense chosen is predicated on commander's intent and METT-T.

(1) Position Defense. The type of defense in which the bulk of the defending force is disposed in selected tactical localities where the decisive battle is to be fought. Principal reliance is placed on the ability of the forces in the defended localities to maintain their positions and to control the terrain between them. The reserve is used to add depth, to block, or restore the battle position by counterattack. (Joint Pub 1-02.) The position defense is usually characterized by defending key terrain. In the urban environment, this equates to the physical occupation of key public buildings, avenues of approach, transportation centers, industrial parks, etc. This defense focuses combat power to repel the attacker's advance and/or penetration while retaining the terrain. This type of defense is common for battalion-sized units and below.

The commander conducting a position defense organizes forces utilizes the same fundamentals that apply to any defense. Forces can be assigned sectors to defend and/or battle positions or strongpoints based on the commander's analysis of METT-T. Reserves are identified and located to respond quickly to exploit success.

.3

(2) Mobile Defense. Defense of an area or position in which maneuver is used with organization of fire and utilization of terrain to seize the initiative from the enemy. (Joint Pub 1-02.) The mobile defense is characterized by its focus on the enemy, not on terrain. Terrain is still important. However, this defense is organized with a mobility capability to exploit enemy weakness. In order to conduct the mobile defense, the defender must have enough forces to defend the built-up area, and to rapidly conduct an enveloping or flanking attack.

As in the position defense, the sector, battle position, strongpoint, or any combinations thereof are also used in shaping the battlespace by canalizing the enemy into a designated location where he can be destroyed.

b. Defensive Options

(1) Defense Outside the City. The commander may defend from outside the city. This option is often selected when the terrain surrounding an urban area offers an advantage to a defender and when his defending force is of sufficient strength (See Cherbourg paragraph in Chapter 1). The presence of a large, friendly, civilian population, or buildings of historic significance or specific cultural value, would also contribute to deciding on an outside defense. A situation in which the built-up area has a preponderance of building structures that offer little protection to defenders or terrain that favors the attacker.

A commander organizes his defenses around strongpoints selected on the outer edges of the urban area when these may be suitable structures for siting weapons to gain maximum firepower range. The organization of an outer-city defense is also heavily influenced by the number and type of weapons with which the hostile force is armed. A defense against a predominantly infantry threat would differ from that planned to defeat an armored threat. The decision to defend or concentrate assets solely on the outer edge of a city are weighed against the enemy's ability to mass fires, infiltrate, or bypass strongpoints.

(2) Defense Inside the City. The commander may decide that his best defense is to defend within the city. This option is usually chosen when hostile forces possess strong, accurate, long-range fires and the defender wishes to minimize their effectiveness (See Quang Tri City example in Chapter 1). The strength of the defending force relative to the attacker may also lead to this option as an economy-of-force measure.

Structures on the outer edge of an urban area that obstruct fields of fire or that are likely to be used by an attacker should be prepared for demolition or demolished ahead of time. All defensive positions should be supplemented or covered by direct-fire weapons. Indirect fires should be planned for most likely avenues of approach. Additionally, close air support should be planned and integrated with the defense.

Security forces are placed outside the defended area to gain early enemy contact, inflict maximum casualties, and attempt to deceive the enemy as to the location of the friendly positions.

(3) Defense of a Key Sector. This option is usually employed when only a portion of an urban area be held or when strength is insufficient to defend an entire area (See the Japanese defense of Manila in Chapter 1). This option is often employed in situations where the mission requires clearing and holding a key sector, or sectors, in order to facilitate other operations.

(4) Entrapment and Ambush. This option is preferred when the defender lacks requisite strength to man organized, in-depth, defensive positions (See the highly successful Egyptian defense of Suez City example in Chapter 1). It requires the preparation of a series of blocking and/or ambush positions along major avenues of approach through an urban area. The defending force uses mines and obstacles to block alternate or possible detour routes in order to canalize the attack force into prepared ambush sites. The defender employs security forces forward of the main battle area and establishes inner or outer defensive strongpoints with extensive obstacles designed to assist in canalize the enemy into kill zones.

The entrapment and ambush option requires detailed planning and rehearsal by all echelons. Alternate plans should be prepared in the event the ambush is only partially successful or if the enemy fails to enter prepared ambush routes. Alternate ambush positions or counterattacks are planned.

Ambush planning in an urban environment does not differ significantly from planning in other environments. However, the urban environment offers an abundance of kill zones that afford the enemy little or no opportunity to escape. Ambush positions are established in selected buildings in the kill zone which offer optimum fields of fire for all weapons. Indirect fires and close air support should be planned to obtain maximum effect in the kill zone. The kill zone should also be covered with targets for indirect fire and CAS. Entrances into buildings not used as firing points should be blocked to entrap hostile forces in the kill zones.

(5) Defense in Depth. Defense in depth should be incorporated into any defensive plan. Units should be given enough terrain to allow for primary, supplementary, and alternate fighting positions as well as successive battle positions or strongpoints. This allows enough room for multiple ambush/entrapment kill zones and the positioning of a reserve.

c. Defensive Organization. The defensive battlespace is divided into three areas: the *security area*, *main battle area*, and *rear area*. Depending on the mission of the unit, forces defending a built-up area may have missions in any one or more of these areas. (Figure 3-5 on page 3-12).

.3

(1) Security Area. The defensive battle begins with a combined-arms force conducting security operations well forward of the main battle area. Security area missions can include screen , guard, and cover. Based on their assigned mission, security forces use all

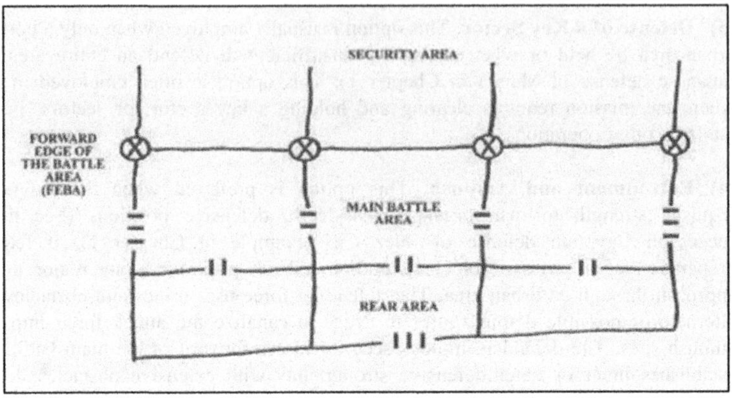

available combat power to gain time, destroy the enemy, or slow his momentum. The initial battle is fought using aviation assets and indirect fire weapons.

(a) Value of Security Area Operations. Security area operations complement to the defense by:

- Alerting the defense to the strength, location, and general direction of the enemy's main and supporting attacks
- Delaying enemy first-echelon detachments
- Initiating early engagement of enemy forces
- Deceiving the enemy as to the true location of the defender's main forces and main effort.

Figure 3-5. Organization of the Battlefield

(b) Withdrawal of Security Area Forces. The urban environment may complicate battle handover from the security force to the main battle area force. However, this transition must be accomplished smoothly to prevent the enemy from gaining momentum. The withdrawal of the security area forces must not result in an easing of pressure on the enemy.

(2) Main Battle Area. The decisive battle is fought in the main battle area. However, the

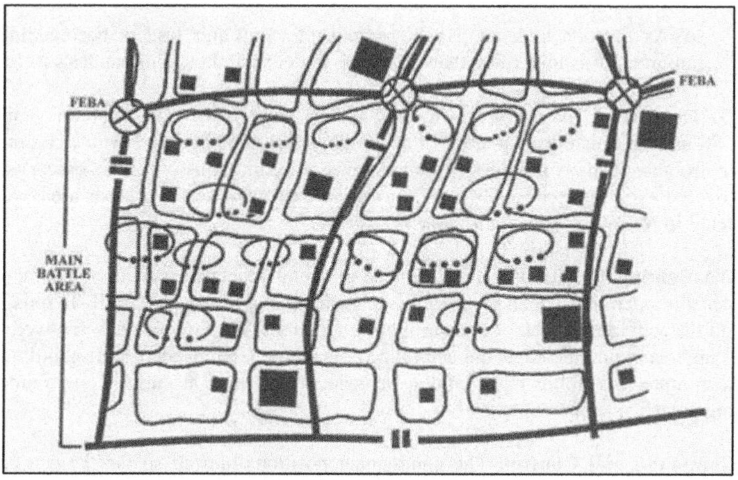

commander may deploy units on the forward edges of the city or in battle positions in depth. The defensive scheme should include forces to defend along the flanks. The commander normally employs security forces to the front to provide early warning and to deny the enemy intelligence on the unit's defensive dispositions (counterreconnaissance) (Figure 3-6).

(a) The size and location of battle positions depend on METT-T and the type of defense prepared. A sophisticated defense will often include a combination of both position and mobile defense but will always be planned in depth.

(b) A unit assigned battle positions on the forward edge of a city or town should:

- Provide early warning of the enemy's advance
- Engage the enemy at long range
- Deceive the enemy as to the true location of the defense.

Figure 3-6. Main Battle Area

(c) The defender employs all available fires to destroy and suppress the direct-fire weapons that support the ground attack. Depending on the concept of operations, weapons may engage targets at their maximum effective range or engage simultaneously for an ambush.

(d) As the attack develops, the actions of small-unit leaders become increasingly important. It is imperitive that all leaders understand their commander's intent.

(3) Rear Area. The rear area is located behind the main battle area and it contains the CSS and administrative support. Protection of these units is vital. Fortunately, the urban environment usually provides sufficient cover and concealment for CSS assets as well as covered and protected routes to the forward areas. Rear area operations are covered in detail in MCWP 3-41.1, *Rear Area Operations*.

3104. Warfighting Functions. The defensive planning sequence remains the same as in any other defensive operation. Defense planning considerations, based on METT-T, must take into account all the activities in each of the warfighting functions. The complexities involved in defense of a built-up area require detailed and centralized planning. Commanders and subordinate leaders must incorporate the following planning considerations for an urban environment when conducting a defensive operation.

a. Command and Control. The commander positions himself so that he can control the action. In an urban environment, this can be more difficult because of the close nature of fighting, prolific obstacles, poor visibility, and difficulty in communication.

Graphic control measures oriented on prominent terrain features are also used in planning and conducting combat in built-up areas. Prominent streets are ideal for use as phase lines and boundries. These and other control measures assist the commander in controlling manuever and fires throughout the battlespace.

COCs should be located in secure locations. (These can be inside or underground when possible.) (See Chapter 2, Section IV, Paragraph 2402.)

Commanders consider the effects of built-up areas on communications when they allocate time to establish communications. LOS limitations affect both visual and radio communications. Wire laid at street level is easily damaged by rubble and vehicle traffic. Also, the noise of urban combat is much louder than in other areas, making sound signals difficult to hear. Therefore, the time needed to establish an effective communications system may be greater than in more conventional terrain.

Wire is the primary means of communication for controlling the defense of a city and for enforcing security. However, wire can be compromised if interdicted by the enemy. Radio communication in built-up areas is normally degraded by structures and a high concentration of electrical power lines. Nonetheless, radio is an alternate means of communication. Messengers can be used as another means. Visual signals may also be used but are often not effective because of the screening effects of buildings, walls, and so forth. Signals must be planned, widely disseminated, and understood by all assigned and attached units. Commanders consider the following techniques when planning for communications:

(1) If possible, lay wire through buildings for maximum protection.

(2) Use existing telephone systems. (Remember, telephones are not secure.)

(3) Emplace radios and retransmission sites in buildings on top floors or atop buildings that are higher than surrounding structures. Ensure that antennas are placed on the side of buildings where the radio signal can be directed at friendly units. This will enhance LOS radio communications.

(4) Use messengers at all levels because they are the most reliable and secure means of communications.

b. Intelligence. The discussion in Chapter 2 pertains equally to a defensive operation. Intelligence gathering for defensive operations is not limited to studying only the enemy. Commanders must emphasize obtaining and using all intelligence. Commanders must emphasize the utility of the urban infrastructure to enhance his intelligence gathering and dissemination effort. The defender usually has the advantage of being familiar with the intricacies of the urban terrain that he is defending. For example, the:

- The sanitation or public works department will have knowledge of the subterranean sewer lines
- Civil engineers will have extensive knowledge of water and power distribution nodes
- Law enforcement representatives may have access to lists of people sympathetic to the attacker
- Local journalists may agree to use local sources to assist in intelligence gathering effort.

c. Manuever

(1) Positioning of Forces. Finding fighting positions in urban areas that provide both good fields of fire and cover is often difficult. The forward edges of a town usually offer the best fields of fire but can be easily targeted by enemy observation and supporting fire. These areas often contain residential buildings constructed of light material. Factories, civic buildings, and other heavy structures that provide adequate cover and are more suitable for a defense are more likely to be found deeper within the town, but they have limited fields of fire.

Because the forward edge of a town is the obvious position for the defender, it should generally be avoided. However, the defender may choose to use the edge of the town if the terrain limits the enemy's ability for engagement or if strongly constructed buildings that give defenders protection are available. The mission of such a force is to provide early warning of the enemy's advance, to engage the enemy at long range, and to deceive the enemy as to the true location of the defense. This force should withdraw in time to avoid decisive engagement. If there is limited observation from the forward edge, a force should be positioned on more favorable terrain to gain better observation and to engage the enemy at long range.

Reconnaissance by all defending elements should help select routes for use by defenders moving between positions. Movement is crucial in fighting in built-up areas. Early selection and marking of routes adds to the defender's advantages.

The commander covers probable landing and drop zones such as parks, stadiums, or large rooftops and heliports with obstacles or fire to prevent air assault. (See Appendix F.) In a large built-up area, a unit is given a sector to defend and normally establishes a series of defensive positions. Although mutual support between positions should be maintained, built-up terrain often allows the enemy to infiltrate between positions. Therefore, the defender must identify the following:

- Positions that allow suppressive fires on infiltrating routes
- Covered and concealed routes, such as subways and sewers, for friendly forces to use to move between positions
- Structures that dominate large areas
- Areas such as parks, boulevards, rivers, highways, and railroads where antiarmor weapons have longer fields of fire
- Firing positions for mortars
- Command locations that offer cover, concealment, and ease of command and control
- Protected storage areas for supplies.

Buildings that enhance the general plan of defense are selected. Mutual support between these positions is vital to prevent the attacker from maneuvering and outflanking the defensive position. Buildings chosen for occupation as defensive positions should:

- Offer good protection
- Have strong floors to keep the structure from collapsing under the weight of debris

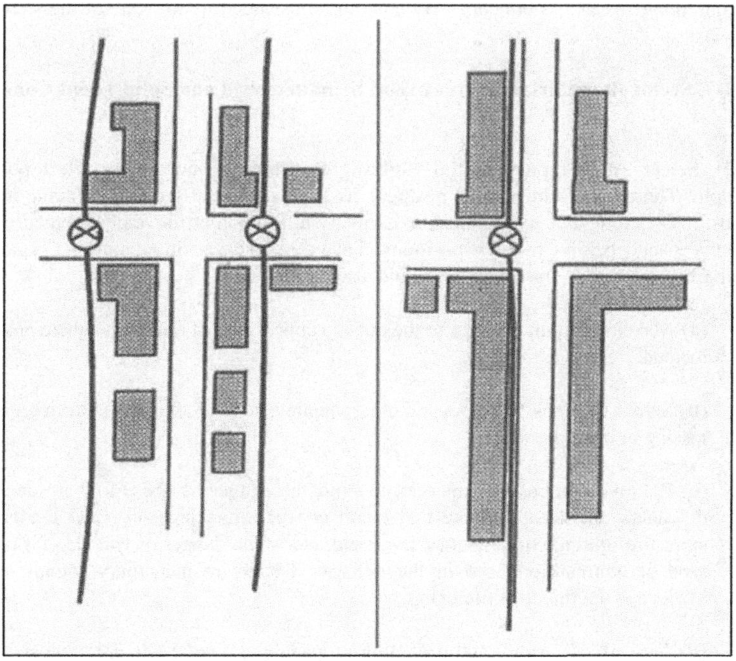

- Have thick walls
- Be constructed of nonflammable materials
- Be strategically located (corner buildings and prominent structures)
- Be adjacent to streets, alleys, vacant lots, and park sites (these buildings usually provide better fields of fire and are more easily tied in with other buildings)
- Provide for 360-degree protection and fields of fire
- Be stocked with adequate supplies as positions may become isolated and cut off for extended periods.

(2) Assign Sectors of Responsibility. Boundaries define sectors of responsibility. Sectors include areas where units may fire and maneuver without interference or coordination with other units. Responsibility for primary avenues of approach should never be split among units. In areas of semi-detached construction, where observation and movement are less restricted, boundaries should be established parallel to alleys or streets so that both sides of a street will be in a single sector. Where buildings present a solid

.3

front along streets, a boundary may have to be extended to one side of the street (Figure 3-7).

Figure 3-7. Sector Boundaries in Areas With Semi-detached and Solid-Front Construction

(3) Select and Prepare Initial Fighting Positions. Commanders select positions in depth. The unit should prepare positions as soon as they arrive and continue improving positions as long as they are occupied. Enemy infiltration or movement sometimes occurs between and behind friendly positions. Therefore, each position must be organized for all-around defense. The defender should also:

(a) Make minimum changes to the outside appearance of buildings where positions are located.

(b) Screen or block windows and other openings to keep the enemy from seeing in and tossing in hand grenades.

(c) Remove combustible material to limit the danger of fire. Fires are dangerous to defenders and create smoke that could conceal attacking forces. Defenders should store firefighting equipment (water, sand, etc.). The danger of fire also influences the type of ammunition used in the defense. Tracers or incendiary rounds should be avoided if the threat of fire exists.

(d) Turn off electricity and gas. Both propane and natural gas are explosive. Natural gas is also poisonous and is not filtered out by a protective mask. Propane gas, although not poisonous, is heavier than air. If it leaks into an enclosed area, it displaces the oxygen and can cause suffocation. Gas mains and electricity should be shut off at the main distribution nodes.

(e) Battle positions should be located so as not to establish a pattern. The unit should avoid obvious firing locations like church steeples.

(f) The unit should rehearse movements between positions.

(g) Camouflage positions.

(h) Reinforce positions with available materials, such as beds, furniture, etc.

(i) Block stairwells and doors with wire or other material to prevent or delay enemy movement. Create holes between floors and rooms to allow for covered movement within a building.

(j) Prepare range cards, fire plans, and sector sketches.

(k) Emplace machine guns in basements windows, where the guns can provide grazing fires. When basements are not used, access to them should be sealed to prevent enemy entry.

(l) Stockpile V (ammunition) and class VIII (medical supplies) items.

(4) Clear Fields of Fire. Commanders may need to rubble certain buildings and structures to provide greater protection and fields of fire. If the ceiling of a lower story room can support the weight of the rubble, collapsing the top floor of a building before the battle starts may afford better protection from indirect fires. Rubbling an entire building can increase the fields of fire and create an obstacle to enemy movement. Rubbling buildings too soon (or rubbling too many) may give away defensive locations and may destroy cover from direct fire. Rubbled buildings may also interfere with planned routes of withdrawal or counterattack.

(5) Improve Fighting Positions. When time permits, all positions, including supplementary and alternate positions, should be reinforced with sandbags and provided

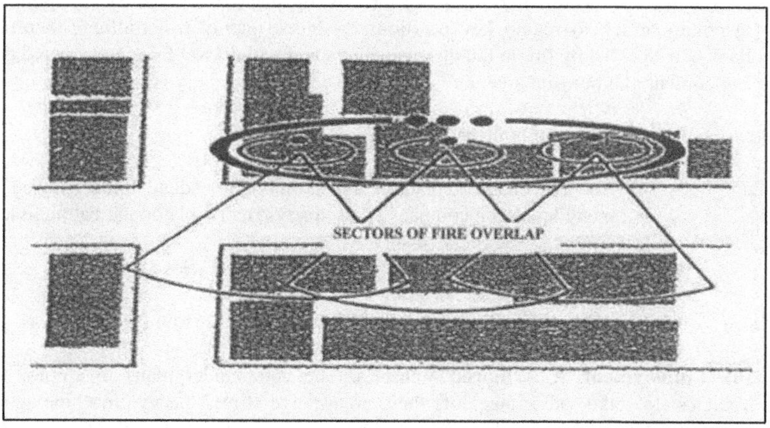

with overhead cover. Obtain support from attached engineers. (See Appendix D).

(6) Employment of Infantry

(a) The infantry battalion is normally assigned a sector to defend. However, depending on METT-T, it may be assigned to defend from a battle position or strongpoint. The battalion will usually assign its companies to either a battle position, a strongpoint, or a sector. These battle positions, strongpoints, or sectors are placed along avenues of approach to block or restrict the enemy's movement.

Depending on the type of built-up area, a company may be employed on the forward edge of the flanks of the area. This forces the enemy to deploy early without decisive engagement because it deceives the enemy as to the true location of the main defense. Other companies may then be employed in a series of battle positions and/or strongpoints in the center of the city or town. In all cases, mutual support between positions is vital. Companies and platoons should have designated alternate and supplementary positions.

(b) Once the rifle platoon commander has received his battle position or sector, he then selects the positions for his squads and crew-served weapons. Squad positions within the built-up area may be separated by rooms within buildings or be deployed in different buildings. Squad positions must be mutually supporting and allow for overlapping sectors of fire (Figure 3-8).

Figure 3-8. Sectors of Fire

(7) Employment of the Reserve. The commander's defensive plan always considers the employment of a reserve. The reserve force is organized and prepared to exploit success, to counterattack to regain key positions, to block enemy penetrations, to protect the flanks, or to assist by fire in the disengagement and withdrawal from endangered positions. For combat in a built-up area, a reserve force:

- Must be as mobile as possible
- May be a mechanized platoon, a platoon (-), or a squad at mechanized company level or a company, a company (-), or a platoon at battalion level
- May be supported by tanks or other armored vehicles
- Should be centrally located to facilitate manuever to threatened areas.

(8) Employment of Armored Vehicles. The commander plans to employ armored vehicles to take advantage of their long-range fires, heavy machine gun fires, self-generating smoke, and mobility. However, built-up areas restrict the mobility of armored vehicles and make them vulnerable to the antiarmor weapons of the enemy. Both the LAV and AAV are lightly armored and can be penetrated by heavy machine gun and antiarmor fires. Armored vehicles provide the commander with a mobile force to respond quickly to enemy threats on different avenues of approach.

(a) When armored vehicles are employed in the defense of a city, infantry should be positioned to provide them with security against close antitank fires and to detect targets for the armored vehicles. Armored vehicles should be assigned *primary*, *alternate*, and *supplementary* positions, as well as *primary* and *alternate* sectors.

(b) Armored vehicles should be located on likely avenues of approach to take advantage of their long-range fires. They may be:

- Positioned on the edge of the city in mutually supporting positions
- Positioned on key terrain on the flanks of towns and villages
- Used to cover barricades and obstacles by fire
- Placed with the reserve.

(c) Armored vehicles are normally employed as a platoon. However, sections of armored vehicles may be employed with rifle platoons or squads. This provides the armored vehicles with the close security of the infantry.

(9) Employment of Helicopterborne Forces

(a) Helicopterborne Task Force in the Defense. A helicopterborne force can defend against an infantry-heavy threat by utilizing its mobility to achieve a maneuver advantage. This force can be key in the execution of a mobile defense in both the security area and main battle area.

(b) Security Area. The helicopterborne force may be able to conduct security force operations for a larger force. Helicopters can position combat power and combat support quickly during rapidly changing situations.

(c) Main Battle Area. The mobility advantage that the helicopterborne force has over enemy infantry-heavy units may allow it to defend in greater depth in a large city. The helicopterborne force focuses on the destruction of advancing enemy forces and fights a series of battles in depth, it can attack the enemy from the front, flanks, and rear. Battle positions are selected and prepared throughout the main battle area along likely avenues of approach. Primary and alternate LZs and pick-up zones should be selected.

d. Fires. Fire support in the defense is as important as it is in the offense and has many of the same characteristics. The proximity of buildings to targets, minimum range restrictions, and repositioning requirements are all factors in the utilization of fire support in the defense. Additionally indirect fires are planned on top of and immediately around defensive positions for close support. Some of the indirect fire support considerations for the defense include:

(1) Historically, artillery fire has been used in some unique and innovative ways in an urban environment. Artillery fire support may be used in the direct- or indirect-fire mode. Artillery fire should be used to:

- Suppress and blind enemy overwatch elements

.3

- Disrupt or destroy an assault
- Provide counterfire
- Support counterattacks
- Provide direct fire when necessary
- Provide illumination in hours of darkness.

(2) Mortars at the infantry battalion and rifle company level are employed to maximize the effect of their high-angle fires. Mortars are one of the most effective indirect fire weapons in the urban environment. They should be used to engage:

- Enemy overwatch positions
- Enemy infantry before they seize a foothold
- Targets on rooftops
- Enemy reinforcements within range.

(3) Final protective fires are planned to stop dismounted assaults in front of the defensive positions. Fires within the city are planned along likely routes of advance to delay, disrupt, and destroy the enemy as he attempts to deepen a penetration.

(4) Commanders establish priorities of fire on the basis of enemy avenues of approach and threat systems that present the greatest danger to the defense. For example, during the attacker's initial advance, tanks, IFVs, and overwatching elements are the greatest threat to the defense. ATGMs should concentrate on destroying tanks first, then IFVs. Indirect fires should suppress and destroy enemy ATGMs and overwatch positions and/or elements. If enemy formations secure a foothold, priority is shifted to the destruction of enemy forces within the penetration.

(5) As the enemy attack progresses into the city, fires are increased to separate enemy infantry from its supporting armored vehicles. Friendly artillery concentrates on attacking infantry, conducting counterfire missions, and destroying reinforcements.

(6) Ample consideration must also be given to the rubbling effects of supporting arms. These effects often create massive rubbling of structures and buildings, making obstacles to both friendly and enemy movement.

(7) Mortars may be initially positioned forward in support of security forces. After their withdrawal, mortars are repositioned where they can support the in the main battle area.

(8) The commander assigns final protective fires and machine gun final protective lines (FPLs). Machine gun positions provide interlocking grazing fire and mutual support between adjacent units. FPLs are fired on planned signals. Proposed FPLs must be "walked out" to determine the extent of grazing fire available and to locate dead space. Dead space can be covered by:

- Rifle and light machine gun fire
- Grenade launchers
- Mines and boobytraps
- Indirect fires.

(9) Long-range antiarmor weapons are employed in the forward areas of the main battle area and the security area. Antiarmor weapons must be carefully positioned to prevent terrain and obstacles from reducing their range. Dismounted and manpacked antiarmor weapons should be positioned in buildings and along routes where engagement distances are reduced but where surprise and ambush are ideal.

e. Logistics. Just as in the offense, the defense utilizes a greater amount of class V and class VIII supplies in the urban environment. The defender needs to plan for multiple routes and means of supply. These can include:

- Using the city's existing transportation infrastructure to move supplies
- Prepositioning supplies in the defense in depth
- Using helicopter assets for rapid resupply
- Establishing and protecting supply routes
- Plan for utilization of existing repair and maintenance facilities
- Identify potable water sources

f. Force Protection

(1) Employment of Engineers. Engineers are employed at the battalion level or attached to companies and integrated down to squad level. Commanders must consider and assign the priority of work for engineer tasks that enhance survivability, mobility, and countermobility. Tasks that engineers can accomplish in the defense of a built-up area include:

- Constructing obstacles and rubbling for countermobility

.3

- Clearing fields of fire
- Laying mines
- Preparing routes to the rear
- Preparing fighting positions.

(a) Employment of Obstacles. Obstacles are used to shape the battlespace. (See FM 5-102, *Countermobility*, and FM 20-32, *Mine/Countermine Operations*, for details.) The four types of obstacles are:

- **Disrupting.** Disrupting obstacles are used to disrupt the enemy's timetable and force him into conducting a time-costly bypass or breach, or cause his march columns to become staggered or "telescoped."

- **Turning.** Turning obstacles are used to canalize the enemy into engagement areas or kill zones where combined-arms fires can be concentrated and flanking shots are increased.

- **Fixing.** Fixing obstacles increase the time the enemy can be fired on in an engagement or target area and to cause increased confusion.

- **Blocking.** Blocking obstacles prevent or delay the enemy from moving past a certain point.

(b) Obstacle Belts. In defensive obstacle planning, obstacles may be emplaced in a series of belts. This enhances the overall effectiveness of the obstacle plan with successive belts designed for specific purposes. The following example lays out a three-belt obstacle plan.

1 **First Obstacle Belt.** The first obstacle belt is enplaced at the nearest buildings across from and parallel to the main defensive position. This belt consists of wire and improvised barriers (including inside buildings, in subterranean avenues of approach, and outside in open areas), danger areas, and dead space. These barriers and obstacles should be heavily boobytrapped and covered by fires. This belt is designed to impede enemy movement, breaks up and disorganize attack formations, and inflicts casualties.

2 **Second Obstacle Belt.** The second obstacle belt is placed between the first belt and the main defensive position buildings, but out of hand grenade range. It is designed to impede movement, canalize the enemy into the best fields of fire, break up attack formations, and inflict casualties. This belt is not meant to stop the enemy permanently. It consists mainly of wire obstacles, improvised barriers, road

craters, and minefields. It should be heavily boobytrapped. Triple-strand concertina wire is placed along the machine gun FPL to delay the enemy in the kill zone.

3 **Third Obstacle Belt.** The third obstacle belt is the defensive position's denial belt. It consists of wire obstacles placed around and through the defensive buildings, close-in minefields, and subterranean accesses. It is designed to impede the enemy's ability to gain a foothold in the defensive area. It should be heavily boobytrapped, and integrated with mines.

(c) Emplace Obstacles and Mines. While principles for employing mines and obstacles do not change in the defense of a urban area, techniques for their employment change. For example, burying and concealing mines in streets is usually not done. To save time and resources in preparing the defense, commanders must emphasize using all available materials (automobiles, railcars, rubble) to create obstacles. Civilian construction equipment and materials should be utilized. Obstacles must be integrated into buildings and rubble areas to increase their effectiveness and to canalize the enemy. The family of scatterable mines (FASCAM) may be effective on the outskirts of a city or in parks and open areas. (See Appendix F for further discussion on obstacles, mines, and demolitions.)

(2) Air Defense. Air defense (Stinger) weapons are positioned on rooftops and parking garages in order to obtain LOS coverage for the units they are protecting. Stinger teams can be assigned the mission of protecting specific positions or placed in general support (GS) of the battalion.

(3) Fire Hazards. All cities are vulnerable to destruction by fire, especially those with many wooden structures. The defender's detailed knowledge of the terrain permits him to avoid areas that are likely to be fire hazards. The defender can deliberately set fires:

- To disrupt and disorganize the attackers
- To canalize the attackers into more favorable engagement areas
- To obscure the attackers' observation.

Section II

Infantry Battalion Defense of a Built-Up Area

This section addresses some of the missions a task-organized and/or reinforced infantry battalion could be assigned. These missions are normally conducted as part of a larger joint operation.

3201. Defense in Sector. An infantry battalion is normally given the mission of defending a sector in a city (Figure 3-9). The interior buildings in the battalion sector form battle positions for the battalion's fixed defense. The size of a sector varies, but may cover the outlying structures and can be used to provide early warning and to delay the enemy.

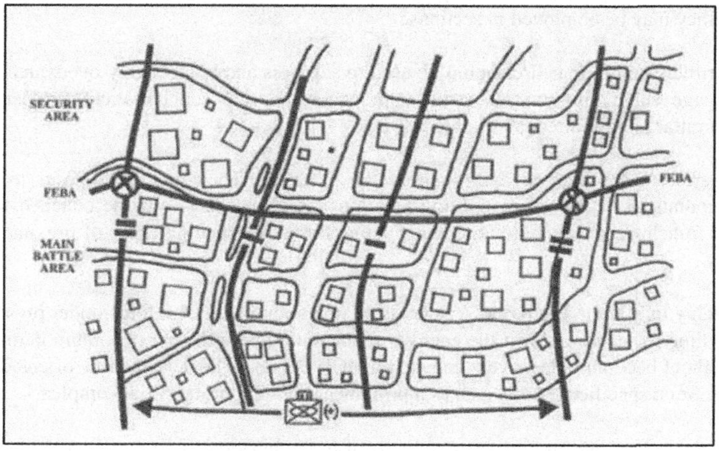

Figure 3-9. Defense of a Built-Up Sector

a. Defensive operations begin with the reconnoitering of the assigned sector to learn the terrain and to orient toward the enemy's expected advance. The main defensive positions should be formed on key terrain features, such as prominent buildings, road junctions, and positions which offer good fields of fire. Long-range antiarmor weapon systems should be deployed to engage the enemy as early as possible.

b. Depending on the situation, a reinforced rifle company may be used as a security force. The security force would operate forward of the FEBA and, depending on the mission, cause the enemy to deploy without without becoming decisively engaged. This can be done through maximum use of ambushes and obstacles while using covered and concealed routes. These routes are used for disengagement. Covered and concealed routes can be artificially made by making holes through buildings. The security force also attempts to detect the location of the enemy's main attack. Once the security force withdraws to the main battle area, it can be used as a reserve to reinforce or counterattack.

c. Based on METT-T considerations, the defense in sector (behind the FEBA in the main battle area) may consist of either company sectors, strongpoints, or battle positions. Strongpoints located on or covering decisive terrain are extremely effective in the defense.

d. Antiarmor vehicles and weapons should be used to engage enemy armored vehicles and/or cover obstacles with fire on likely avenues of approach.

e. Tanks are best used to engage enemy tanks, cover obstacles by fire, and engage in counterattacks. They should be employed in platoons where possible. However, in congested areas they may be employed in sections.

f. Artillery and mortar fire should be used to suppress and blind enemy overwatch elements, to engage enemy infantry on street approaches, provide counterbattery fire, and support counterattacks with direct and indirect fire.

g. Engineers are integrated throughout the defense. They provide support to forces in strongpoints to help prepare fighting positions. Engineers may also be attached to security forces to help in laying mines and constructing obstacles, clearing fields of fire, and preparing routes.

3202. Delay in a Built-Up Area. A delay is an operation in which a force under pressure trades space for time by slowing down the enemy's momentum and inflicting maximum damage on the enemy without becoming decisively engaged (MCWP 3-1). The delay can be oriented either on the enemy or on specified terrain such as a prominent building or industrial complex.

a. A delay in a built-up area may consist of a combination of ambushes and alternating or successive battle positions (Figures 3-10A and 3-10B on page 3-28).

(1) Ambushes are normally planned on observed obstacles. They require decentralized control and execution. Ambushes can be combined with limited-objective attacks on the enemy's flanks. They are most effective at the edge of open spaces, parks, wide streets, and so on. These attacks may be executed by dismounted infantry supported by tanks, LAVs/AAVs, or HMMWV-mounted machine guns and antitank vehicles.

(2) Battle positions should be placed where heavy weapons, antiarmor weapons, and machine guns will have the best fields of fire. Such locations are normally found at major street intersections, parks, and the edge of open residential areas. Battle positions should be carefully and deliberately prepared, reinforced by obstacles, and supported by indirect fire weapons. Battle positions are designed to inflict maximum casualties on the enemy and cause him to deploy for a deliberate attack.

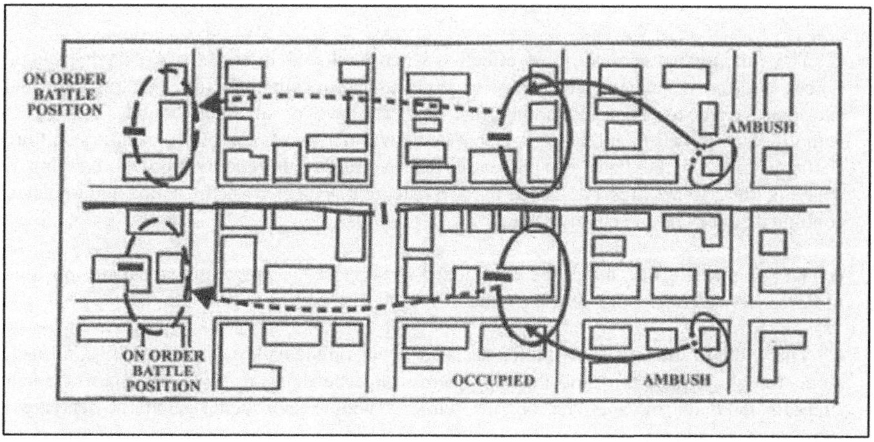

b. Tanks, LAVs/AAVs, and antiarmor weapons should prepare primary and alternate positions to improve their surivability and flexibility.

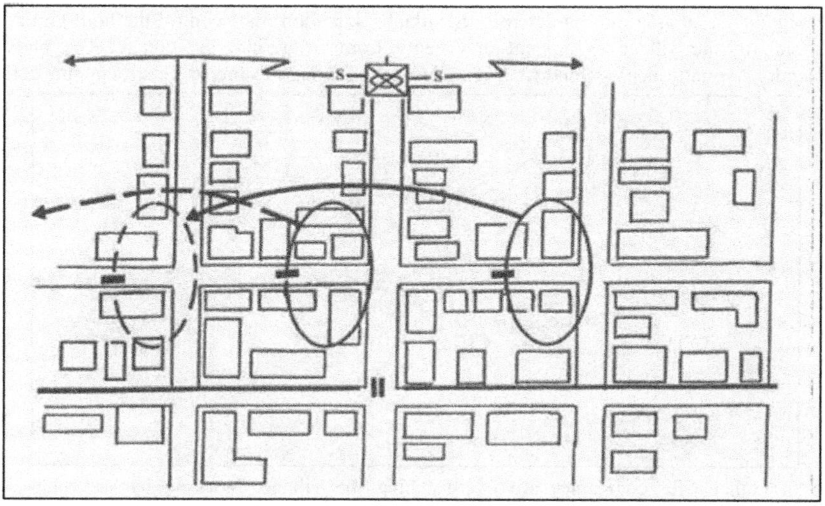

Figure 3-10A. Delay in a Built-Up Area (Successive Bounds)

Figure 3-10B. Delay in a Built-Up Area (Alternate Bounds)

c. The infantry battalion is most effective when deployed in successive delaying actions. Forces alternately engage the enemy to the maximum extent possible without becoming decisively engaged, then move on order in successive or alternate bounds. Smoke and demolitions are used to aid disengagement and cover the movement of the withdrawing force. Security units on the flank can be employed to prevent the enemy from outflanking the delaying unit. A small reserve can be used to react to unexpected enemy action and to conduct continued attacks on the enemy's flank.

d. The engineer effort should be centralized to support the preparation of battle positions, obstacle emplacement, and clearing routes.

e. The width of the infantry battalion sector depends on the nature of the buildings, obstacles along the street, and the time that the enemy must be delayed. Normally, narrow sectors supports delaying by alternate bounds while a wide sector better supports delaying by successive bounds.

3203. Defense of a Village. An infantry battalion assigned a defensive sector that includes a village may decide to incorporate the village as a strongpoint. This use is most common where the village stands astride a high-speed avenue of approach or lies between two difficult obstacles. To incorporate such an area into its defense, the infantry battalion must control the high ground on either side of the village to prevent the enemy from firing into the village. The battalion commander normally deploys one rifle company in the village to prepare a strongpoint defense

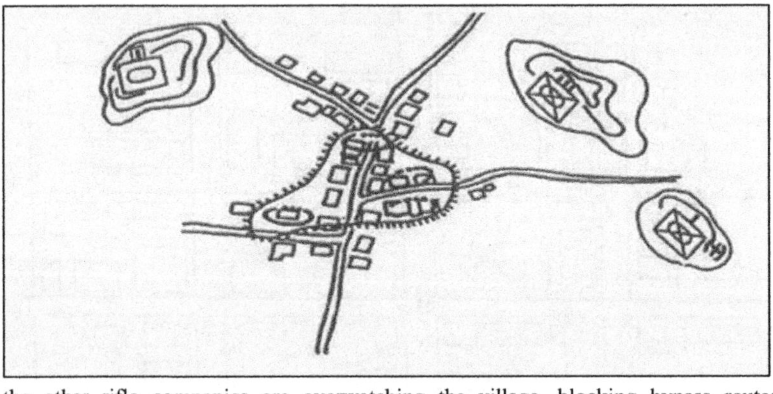

while the other rifle companies are overwatching the village, blocking bypass routes, and protecting the strongpoint's flanks. The strongpoint defense should provide a location where the enemy can be stopped and counterattacks can be launched (Figure 3-11).

Figure 3-11. Defense of a Village

a. Armored Vehicles and antiarmor weapons should be employed where the maneuver room is the greatest. This is usually on key terrain to the flanks of the village. As the security force withdraws and rifle companies become engaged, armored vehicles and weapons support by fire.

b. Although the infantry battalion's disposition should prevent large enemy forces from threatening the rear and flanks of the village, the danger of small-unit enemy infiltration requires the village be prepared for all-around defense.

c. Engineers should be assigned to the rifle company in the village strongpoint to provide continuous engineer support in preparation for the enemy's attack. The priority of engineer support should go to the strongpoint.

Section III

Rifle Company Operations

This section addresses some of the missions the infantry battalion could assign to the rifle company. For the purpose of this publication, the rifle company is a task-organized and/or reinforced unit. Assigned missions are normally part of a larger battalion operation.

3301. Defense of a Village. Once the commander has completed his reconnaissance of the village and the surrounding terrain, he develops his plan for the defense (Figure 3-12). The commander must decide whether to defend on the leading edge of the village or within the village.

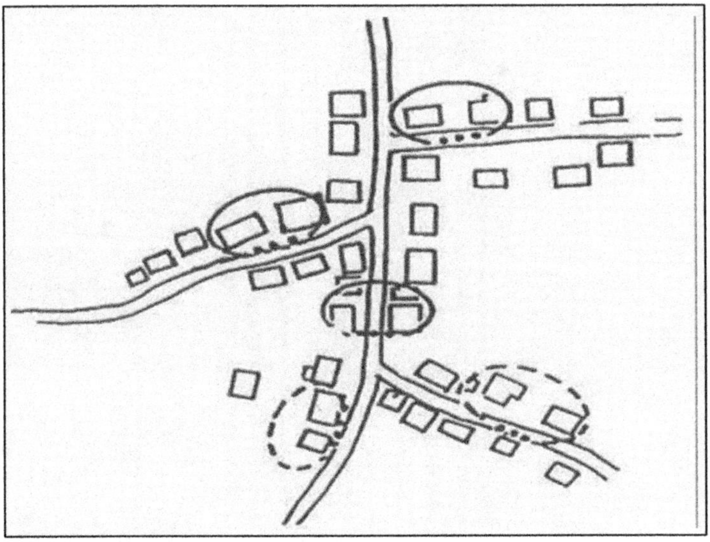

Figure 3-12. Defense of a Village

a. Several factors influence his decision. First, he must know the type of enemy that he will defend against. If the threat is mainly dismounted infantry, the greatest danger is in allowing them to gain a foothold in the town. If the threat is armor or motorized infantry, the greatest danger is that the company's defensive positions can be destroyed by direct fire. The commander must also consider the terrain forward and to the flanks of the village from which the enemy can direct fires against his positions.

b. The commander places platoons in a small group of buildings where each platoon leader establishes mutually supporting squad-sized positions. This positioning increases the area that the rifle platoon can control and hampers the enemy's ability to isolate or bypass a platoon. A rifle platoon can be responsible for the road through a village. The rest of the rifle company is positioned to provide all-around security and defense in depth.

c. Mortar positions must be protected from direct fire and allow for overhead clearance.

d. Combat vehicles should be assigned primary, alternate, and supplementary positions as well as primary and secondary sectors of fire. They should be positioned in defilade behind rubble and walls or inside buildings for cover, concealment, and movement into and out of the area.

e. If the company is reinforced with antiarmor weapons, the reinforcements should be positioned on high ground in and around the town to attain good fields of fire.

f. If the company is reinforced with tanks, the commander could place the tanks along the leading edge complement the other antiarmor weapons. The tank platoon leader selects exact firing positions and assign sectors of fire. If attacked by enemy infantry, the tanks should move to alternate positions protected by infantry. The alternate positions should allow the tanks to engage to the front and flanks with as little movement as possible. Tanks can be used as a mobile reserve after withdrawal from the leading edge of the town.

g. Obstacles are easily constructed in a built-up area. Commanders must stop enemy vehicles without interfering with his own movement in the village. Command detonating cratering charges at key street locations may be used to hamper enemy mobility. Mines should be laid on the outskirts of the town and along routes that the company will not use.

h. Engineers assist rifle platoons to prepare defensive positions. Engineers use explosives to make firing ports, mouseholes, and demolition obstacles. They also assist in the emplacement and recording of the minefields.

i. Ammunition expenditure is increased when fighting in an urban environment. Ammunition should be stockpiled at each squad position.

Other factors that the company commander may consider are:

- Resupply
- MEDEVAC
- Communications
- Firefighting

.3

- Sleep and alert plans
- Security
- Limited visibility
- Control of noncombatants.

j. The company should install a wire net and develop a plan for pyrotechnic signals to ensure adequate communications. Backup wire should be laid in case primary lines are cut by vehicles, fires, or the enemy. The commander also plans for the use of messengers .

3302. Delay in a Built-Up Area. A rifle company may be assigned the mission to delay in a built-up area.(Figure 3-13). To perform this mission, a rifle company can destroy enemy reconnaissance elements forward of the outskirts of the town, prevent their penetration into the built-up areas, and gain and maintain contact with the enemy to determine the strength and

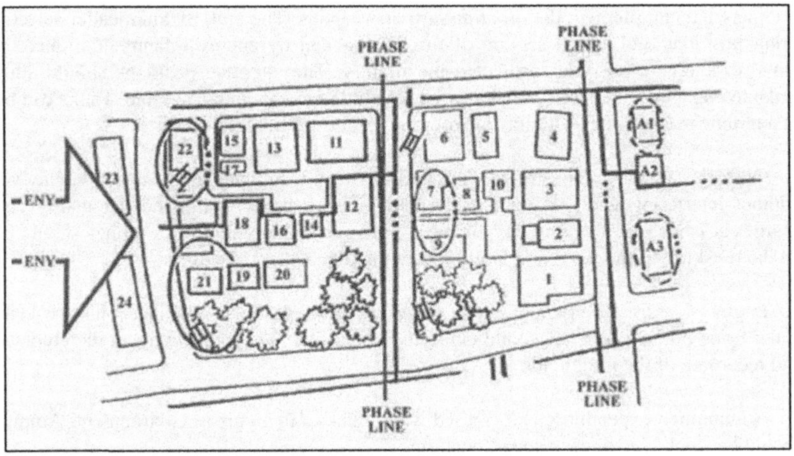

location of the main attack.

Figure 3-13. Delay in a Built-Up Area

a. Platoons delay by using patrols, OPs, and ambushes. Following each action, a platoon will disengage and withdraw.

b. The company's sector should be prepared with obstacles to enhance the delay. Engineers should prepare obstacles on anticipated enemy main routes of advance.

c. Antiarmor weapons may be positioned on the outskirts of the town to destroy the enemy at maximum range. They should be located in defilade positions or in prepared shelters. They fire at visible targets and then fall back or proceed to alternate positions.

d. Tanks in support may be used to engage enemy tanks, provide reinforcing fires, aid the disengagement of platoons, and cover obstacles by fire.

3303. Defense of a City Block. The rifle company may be assigned to defend a city block. The company actions should be coordinated with those of the security forces charged with delaying to the front of the company's position. The defense should take advantage of the protection of buildings that dominate the roads.

a. A well-organized company defense:

- Stops the attack of the enemy on the roads by using obstacles and enfilade fire
- Destroys the enemy by ambush and direct fire from prepared positions
- Ejects the enemy from footholds or fixes them for the battalion's counterattack.

b. The defensive forces should prepare ambushes on the avenues of approach, cover the obstacles by fire, and build a defense inside the buildings (Appendix D). The reserve can be tasked:

- To reinforce the fires of the defense
- To react to a danger on the flank
- To counterattack to throw the enemy from a foothold.

c. Engineers should be employed in GS to construct obstacles, prepare access routes, and assist in preparing defensive positions.

d. Tanks attached to the rifle company may be used to direct-fire support, engage enemy tanks, and support a counterattack.

3304. Defense of a Traffic Circle or Key Intersection. A rifle company may be assigned the mission to defend a key traffic circle or road junction (Figure 3-14).

a. The commander analyzes enemy avenues of approach and buildings that dominate those avenues. He plans to focus all possible firepower on the traffic circle and its approaches. He also plans for the all-around defense of the buildings that dominate the traffic circle to prevent encirclement. The commander prepares as many covered and concealed routes between these

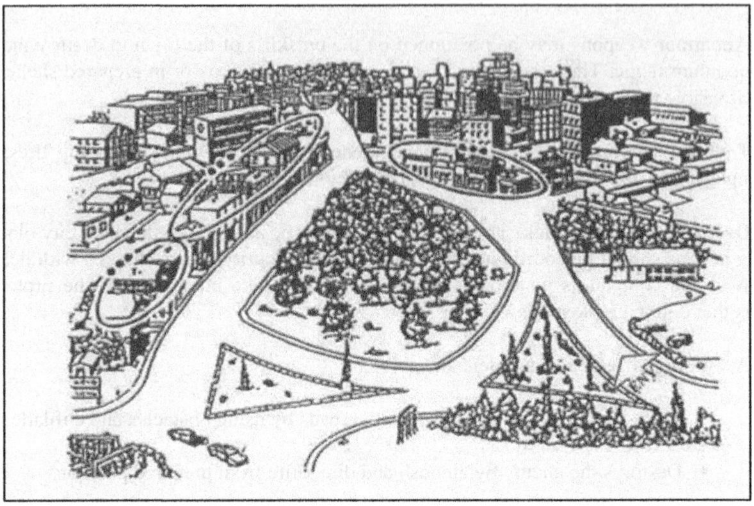

buildings as possible. These routes make it easier to mass or shift fires and to execute counterattacks.

Figure 3-14. Defense of a Traffic Circle

b. Obstacles can be used to deny the enemy the use of the traffic circle. In this case, obstacle planning should consider whether friendly forces require access to the traffic circle. Antiarmor weapons can be employed across the traffic circle if designated fields of fire exist. Tanks can be used to engage enemy tanks and provide direct-fire for counterattacks.

Section IV

Rifle Platoon Operations

This section discusses some missions the rifle company might assign to a rifle platoon.

3401. Defense of a Strongpoint. A rifle platoon may be given the mission to provide or conduct the strongpoint defense of a building, part of a building, or a group of small buildings (Figure 3-15). The rifle platoon's defense is normally integrated into the rifle company's defense. The rifle platoon must keep the enemy from gaining a foothold in its assigned buildings. It makes the best use of its weapons and supporting fires, organizes all-around defense, and counterattacks or calls for a rifle company counterattack to eject an enemy that has a foothold. The platoon leader analyzes his defensive sector to recommend to the company commander the best use of obstacles and supporting fires. (See Appendices B and D.)

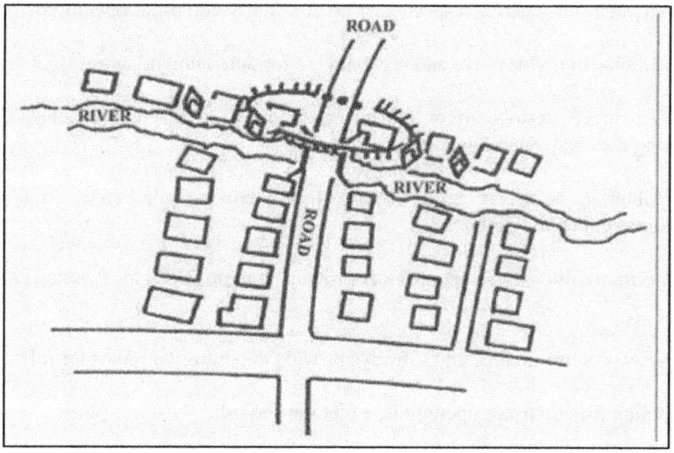

Figure 3-15. Defense of a Strongpoint

a. The rifle platoon should be organized into a series of firing positions located to cover avenues of approach, cover obstacles, and provide mutual support. Snipers may be located on the upper floors of the buildings. Unengaged units should be ready to counterattack, fight fires, or reinforce other elements of the rifle platoon.

b. Depending on the length of the mission, the rifle platoon should have available:

- Pioneer equipment (axes, shovels, hammers, picket pounders)

- Barrier material (barbed wire, sandbags)
- Munitions (especially grenades)
- Food and water
- Medical supplies
- Firefighting equipment.

3402. Defense Against Armor. Urbanized terrain is well suited to an infantry's defense against mechanized infantry and armored forces. Armored forces will usually avoid built-up areas but may be forced to pass through them.

a. Built-up areas have certain aspects that favor infantry antiarmor operations.

(1) Buildings provide cover and concealment for defending infantry.

(2) The streets restrict armor maneuver, fields of fire, and communications, thereby reducing the enemy's ability to reinforce.

(3) Rubble in the streets can be used to block enemy vehicles, conceal mines, and cover and conceal defending infantry.

(4) Rooftops, alleys, and upper floors provide good positions for flanking and overhead fire.

(5) Sewers, storm drains, and subways provide underground routes for infantry forces.

b. To prepare for antiarmor operations, a platoon should:

(1) Choose a Good Engagement Area. The best way for infantrymen to engage tanks is one at a time so that they can destroy one tank without being vulnerable to the fires of another. Good engagement areas include narrow streets, turns in the road, "T" intersections, bridges, tunnels, split-level roads, and rubbled areas. Less obvious locations can also be created by using demolitions or mines to create obstacles.

(2) Select Good Weapons Positions. The best weapons positions are places where the tank is weakest (tank deadspace) and the infantry is most protected. Forcing tanks to button up greatly reduces their ability to see and fire on close-in street and overhead targets. Figure 3-16 (on page 3-38) shows close-in tank weapon vision and engagement limitations (tank buttoned up) at street and overhead levels.

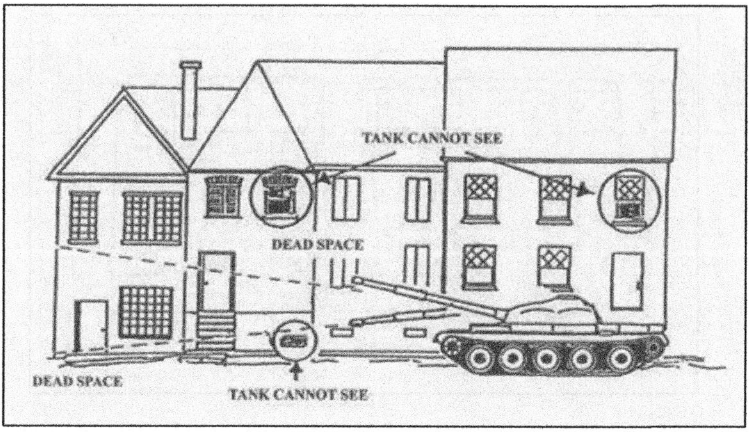

Figure 3-16. Close-in Tank Limitations at Street and Overhead Levels

(3) Assign Target Reference Points and Select Method of Engagement. After selecting the weapons positions, target reference points should be assigned to ensure adequate coverage and control fires. The target reference points should be clearly visible through the gunner's sights and should be relatively resistant to battle damage (for example, large buildings or bridge abutments, but not trees or cars). The three types of antiarmor engagement techniques are frontal, crossfire, and depth. Frontal fire is the least preferred because it exposes the gunner and it attacks the tank where the armor is the thickest. (For more information on target engagement techniques, see MCWP 3-15.5 *Antiarmor Operations*.)

Infantry forces should fire at tanks on the flanks and rear from ground level or at the top of tanks from an elevated position. A suitable antiarmor defense might be set up as shown in Figure 3-17. For a safe engagement from an elevated position, Marines should allow the tank to approach to a range three times the elevation of the weapons. However, engagement at longer ranges risks counterfire because the attacking weapon's position will not be in the tank's overhead dead space.

(4) Coordinate Target Engagement. Tanks are more vulnerable to infantry operations when they are buttoned up. The first task is to compel the tanks to button up by using direct and indirect fire. The next task is to coordinate the fires of the antitank weapons so that all targets in the engagement area are attacked simultaneously (Figure 3-18).

Figure 3-17. A Rifle Platoon's Antiarmor Defense

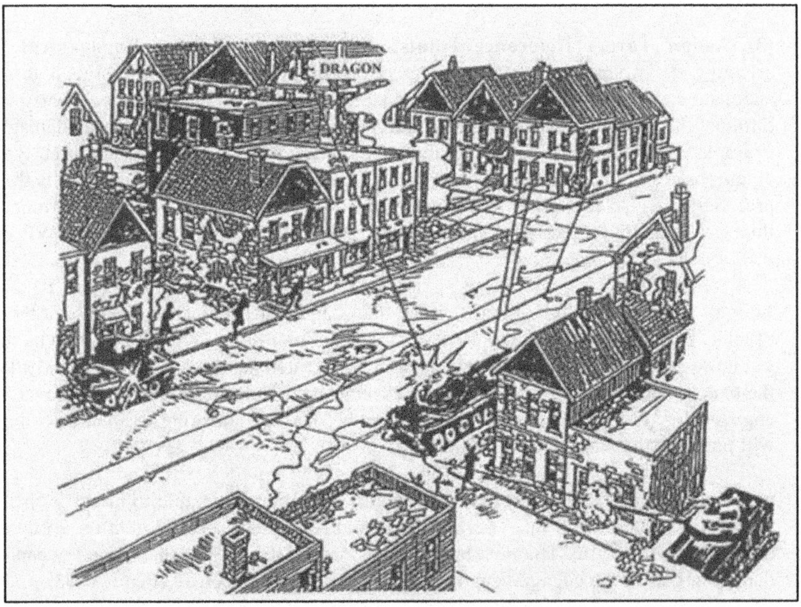

Figure 3-18. Coordinated Antiarmor Ambush

.3

c. Infantry often accompany armored vehicles in built-up areas to provide them with close-in protection. Antiarmor weapons should be supported by an effective antipersonnel defense.

d. At a planned signal (for example, the detonation of a mine), all targets in the engagement area are attcked at the same time. If that is not possible, then the most dangerous target should be engaged first. Although tanks present the greatest threat, BMPs are also dangerous because their infantry can dismount and destroy friendly antiarmor positions. Rubble and mines should be used to reduce target mobility so that a target's time in the engagement area is increased.

3403. Conduct of an Armored Ambush. A rifle company can use an attached tank platoon to conduct an armored ambush (Figure 3-19). The tank platoon should be reinforced with infantry.

Figure 3-19. Armored Ambush

a. An armored ambush requires maneuver on obstacle-free roads. Obstacles outside the ambush area can be used to canalize and delay the enemy.

b. Ambushing tanks should hide approximately 1,000 meters from the expected enemy avenue of approach. A security post, located at a chokepoint, observes and reports the approach, speed, security posture, and activity of the enemy. When the enemy is reported at a trigger point or target reference point, the tank platoon leader orders the ambush.

c. The tanks move quickly from their hiding positions to firing positions, taking advantage of all available concealment. They attempt to fire flank shots on the approaching enemy at an average range of 300 to 400 meters. Firing at these ranges does not expose tanks to enemy infantry. After the engagement, the tank breaks contact and moves to a rally point with close security provided by infantry. They may then move to a new ambush site.

Chapter 4

Combat Support

Combat support units provide critical combat functions in conjunction with combat arms units to secure victory (MCRP 5-2A).

4001. Introduction. Success against a determined enemy is achieved in part by generating decisive combat power through the application of *combined arms*. This is especially true in MOUT. The MAGTF commander must effectively integrate all the capabilities of the command element (CE), GCE, ACE, and CSSE to win the battle in the shortest amount of time with the fewest casualties. The MOUT environment requires special considerations pertaining to combat support forces and capabilities.

Section I

Fire Support

4101. Indirect and Direct Fire Support

a. General. MOUT operations present unique fire support considerations. On urbanized terrain, buildings provide excellent cover and concealment to the enemy while limiting friendly observation and targeting efforts. Targets are generally exposed for brief periods of time and are often in close proximity to friendly forces. Observers will experience difficulty in finding OPs with adequate fields of observation. Terrain masking by tall buildings may restrict the delivery of indirect fires. Collateral damage and rubble effects must be considered during ammunition selection. The importance of effective communications, ROE, control measures, and procedures to prevent fratricide is magnified.

b. Offensive Support. In the offense, fire support plans should include fires to isolate the objective area, support the assault, and support the clearing action. Fires are delivered to isolate and fix the enemy and deny him the use of avenues of approach into and out of the built-up area. Most fires are normally planned and executed at the GCE or higher echelons. Fires are employed to rupture the enemy's established defenses and screen friendly maneuver, in order to maintain the momentum of the attack. Fire support is also allocated to units involved in clearing operations. Fire support plans should incorporate the employment of aerial observers and UAVs to compensate for restrictions to observation and to assist in the delivery of deep fires. Procedures for designating the forward line of own troops (FLOT), marking targets, shifting fires, and communicating in the urban environment should also be considered.

c. Defensive Support. In the defense, fire support plans address fires to disrupt and slow the enemy attack. Fires are delivered at maximum ranges along avenues of approach to separate armor and infantry forces, to canalize the enemy into killing zones, and to deny the enemy use of key facilities in the city, such as communications and transportation facilities. Most fires are planned and executed at the GCE or higher echelon. Defensive fires planned at lower echelons support fighting within the built-up area. Defensive fire support plans identify fire support coordination procedures necessary to execute the delivery of fire support.

4102. Artillery

a. Employment of Artillery. The mission of artillery does not change in MOUT. However, the method of employment is tailored to the unique requirements posed by the urban battlespace. MCWP 3-16.3, *Marine Artillery Support,* provides a detailed discussion on the employment of artillery. Considerations for MOUT include:

(1) Organization for Combat. Standard command relationships for tactical missions are used to organize for combat. During the initial phases of offensive MOUT, particularly isolation of the objective area and the initial assault to secure a foothold, control of artillery is more centralized than in other types of offensive operations. Centralized control provides better responsiveness in shifting and massing fires. As the attack progresses into the clearance phase, control of artillery becomes more decentralized with direct support (DS) artillery providing direct and indirect fires. In the defense, control of artillery should be sufficiently centralized to enable the rapid massing of fires.

(2) Movement and Positioning. Key considerations for the positioning of artillery in MOUT include:

- Mobility restrictions
- Limited availability of firing positions
- Masking of fires
- Security
- Enemy counterbattery fires

(a) Offense. Artillery is initially positioned outside or in the outskirts of the city. Artillery may displace to suitable firing positions within the city as the attack progresses. Positions should be selected that minimize masking, provide several routes for escape, and afford as much cover and concealment as possible. Potential firing positions can be found in parks, racetracks, sports fields, school yards, stadiums, parking lots, and similar areas. Special techniques for the emplacement of howitzers may be required if the ground is not suitable for normal emplacement. Some artillery positions may be maintained outside the city (see appendix B). Artillery generally makes few displacements in MOUT because it requires extensive route reconnaissance and security. If and when artillery is moved, it should be given priority of movement during displacements. Movement is conducted during night or periods of reduced visibility.

(b) Defense. Artillery is positioned on the friendly outskirts of the built-up area. Artillery displaces along predetermined routes to avoid traffic congestion.

(3) Delivery of Fires. The commander, in coordination with the fire support coordinator (FSC), considers the following when planning fire support in a built-up area:

(a) Target acquisition may be more difficult because of the increased cover and concealment afforded by the terrain. Ground observation is limited; therefore, FOs should be placed near tops of buildings.

(b) For targets in defilade, fire may be adjusted laterally so that the rounds impact on the street perpendicular to the observer's position. Airburst rounds are more suitable for this adjustment. Adjustments may have to be made by sound. When rounds impact on a perpendicular street, they are adjusted for range. When the range is correct, a lateral shift is made onto the target and the observer fires for effect.

(c) Special consideration must be given to shell and fuze combinations when effects of munitions are influenced by buildings. These include:

- Careful use of variable time is required to avoid premature arming.
- HE ammunition with impact or delay fuzes may create unwanted rubble.
- The close proximity of enemy and friendly forces requires careful coordination.
- White phosphorus (WP) may create unwanted fires and smoke.
- Fuze delay should be used to penetrate fortifications.
- Illumination rounds can be effective. However, friendly positions should remain in shadows and enemy positions should be highlighted. Tall buildings may mask the effects of illumination rounds.
- HE/variable time, HE/time fuze, and improved conventional munitions are effective for clearing enemy positions, observers, and antennas from rooftops.
- Swirling winds may degrade smoke operations.
- FASCAM may be used to impede enemy movements. FASCAM effectiveness is reduced when delivered on a hard surface.

(4) Targeting is difficult on urbanized terrain because the enemy has many covered and concealed positions and movement lanes. The enemy may be on rooftops and in buildings and may use sewer and subway systems. Targets should be planned on rooftops to clear enemy FOs, communications, and radar equipment. Targets should also be planned on major roads, at road intersections, and on known or likely enemy fortifications. Employing artillery in the direct-fire mode to destroy fortifications should be considered. Also, restrictive fire support coordination measures (such as a restrictive fire area or no fire area) may be imposed to protect civilians and critical installations.

(5) The 155-mm howitzer can be effective in neutralizing concrete targets by direct fire. A 155-mm round can penetrate 36 inches of concrete at ranges up to 2,200 meters. Restrictions may be placed on the type of ammunition used in order to reduce rubbling on friendly avenues of approach.

(6) FOs must be able to determine the location and size of dead space. Dead space is the area in which indirect fires cannot reach the street level because of buildings. This area is a safe haven for the enemy because he is protected from indirect fires. For low-angle artillery, the dead space is about five times the height of the building. For mortars and high-angle artillery, the dead space is about one-half the height of the building.

(7) Aerial observers and UAVs are effective in providing observation behind buildings immediately to the front of friendly forces. They are extremely helpful in adjusting fire. Aerial observers can also relay calls for fire when communications are degraded because of power lines or building masking.

(8) Weapons locating radars are effective in locating enemy artillery and mortar targets in an urban environment because of their high percentage of high-angle fires. However, LOS must be maintained in order to exploit this advantage. Radars should not be placed in the midst of a city due to masking. They are best located on the outskirts of the city.

(9) The use of airburst fires is an effective means of clearing snipers and manpads from rooftops. HE shells with delay fuzes may be effective against enemy troops in the upper floors of buildings, however, because of the overhead cover provided by the building, such shells have little effect on the enemy in the lower floors. (The planning and employment of artillery in offensive and defensive operations is also addressed in Chapters 2 and 3.)

b. Local Security. Artillery crews and howitzers require close security when used in the direct-fire mode. Artillery is particularly vulnerable when units are employed in isolated firing positions.

c. Communications and Information Systems (CIS). CIS must be able to satisfy the unique C2 requirements for artillery in MOUT. CIS should provide the commander with the ability to collect, process, analyze, and exchange information. Requirements such as massing fires and call for fire must be carefully planned and adapted to meet the requirements of

MOUT. For example, small units conducting independent operations must have the communications capabilities and maps to allow them to call for fire.

4103. Mortars

a. Employment of Mortars. Mortars are often the primary indirect-fire weapon for forward units in the assault or defense of a built-up area. Their high rate of fire, short minimum range, and high trajectory give mortar units the ability to fire in the tight confines of the city. Mortars can provide obscuration, neutralization, suppression, or illumination fires.

(1) Organization for Combat. Normally, 81-mm mortars are employed in GS of the battalion. The 60-mm mortar is normally used at the discretion of the rifle company commander.

(2) Movement and Positioning. 81-mm mortars employed in GS normally occupy a position within or near the infantry battalion reserve. Key considerations for the selection of positions include:

- The minimum range of the weapon
- Mask and overhead clearance
- Terrain suitable for setting base plates
- Dispersion, and accessibility.

Special techniques can be used to position and lay the 81-mm mortar. These techniques include placing buffers under base plates, using curbs and sandbags to anchor or brace mortars, and filling cans with dirt or using sandbags to emplace aiming posts. Fixed objects may be used as distant aiming points. When the depth of the defensive position is shallow, or when suitable firing positions are not available, mortars may have to be positioned behind the reserve. Positioning of mortars behind buildings greatly enhances survivability. Displacement is often executed by section. Mortars are usually not placed on top of buildings because the location lacks cover and concealment and recoil/concussion could collapse roofs.

(3) Delivery of Fires. Mortar fires are effective in the offense as well as in the defense. The indirect fires are extremely responsive. Mortars are well suited for combat in built-up areas because of their high rate of fire, steep angle of fall, and short minimum range. Mortar fires can be used to inhibit enemy fires and movement, allowing friendly forces to maneuver to a position of advantage. Effectively integrating mortar fires with dismounted maneuver is key to successful combat in a built-up area.

.3

(a) HE Ammunition. HE is used more than any other type of ammunition. Although mortar fires are often targeted against roads and other open areas, the natural dispersion of indirect fires will result in many hits on buildings. Leaders must use care when planning mortar fires to minimize collateral damage. Considerations when using HE include:

- HE ammunition is particularly effective when used on lightly built structures within cities. However, it does not perform well against the reinforced concrete found in larger urban areas.

- Point-detonating fuzes should be used. The use of proximity fuzes should be avoided because the nature of built-up areas causes proximity fuzes to function prematurely. However, proximity fuzes are useful in attacking targets such as OPs on tops of buildings.

- During World War II, the Middle East conflicts, and most recently in Bosnia, mortar HE fires have been used extensively to deny the use of streets, parks, and plazas to enemy and civilian personnel.

(b) Incendiary Fires. Employment of these munitions is planned in detail to prevent the burning of buildings and surrounding terrain that may create obstacles to friendly forces.

(c) Illumination. In the offense, illumination rounds should burst above the objective to put enemy troops in the light. Buildings reduce the effectiveness of the illumination by creating shadows. If the illumination is behind the objective, the enemy troops would be in the shadows rather than in the light. In the defense, illumination is planned to burst behind friendly forces to put them in the shadows and place the enemy forces in the light. Continuous illumination requires close coordination between the FO and the fire direction center (FDC) to produce the proper effect.

(d) Other Considerations. When planning the use of mortars, commanders must consider the following:

- FOs should be positioned on tops of buildings so target acquisition and adjustments in fire can best be accomplished.

- Commanders must understand ammunition effects to correctly estimate the number of volleys needed for the specific target coverage. Also, the effects of using white or red phosphorus may create unwanted smoke screens or limit visibility, which could interfere with the tactical plan.

- Mortar sections should plan to provide their own security.

- Commanders must give consideration to where, when, and how mortars are to displace in order to maintain immediate indirect fire support. Combat in built-up areas may adversely affects the ability of mortars to displace because of rubbling.

b. Communications. An increased use of wire, messenger, and visual signals will be required. However, wire should be the primary means of communication between the FOs and the FDC because elements are close to each other. Also, frequency-modulated (FM) radio transmissions in built-up areas are likely to be erratic. Remote siting of antennas to upper floors or roofs may improve communications and enhance operator survivability. The use of radio retransmissions can also be effective. Both in the offense and defense, commanders consider using existing civilian systems to supplement the unit's capability.

c. Magnetic Interference. Anything that uses or generates electricity generates its own magnetic field. In an urban environment, all magnetic instruments are affected by surrounding structural steel, electrical cables, and automobiles. Minimum distance restrictions for the use of the M-2 aiming circle will be difficult to overcome. However, an azimuth may be obtained from a distant aiming point.

4104. Naval Surface Fire Support (NSFS) and Naval Gunfire (NGF)

a. General. When NSFS ships are available, naval surface fires and NGF can provide some fire support.

b. Employment of Naval Surface Fire Support. The urban environment severely limits NSFS, and NGF in particular, due to its flat trajectory. NSFS can be used to supplement existing fire support but should not be planned as the primary fire support asset for MOUT. NSFS is most effective when firing on targets to isolate the battlefield and engage targets located on the seaward side of a city. Only when the ROE allows extensive collateral damage can the full effect of NSFS be brought to bear on units occupying a city.

(1) Organization for Combat. NSFS is best employed under centralized control because of the limited use of NSFS within the confines of the city.

(2) Positioning. Many of the fires delivered by NSFS ships require a careful orientation of the gun target line and may limit the size of a fire support area.

(3) Delivery of Fires. NGF can deliver a variety of ammunition in MOUT. NGF can deliver HE, incendiary, and illuminating projectiles. The NGF high-capacity projectile is very effective in producing fragmenting and penetration effects. Offensive and defensive fire support plans can incorporate NSFS fires on bridges, roads, reinforcements, artillery positions, and command and control centers. NSFS can be effectively delivered on targets along the outskirts of the city to rupture established defenses. These fires are particularly effective on buildings and on targets with a linear axis such as streets and airfields. Because of its flat trajectory, NGF employment is limited on targets within the city. Firing reduced charges at high angles does enable NGF to hit more targets but will reduce its range and accuracy.

4105. Aviation

a. Employment of Aviation. Both fixed- and rotary-wing aircraft can provide fire support in MOUT. The aviation combat element (ACE) provides the MAGTF commander with a variety of options in ordnance delivery. Aerial delivered ordnance is particularly effective against hard targets. Cluster bomb units, rockets, cannons, laser-guided bombs, and electro-optically guided missiles can be used against a variety of targets. However, the presence of civilians, the requirement to preserve key facilities, and the residual effects caused by rubble and collateral damage must be taken into consideration when planning for and using aviation in an urban environment. (See MCWP 3-24 *Assault Support Helicopter Tactical Manual* (under development), MCWP 3-23.1 *Close Air Support* (under development),and NWP 55-3-AH1, *AH1 Tactical Manual,* for greater detail on MOUT considerations for aviation.)

(1) Fixed-Wing Aircraft

(a) Offensive/Defensive Operations. During offensive operations, fixed-wing aircraft can isolate, interdict, and reduce enemy strongpoints throughout the battlespace. During defensive operations, they attrite enemy forces as they approach, attack forces attempting to bypass, or canalize the enemy to take avenues of approach that are advantageous to the defender.

(b) Other Considerations

1 Fixed-Wing Attack of a Built-Up Area. Air strikes can reduce or destroy vital supplies and facilities supporting the enemy. The resulting shock, concussion, and psychological effects of such attacks can reduce the efficiency and fighting spirit of enemy forces.

2 Rubble and Debris. The rubble and debris resulting from air attacks may increase the defender's cover and create obstacles to the movement of friendly forces.

3 Proximity of Friendly Forces. The proximity of opposing forces to friendly forces increases the potential for fratricide. This may require the use of precision guided munitions (PGMs) or the temporary disengagement of friendly forces in contact.

4 Locations of Indigenous Civilians or Key Facilities. The use of aerial delivered ordnance may be restricted by the ROE because of the presence of civilians or the requirement to preserve key facilities within a city.

5 **Limited Ground Observation.** Limited ground observation may require the use of forward air controllers (airborne) (FAC(A)).

(2) **Rotary-Wing Aircraft.** Attack helicopters provide a distinct advantage to the commander in MOUT. They can rapidly maneuver within the built-up area to attack a variety of targets. Urban structures offer ideal cover for attack helicopters. The ability to deliver PGMs, rockets, and 20 mm cannon fire makes attack helicopters valuable assets in reducing strongpoints, breaking up enemy concentrations, and providing suppressive fires for attacking forces. Attack helicopters also provide a highly manueverable antiarmor capability.

(a) Attack Helicopter Fire Support During Offensive/Defensive Operations

1 Provide close air support in support of the commander's scheme of maneuver

2 Provide suppression and/or destruction of enemy forces within buildings and strongpoints

3 Conduct counterattacks

4 Isolate the built-up area or objectives within the built-up area

5 Conduct armed reconnaissance and patrols.

6 Conduct rapid concentration of fires.

4106. Armored Vehicles

a. Employment of Armored Vehicles. Tanks, LAVs, and AAVs are primarily mobility assets that function best in an unrestricted, mobile environment. However, these vehicles can play an important role inside the city by using their firepower in support of offensive and defensive operations. LAVs and AAVs highly are susceptible to armor-piercing rounds and manpacked antiarmor weapons. All vehicles are susceptible to top-attack munitions. Some of the fire support roles armored vehicles may provide include:

- Suppression and/or destruction of enemy forces within buildings and strongpoints (tanks/LAVs/AAVs)
- Provide machinegun suppression fires
- Reserve or counterattack suppression fires (tanks and/or AAVs)
- Creating entry/exit points in buildings
- Isolation of the built-up area or objectives within the built-up area (tanks/LAVs/AAVs)

.3

- Artillery-towing vehicle (LAVs/AAVs)
- Antiarmor fires
- Breaching obstacles in a direct fire mode.

b. Special Consideration. The existence of waterways, canals, and rivers in and around urban areas can provide avenues of ingress into designated objective areas. If used in conjunction with supporting fires, smoke, and/or deception operations, AAVs can capitalize on their amphibious capability and make use of waterways previously viewed as obstacles.

4107. Antiarmor Weapons. Antiarmor weapons have more utility in the defense than in the offense. Their primary task is defeating enemy armor. Secondary tasks may involve blasting holes in structures and fortified positions. As with other weapons systems, antiarmor weapons are normally employed in a decentralized fashion at the small unit level. Enemy armor should be engaged in predetermined locations that afford the most advantage to the defender. Weapon positions should be selected that provide maximum range fires. TOWs and Dragons may be fired from an enclosed area if it is a sturdy ventilated structure of appropriate size. For a more detailed discussion of antiarmor weapons, see MCWP 3-15.5, *Antiarmor Operations*.

4108. Employment of Snipers

a. General. Snipers are highly trained in field skills and marksmanship and can deliver long-range precision fire at selected targets from concealed positions. The method by which snipers are employed in urban warfare will be governed by the nature of the terrain/weather, number of snipers available, and whether the enemy employs snipers. A sniper can play an important role in MOUT. Employment of snipers influences the action by:

- Creating adverse psychological affects on the enemy by negating feelings of security
- Inflicting casualties on enemy leadership
- Enabling the infantry to seem to cover a large area, regardless of whether the terrain is physically occupied or not
- Detailing enemy positions and activity through the use of advanced optical equipment and observation techniques
- Controlling fire support
- Conducting reconnaissance in conjunction with their sniper role
- Disrupting enemy movement, observation, and infiltration, and negating the possibility of surprise.

b. Offensive Employment. Snipers should be used to gain and maintain contact with the enemy. This enhances security and prevents surprise by keeping constant and unrelenting pressure on the enemy. Some of the considerations for sniper employment include:

(1) As the "eyes" of a commander, a sniper increases the commander's flexibility by gathering and transmitting information on fire support targets.

(2) Snipers attack those targets that affect the enemy's ability to wage battle and those that will cause the maximum amount of confusion on the battlefield. Potential targets for a sniper are:

- Officers and noncommissioned officers
- Enemy snipers and marksmen
- Engineer personnel
- Personnel manning crew-served weapons
- Communications personnel and equipment
- Fire support observers and/or equipment
- Commanders of armored vehicles.

c. Combat on Urbanized Terrain

(1) Snipers can be employed to operate for extended periods of time from hidden positions to dominate and establish a "no-man's land," screen flanks, protect the rear, and deny the enemy access to certain areas or avenues of approach. Snipers can operate with the covering element to deliver accurate fires in support of search elements. Optical devices enhances their ability to detect movement and engage it instantly.

(2) Snipers can also support (by fire) infantry movement across streets. They can provide immediate precision fire on enemy machine gun nests and enemy snipers. The best countersniper weapon is the sniper. Snipers are assigned the following supporting missions:

- Delaying and inflicting casualties on the advancing enemy
- Observing and reporting potential targets
- Covering (by fire) avenues of approach and obstacles

- Harassing the enemy and causing him to deploy prematurely
- Directing supporting arms fire.

The urban environment provides an abundance of secure firing positions that are not highly accessible to countersniping.

d. Defensive Employment. In defensive operations, the sniper is best employed forward of the FEBA to provide early warning of enemy approach, disrupt it, and if possible, cause the enemy to deploy prematurely. Snipers are employed to:

- Prevent enemy infiltration
- Detect and destroy infiltrators
- Protect unit at the FEBA from surprise attacks
- Protect patrols from ambush
- Screen the flanks and rear of the defensive positions.

More detailed information regarding the employment of snipers can be found in Marine Corps Warfighting Publication (MCWP) 3-15.3, *Scout Sniping* (under development).

Section II

Other Combat Support

4201. Engineers

a. General. The nature of the urban battlespace requires extensive use of engineers. In both the offense and defense, the proper employment of engineers can be decisive. Engineers must also be prepared to repair and maintain the internal support facilities of the city in order to support further operations. MAGTF commanders plan to use all available engineer assets, including engineers in the GCE, ACE, and CSSE, as well as any naval construction forces (NCFs).

b. Engineer Organization. Engineers are almost always task organized. The compartmentalized, closed-in nature of the city, which limits observation, mobility, and communications, will dictate the best organization for the mission. Teams should be formed to perform specific tasks for small units and must be positioned for quick employment. In MOUT, engineer support is key to maintaining the mometum of the attack. Engineer teams will usually move with the units they are supporting. Two-man demolition teams may be assigned to support a rifle platoon or squad to breach obstacles or to neutralize booby traps and improvised explosive devices (IEDs).

Commanders must carefully analyze mission requirements, the alignment of enemy forces, the condition of the battlespace, and available assets, to ensure an organization that will provide optimum engineer support. Most of the actual engineer work is done at the small unit level.

c. Engineer Tasks

(1) Offense. Urban combat is by nature a very slow, deliberate affair. Even if the enemy has not prepared the area for a defense, it will take time to secure the city. A well-constructed urban defense will slow or stop any attack unless impediments to movement can be quickly removed or bypassed. Properly organized and positioned engineers can reduce delays by performing a variety of tasks which include:

(a) Engineer reconnaissance to determine the level of effort required to penetrate outer and inner defenses, clear obstacles, and identify bypass routes.

(b) Bridge repair and construction on critical avenues of approach. Bridge construction may also open additional avenues of approach that the enemy is not prepared to defend.

(c) Breaching obstacles both outside and inside the city. This includes breaching minefields and neutralizing booby traps and IEDs.

(d) Assault breaching into buildings.

(e) Reducing enemy strongpoints using demolitions and heavy equipment.

(f) Preparing helicopter landing zones (HLZs).

(g) Assist in the preparation of defensive positions in the event of a counterattack.

(2) Defense. An urban area can be made nearly impregnable with proper organization of the battlespace and skillful preparation by engineers. Obstacles in outlying areas can delay/disrupt the enemy and canalize him into engagement areas. The streets and alleys inside a city can also be use to canalize enemy forces into engagement areas and kill zones. In preparing the defense of an urban area, engineers perform several critical tasks to include:

(a) Engineer preparation of the urban battlespace. This should begin well outside the city with complex obstacles in support of security forces. Selective demolition of bridges and roads and other countermobility measures can force the enemy into taking avenues of approach that favor the defense.

(b) Constructing obstacles forward of and between defensive positions to block, disrupt, fix, and turn enemy attacks. A majority of the obstacles will be directed at interfering with the movement of personnel and vehicles, but some effort must be directed at preventing vertical and subterranean movement. Material designed especially for obstacles and barricades, such as barbed wire and tetrahedron(s), is necessary, but improvised barriers can be just as effective and are readily available. Vehicles and building rubble are the most common improvised barriers.

(c) Enhancing mobility by opening and/or marking preplanned routes into attack positions for counterattacks or alternate positions to protect a flank.

(d) Providing technical advice on which buildings are more suitable for use as strongpoints or which rooftops are strong enough for helicopter landings can be vital. Any reinforcement of building roofs should be supervised by the engineers.

(e) Preparing tank positions that offer them cover and concealment.

(f) Assisting infantry units in clearing clear fields of fire, especially for longer range weapons like the TOW.

(g) Playing a potential role in the maintenance of critical urban services such as power, water, and communications. This function may also be performed by naval construction units.

(h) Assisting in the preparation and maintenance of railroads, airfields, and ports. This function may also be performed by naval construction units.

4202. Assault Support Helicopters

a. Combat Assault Transportation. The mobility provided by helicopters can be a significant advantage in the urban environment. In many modern cities, office buildings may have helipads on their roofs. Other buildings, such as parking garages, are usually strong enough to support the weight of a helicopter. The delivery of Marines onto a building can also be accomplished by rappelling, fastroping, or jumping out of the helicopter as it hovers just above the roof. The rooftops should be inspected to ensure that no obstacles exist, such as electrical wires, telephone poles, antennas, or mines and wire. The use of helicopters in this role is usually predicated on the antiair threat. Helicopterborne forces are highly vulnerable to small arms fire, AAA and manpads.

b. Small-Scale Helicopterborne Assaults. Small units may be helo-lifted onto the rooftop of a key building. Depending on the construction of the roof, rappelling or fastroping Marines from the helicopter may be more advantageous than landing them on the rooftop. Fastroping provides a rapid and safer means of insertion than jumping from a low hover. With practice, Marines can accomplish a fastrope insertion with minimum exposure.

c. Large-Scale Helicopterborne Assaults. For large-scale assaults, rooftop landings are not practical; however, open spaces (parks, parking lots, sports arenas) within the built-up area may be used. Depending on METT-T, helicopterborne forces may assault into an LZ in one wave or scheduled waves.

d. Mobility of Forces and Supplies. In the urban environment, movement of forces by helicopters may become a major requirement. Units engaged in house-to-house fighting normally suffer more casualties than units fighting in open terrain. These casualties must be evacuated and replaced quickly with new forces. At the same time, roads are likely to be crowded with resupply and evacuation vehicles and may also be blocked with craters or rubble. Helicopters provide a means of overcoming these obstacles.

e. Helicopterborne Assault Techniques. Helicopter assaults into enemy-held territory are extremely difficult (Figure 4-1 on page 4-16). One technique is to fly nap-of-the-earth down a broad street or commercial ribbon while attack helicopters and assault helicopter door gunners suppress buildings on either side of the street. Artillery preparations can be incorporated into the helicopter assault plan to suppress rooftop and building antiair threat. Feints and demonstrations in the form of false insertions can deceive the enemy as to the intended objective. Helicopters often deliver forces to the last covered position, short of the fighting, then return without exposure to enemy direct fire. Similar flight techniques can be used for transportation of supplies and casualties.

Figure 4-1. Helicopterborne Assault of a Built-Up Area

f. Combat Roles. Below are some of the roles that assault support helicopters may play in urban warfare:

- Conduct insertion and extraction of forces (See MCRP 3-11.7A, *Helicopter insertions and Extractions* [under development])
- Provide command/control and aerial retransmission
- Support CSS operations
- Provide air reconnaissance.

4203. Employment of Reconnaissance Forces. Reconnaissance units traditionally make maximum use of rural camouflage and terrain. Area, zone, and route reconnaissance are conducted from well-concealed vantage points through clandestine insertion by foot, vehicle, rubber boat, or helicopter. These methods of insertion may still apply in village and strip areas for collection of information, sensor employment, conducting OP radio relays, or a combination of missions. In major urban areas, undetected insertions are difficult to accomplish because of the high density civilian population.

The use of reconnaissance assets must be used wisely. Reconnaissance assets must not be risked indiscriminately if the information can be collected by other means. Consideration should be given to flank and rearward employment of ground reconnaissance units, in addition to close (0 to about 2,000 meters), distant (approximately 2,000 - 20,000 meters), and deep (from about 20,000 meters and beyond, plus preassault) missions. Reconnaissance of bypassed pockets of resistance, avenues of approach, canals, bridges, subways, and sewer systems (possibly in conjunction with combat engineers) is feasible. Observation, sniper, and listening posts manned by reconnaissance personnel are also missions that reconnaisance personnel perform. The limited combatant diving capability of division reconnaissance may be employed to advantage in many coastal cities.

4204. Military Police (MP). MPs can play a significant role in the urban environment. They may conduct battlefield circulation control, area security, enemy prisoner of war (EPW) and law and order operations. These operations require continuous coordination with MAGTF units as well as with host nation military and civilian police.

a. MP units support area damage control operations. With the increased possibility of rubbling, MP units report on and block off affected areas and reroute movement to alternate road networks.

b. When conducting EPW operations, MPs collect prisoners as far forward as possible and transport them to the rear. They operate temporary collecting points and holding areas to briefly retain EPWs and civilian internees. EPW operations are important in built-up areas because the rate of capture can be higher than normal.

c. Commanders must realize that MP support may not always be available and that infantry units may have to assume certain MP functions. These functions include:

(1) Providing convoy escort and security for lines of communications.

(2) Controling roads, waterways, and railroad terminals that are critical chokepoints in the main supply routes.

(3) Providing security of critical sites and facilities, including communications centers, government buildings, water and electrical supply facilities, command and control nodes, airfields, storage facilities, and other essential areas.

(4) Provide refugee control in close cooperation with host nation civil authorities.

(5) Collect and escort EPWs.

4205. Communications

a. General. Manmade structures create many challenges for CIS in an urban environment. Other CIS challenges include extensive commercial and communications networks and public service radio nets. Commercial communication networks are commposed of miles of underground protected cable which connect and control telephones and computer networks. Public service radio nets, (including police, fire, civil defense, and taxi) compete with existing communication systems. To communicate on an effective, continuous basis in the urban environment, commanders must minimize communication limitations imposed by this environment and maximize the advantage of existing civil communications capabilities.

b. Radio. Manmade structures will decrease the effectiveness of VHF radio communications. The physical locations of radios are critical for effective communications. Some specific locations to avoid include powerlines, underground areas, and steel bridges.

Very high frequency (VHF) radios provide the primary communication capability at the battalion level. It is imperative that these radios be employed as effectively as possible. Communicators can use the following techniques to enhance VHF communications equipment:

- Relocation of VHF radios to overcome phase cancellation due to numerous dead spots
- Use of the higher end (50 - 75 MHz) of the available frequency spectrum
- Use of retransmission facilities to improve the reliability of communications
- Use of radio remotes.
- Use of field expedient antennas.

In setting up retransmission stations, antennas should be camouflaged to blend in with urban surroundings and to prevent them from being targeted by enemy fires. Antennas can be concealed by such objects as water towers and steeples.

In larger cities with many tall buildings, it may be necessary to construct a terrain profile of the area of operations. A communications plateau can be deduced by calculating the average height of the structures in a zone. Operation of antennas above the median line of this plateau will result in better wave propagation and aid in retransmission.

c. Wire Communications. Wire may be used extensively in urban environments. It offers two primary advantages over radio communications:

- It is not susceptible to interference and absorption suffered by radio systems.
- Security of the overall system is greater because wire systems are essentially point-to-point.

Maximum use should be made of existing facilities for concealment. Sewers, subway tunnels, cable stairwells, elevator shafts, intact buildings, and pole lines offer ideal concealment for military wire systems. Precautions must be taken to lessen the possibility of vehicles tearing lines and disrupting communications. Existing telephone poles and tunnels can be used to keep wire above or below the street level.

Maximum use of existing commercial facilities should be made to supplement or replace military CIS. Telephone exchanges are linked together by underground cable systems and constitute a key resource in the urban communications system. These facilities are normally windowless buildings constructed of steel-reinforced concrete and provide ideal CPs and communications centers. The physical location of these exchanges, along with circuit diagrams and telephone directories, should be determined as soon as possible. Action should be initiated to secure these facilities early. Other facilities that contain extensive internal/external telephone systems may also be used as command centers. These include most municipal buildings, department stores, local police stations, and large business offices.

d. Messengers. Messengers provide a secure and flexible means of communication, particularly in an electronic warfare (EW) environment. However, careful selection of messenger routes is necessary to avoid pockets of enemy resistance. Disorientation in an urban environment occurs more often and more easily than in a rural environment. Marking facilities, transportation networks, and city streets on overlays often facilitates messenger service. Selected primary and alternate routes can often interface with designated logistics/resupply routes. Early restoration of street signs and recognizable landmarks are is essential.

e. Telephone Networks. Captured civil or military communications facilities can be utilized to augment organic communications capabilities. A civilian phone system can provide a reliable means of communication. If a telephone encryption device is not available, codes and authentication tables should be used.

Telephone technology differs from place to place, but there are universal concepts and practices in their operation. The most important concept to the military planner is network hierarchy.

Network hierarchy is a concept of organization by which a telephone company can rapidly and expeditiously allocate calls through a particular group of offices. As an ancillary function, it represents a control feature by which traffic can be rerouted to avoid congested areas or damaged facilities. This network hierarchy could play an important role in the planning and execution of the scheme of maneuver and fire support.

4206. Employment of Sensors

a. Sensor systems (seismic, infrared, acoustic, or magnetic) are normally employed in a rural context but also have traditional applications in village and strip area terrain. Use of sensors in urban areas is largely undocumented. In major urban areas, the capability for emplacement by air over select areas is reduced. Emplacement in, under, or around urban structures may also reduce microphone efficiency. Tactical employment will be degraded by such obstacles as concrete and steel structures, high-density population, and traffic.

b. Training in camouflage techniques for microphones and sensor terminals in an urban environment should be conducted. Operators should receive sufficient training to distinguish background-generated clutter from enemy foot, vehicular, and mechanized traffic. Use of and reliance on sensors may be seriously reduced during offensive operations in major urban areas. However, greater use of sensors will occur in subterranean defensive operations. Sensors should continue to be employed as far forward as possible and be centrally controlled.

c. Sensor employment should be coordinated with counterintelligence (CI) efforts for placement, readouts, and recovery. Sensors can also play a role in rear areas to augment rear security operations. Sensors, in themselves, are only component elements of an integrated intelligence system. Their employment is limited only by the imagination.

4207. Employment of Air Defense. Basic air defense doctrine does not change when units operate on urbanized terrain. The employment principles of mix, mass, mobility, and integration all apply to the employment of air defense assets. The commander must consider the following when developing his air defense plan:

a. Enemy air targets such as principal lines of communication and road and rail networks and bridges are often found in and around built-up areas.

b. Good firing positions may be difficult to find and occupy for long-range air defense missile systems in the built-up areas.

c. Movement between positions is normally restricted in built-up areas.

d. Long-range systems can provide air defense cover from positions on or outside of the edge of the city.

e. Radar masking and degraded communications reduce air defense warning time for all units. Air defense control measures must be adjusted to permit responsive air defense within this reduced warning environment.

f. Positioning of air defense weapons in built-up areas may be limited to more open areas without masking, such as parks, fields, and rail yards.

g. Stinger teams provide protection for infantry battalions the same as in any operation. When employed within the built-up area, rooftops normally offer the best firing positions.

h. Heavy machine guns emplaced on rooftops can also provide additional air defense.

Chapter 5

Combat Service Support and Legal Aspects of Combat

During combat in built-up areas, the terrain and the nature of operations create unique demands on the CSS system. Meeting these demands will require innovative techniques and in-depth planning.

Section I

Combat Service Support

5101. Introduction. CSS is "the essential capabilities, functions, activities, and tasks necessary to sustain all elements of operating forces in theater at all levels of war" (Joint Pub 1-02). The CSS organization throughout the MAGTF, down to battalion level, must provide a responsive CSS system that can perform all functions and tasks associated with meeting identified CSS requirements. This responsibility does not change during urban operations. Forward support for combat forces continues to be the basic concept governing CSS operations. No significant changes in doctrine or organization are required. However, the characteristics and nature of urban combat do affect how CSS is provided.

5102. Combat Service Support Resupply, Maintenance/Repair, and Replacement

a. Resupply of Ammunition. Combat in built-up areas is characterized by extremely high ammunition expenditure rates. Not only do individual Marines expend more ammunition, but they also use greater quantities of munitions such as smoke, concussion, and fragmentation grenades; AT4s; claymore mines; demolitions; and so on. The ammunition consumption rate for the first day of combat in a built-up area can be up to four times the normal rate. Even though this rate decreases during succeeding days, consumption remains high. Leaders should plan to meet these high consumption rates. The plan must include how ammunition and demolitions are to be moved forward to the companies. AAVs should be designated for the movement of ammunition if rubble or glass prevents wheeled-vehicle traffic. Marines may need to organize into carrying elements if streets are blocked by rubble.

b. Resupply of Fuel. The amount of bulk fuel needed by forces during combat in built-up areas is greatly reduced. Combat vehicles normally use less fuel in built-up areas because they travel shorter distances and perform less cross-country traveling. Engineer equipment and power generation equipment may use more fuel, but requirements are small. A unit may not use much fuel daily, but when it does need fuel, a problem exists in delivering bulk fuel to the vehicle. In open terrain, a vehicle that has run out of fuel can be recovered later. In built-up areas, the same vehicle is probably going to be lost quickly. Leaders should plan for and provide the means of moving limited amounts of bulk fuel forward to combat units.

c. Maintenance and Repair. Maintenance contact teams must operate well forward to support units fighting in built-up areas. Although some maintenance operations may be consolidated in civilian facilities, many vehicles will have to be repaired near fighting positions. Battle damage assessment (BDA) and repair procedures allow mechanics to analyze, repair, and return damaged vehicles to a serviceable condition. Other considerations:

(1) Combat in built-up areas generates a high demand for tires.

(2) The dust and rough handling that are characteristic of combat in built-up areas also place great strains on communications and night vision devices.

(3) The unit armorers and their small-arms repair kits provide only limited maintenance. S-4s should plan for increased weapons maintenance demands.

d. Replacements. Units conducting combat in built-up areas must expect high casualty rates. Casualty reports must be prepared scrupulously and forwarded via the S-1 to the battalion personnel officer located at the consolidated administration center (CONAD). Other considerations:

(1) The S-4s must plan to expedite the evacuation of wounded from the built-up area. Location of battalion aid stations and evacuation routes must be planned and disseminated to the lowest level. Higher casualty rates should be expected and may require the stockpiling of medical supplies and augmentation of additional HSS personnel.

(2) The personnel officer quickly processes replacements and coordinates their movement forward via the S-1 officer.

(3) Proper accountability of personnel at all levels is required. Timely and accurate personnel accountability and strength reporting support decisionmaking and initiate the replacement cycle. Leaders maintain accountability through the use of battle rosters and by establishing procedures for periodic reporting of numbers to higher commands. During combat, strength reports are provided on request or as significant changes occur.

5103. Critical Classes of Supply. Requirements for supply and resupply will vary from those encountered during operations in other tactical environments. The critical classes of supply listed below will assume greater importance during combat inside a built-up area.

a. Class I (Rations). The process of ordering and moving rations to forward units is complicated by the dispersed nature of combat in built-up areas and its increased caloric demands on Marines. Hot meals should be provided when practical.

(1) Combat in built-up areas not only causes great stress on Marines, but also requires great physical exertion. This combination of stress and exertion quickly causes dehydration. Unless potable water is continuously provided, Marines will seek local sources, which are usually contaminated by petroleum, oils, and lubricants (POL) runoff, sewage, bacteria, or unburied corpses. Marines who are not provided sufficient quantities of potable water become casualties as a result of drinking from contaminated sources or from dehydration. Waterborne contaminants can quickly render entire units combat ineffective.

(2) Water and other liquid supplements such as coffee, tea, or soup that must be forwarded to exposed positions may need to be backpacked at night.

b. Class II (General Supplies). Combat in built-up areas increases wear and tear on combat uniforms and footgear. Supply officers should increase on-hand stocks of uniforms, boots, and individual combat equipment such as protective masks and armored vests. Nuclear, biological, and chemical (NBC) protective suits become damaged quickly when worn in the urban environment. Extra stocks of these and protective mask filters should be kept on hand. (For further information on NBC considerations, see Appendix G.) Limited amounts of other Class II and IV (barrier materials) items may be available locally. These should be gathered and used if authorized and practical. Local shops may provide such items as hand tools, nails, bolts, chains, and light construction equipment, which are useful in preparing a defense or reducing enemy-held positions. The unit's organic wire communications net may be augmented with locally obtained telephone wire and electrical wire.

c. Class III (POL). Bulk fuel may have to be brought forward from fuel tankers by using 5-gallon cans. One man can carry a fuel can long distances, even over rubble, if it is lashed to a pack frame. Supplies of bulk Class III items and some prepackaged POL may be available at local gas stations and garages. These may be contaminated or of poor quality. The S-4s should coordinate to have fuel tests performed.

d. Class IV (Barrier Materials). If a unit is defending a built-up area, the requirements for Class IV materials are less than in other areas. This class of supply is probably the most available locally. After coordinating the effort with higher headquarters, S-4s and supporting engineer officers gather materials for use in strengthening the defense. Cargo trucks, wreckers, or recovery vehicles from maintenance platoons or engineer units can be used to load and move barrier materials. The defense of a built-up area may require concertina wire and/or barbed wire to restrict the enemy infantry's movements. Barriers can be built of abandoned cars and buses, which are dragged into position, turned on their sides, and chained together through the axles.

e. Class V (Munitions). Combat in built-up areas causes ammunition to be expended at extremely high rates. Commanders should plan for early resupply of explosives, grenades, and ammunition for small arms, direct fire, and indirect fire.

(1) In the defense, the S-4 should prestage as much ammunition as practical in dispersed storage areas. These storage areas should be protected, and they should be easily accessible to the forward defensive positions. In the offense, attacking forces should not be overburdened with excessive ammunition. Mobile distribution points may be set up as low as at company level.

(2) Leaders should plan to continuously deliver ammunition to the leading units as they advance. Ammunition may be carried by armored vehicles close behind the advancing forces or by designated carrying elements. Modern ammunition, particularly missiles, is characterized by extensive amounts of packing material. S-4s must remove the ammunition depot overpack before the ammunition is transported forward. Resupply by helicopter (prepackaged slingloads) may be feasible.

(3) Removing the overpack from large amounts of ammunition can be a time-consuming process. It may require being augmented by available Marines. If carrying elements are used to move ammunition forward, a Marine can carry about 75 to 90 pounds by using a pack frame. Bulky and heavier loads can be carried by lashing them to litters and using teams of two to four men. Loads of up to 400 pounds can be carried moderate distances using four-man teams.

Note: DO NOT use medical corpsmen to carry ammunition forward as described above—it is a violation of the Geneva Accords.

f. Class VIII (Medical Supplies). Because of the decentralized nature of combat in built-up areas, medical supplies should be dispersed throughout the unit, not just consolidated with the aid station and the individual corpsman. Marines should carry additional bandages, and units should have additional splints and stretchers.

5104. Health Service Support. The regimental and battalion surgeons are responsible for planning and executing HSS within their respective units. The most critical functions during combat in built-up areas include preventive medicine, trauma treatment, and evacuation. In addition, there should be a plan for the treatment and evacuation of NBC-related casualties that could occur in combat in built-up areas.

a. Combat in built-up areas exposes Marines not only to combat wounds, but also to the diseases endemic to the AO. Commanders must enforce preventive measures against the spread of infectious diseases. The unit surgeon advises the commander on how best to implement the use of prophylactics.

b. Corpsmen should be placed at strongpoints, at battle positions, and in units likely to be fighting in somewhat isolated positions.

c. Battalion aid stations must be placed farther forward than in operations in more open terrain. Protection offered by urban structures will permit this forward location, but the limited range of observation will require that personnel involved in evacuation be aware of their surroundings and the threat.

d. Corpsmen attached to rifle platoons are trained in the treatment of traumatic injuries; however, they can quickly become overwhelmed by the number of casualties needing care. If time and training are available before the conduct of urban operations, the commander should increase first aid training. Immediate first aid support by a buddy to an injured Marine may be the difference between life and death and will greatly assist the corpsmen. The aid station should plan to care for the increased casualties that are inherent to combat in built-up areas, as the incidence of crushing injuries, eye injuries, burns, shrapnel wounds, and fractures increases.

e. The difficulties encountered when evacuating casualties from urbanized terrain are many and require innovative techniques and procedures. The planning for evacuation on urbanized terrain must include special equipment requirements, use of litter teams, use of air MEDEVAC, use of an ambulance shuttle system, communications requirements, and techniques for locating casualties.

(1) Special equipment requirements include ropes, pulleys, skid litters, axes, crowbars, and other tools used to break through barriers.

(2) Although litter teams are labor intensive, they are required for evacuation from buildings, where casualties can occur on any level. Also, rubble in the streets, barricades, and demolition of roads will require a heavy reliance on litter teams.

(3) Communications presents one of the biggest obstacles to casualty evacuation. Urbanized terrain renders LOS radios ineffective. Also, individual Marines normally do not have access to radios. Therefore, a wounded Marine within a building may be difficult to find and evacuate. The unit SOP should contain alternate forms of communication such as colored panels or other forms of markers that can be displayed to hasten rescue when the battle is over. Also, a systematic search of the area after the battle may be required to recover casualties.

f. The use of local medical facilities, hospitals, professional medical help, and medical supplies may be available during combat in large, built-up areas. The commander must adhere to the guidelines established within the theater as to when and how these facilities can be used. If civilians are wounded in the unit's area, the commander is responsible for providing them with aid and protection without disrupting military operations. A commander cannot confiscate civilian medical supplies unless he makes provisions to provide adequate replacements if civilians are wounded.

g. The commander is responsible for the evacuation of deceased personnel to the nearest mortuary affairs collection point, whether they are U.S., allied, or enemy personnel, or civilians. (See Joint Pub 4-06, *JTTP for Mortuary Affairs in Joint Operations.*) Some general considerations for the handling of deceased personnel include:

(1) The theater commander is the approval authority for hasty burial.

(2) The deceased person's personal effects must remain with the body to assist in the identification of the body and to facilitate shipment of personal effects to the next of kin.

.3

Retention of personal items is considered looting and is, therefore, punishable by the Uniform Code of Military Justice (UCMJ).

(3) When operating under NBC conditions, the bodies of deceased personnel should be decontaminated before removal from contaminated areas to prevent further contamination and casualties.

(4) Care must be exercised when handling deceased personnel. Improper handling of deceased personnel can result in a significant decrease in unit and civilian morale.

5105. Personnel Services. Timely and accurate personnel services are just as important during combat in built-up areas as in any other operation.

a. The S-1 plans for all personnel services that support and sustain the morale and fighting spirit of the battalion. Among the most important of these services are:

- Replacement operations
- Strength accounting
- Casualty reporting.

These functions allow the commander to maintain accountability of his forces and to redistribute personnel or change taskings for subordinate units to maintain combat power and tempo.

Chapter 6

Noncombatant Considerations in Urban Operations

Noncombatants can have a significant impact on the conduct of military operations.

Section I

6101. Introduction. Commanders must be well educated in the legal aspects of military operations involving noncombatants. The urban environment contains numerous legal pitfalls in dealing with the civilian population. Adherence to the ROE and the standards of the Geneva Accords will keep the commander on firm legal ground.

6102. Noncombatant Impact. Large concentrations of civilians can greatly impede tactical operations. Noncombatants may have the following effects on military operations:

a. Mobility. Noncombatants civilians, attempting to escape the battlespace, can block military movement. Commanders plan routes to be used by civilians and seek the assistance of the civil police in refugee control.

b. Firepower. The presence of noncombatants can restrict the use of firepower. Areas may be designated no-fire areas to prevent noncombatant casualties. Other areas may be limited to small-arms fire and grenades. The control of fire missions may be complicated by the requirement for positive target identification. Detailed guidance on the use of fires in the presence of noncombatants is promulgated by the MAGTF commander. In the absence of guidance, the general rules of the law of land warfare apply.

c. Security. The presence of noncombatants increases security requirements in an urban environment to preclude:

- Noncombatants entering defensive areas
- Pilferage of equipment
- Sabotage
- Terrorism

d. Obstacle Employment. The presence and movement of Noncombatants will influence the MAGTFs commander's obstacle plan. Minefields may not be allowed on designated refugee routes or, if allowed, must be guarded until the passage of refugees is completed. The use of boobytraps may be curtailed until noncombatants have been evacuated. Commanders need to review and understand current national and international treaties concerning the employment of mines and boobytraps.

6103. Command Authority. The limits of command authority over government officials and noncombatants must be established. The host government's responsibility for its populace and territory can affect the commander's authority in civil-military matters. In less secure areas, where the host government may be only partly effective, the commander may have to assume greater responsibility for noncombatants.

6104. Source Utilization. Operations in highly populated areas may require the diversion of men, time, equipment, and supplies to accomplish humanitarian tasks. If host government agencies collapse, the impact on military resources could be substantial.

6105. Health and Welfare. The disruption of noncombatant health and sanitary services may sharply increases the risk of disease among both noncombatant and military personnel.

6106. Law and Order. Marines may have to augment civilian forces to protect life and property and to restore order. Marines may also have to secure vital government facilities for the host nation. (For more information on how to control civilians violating civil law, see FM 9-15, *Civil Disturbances*.)

6107. Public Affairs Officer and Media Relations. Free media access to units in the field is desirable; however, operational security, existing guidelines, and ROE considerations have priority. All members of the media should have an escort officer. Escort officers may be detailed from line units when public affairs personnel are unavailable. Enforcing established guidelines helps prevent negative publicity which could jeopardize the operation or national and strategic objectives.

6108. Civil Affairs Activities and Psychological Operations (PSYOP).

a. General. CA and PSYOP are mutually supporting and may have essential roles in MOUT. They can be critical force multipliers. CA and PSYOP may offer the possibility of victory without the destruction, suffering, and horror of protracted urban conflict.

(1) CA activities are activities performed by commanders, staffs, department of defense elements and units, and foreign military forces that (1) embrace the relationship between military forces and civil authorities and populations in areas where military forces are present; and (2) involve application of civil affairs functional specialty skills, in areas normally the responsibility of civilian government, which enhance conduct of civil-military operations (Joint Pub 1-02). CA activities are inherently civil-military in nature. They assist in coordinating military efforts to strengthen host government legitimacy, and help prevent or reduce violence by bridging critical gaps between civil and military sectors. MAGTF responsibilities for CA activities are determined during planning. U.S. Army CA units can be requested to support the MAGTF. The Marine Corps Civil Affairs Group (CAG), located in the Marine Corps Forces Reserve, may be activated and assigned to the MAGTF.

(a) CA activities are characterized by applications of functional specialties in areas normally the responsibility of indigenous government or civil authority. CA activities may extend to assumption of government functions required in an occupied territory during or immediately subsequent to hostilities. Refer to Joint Pub 3-57, *Doctrine for Joint Civil Affairs*.

(b) CA support to a MAGTF would normally include procuring local goods and services for military use, reducing civil interference in military operations, and assisting a MAGTF commander in fulfilling moral and legal obligations to the local government and noncombatant populace. See MCWP 3-33.1, *MAGTF Civil Affairs*, for more detailed information.

(2) Psychological operations are planned operations to convey selected information and indicators to foreign audiences to influence their emotions, motives, objective reasoning, and ultimately, their behavior. The purpose of psychological operations is to induce or reinforce foreign attitudes and behavior favorable to the originator's objectives (Joint Pub 1-02).

PSYOPs are planned and conducted by the MAGTF. PSYOPs are integrated and coordinated down to the lowest tactical level. PSYOPs can weaken enemy resistance and reduce its combat effectiveness. PSYOPs can also help prevent civilian interference with military operations.

PSYOPs may include the use of television, radio, posters, leaflets, and loud speakers, to disseminate propaganda and information. Television, including video tapes, is one of the most effective mediums for persuasion. In areas where television is not common, televisions may be distributed to public facilities and selected individuals. See MCWP 3-36.2/FM 33-1, *Psychological Operations*, for more detail.

6109. Refugee Control. The control of refugee movements in combat operations is primarily a host nation function.

The most important principle in refugee control is reliance on host government forces (military/civilian) whenever possible. U.S. forces will have limited refugee control capabilities. If a host government is incapable of performing the function, MAGTFs and MPs may be required to conduct any or all of the following refugee control measures:

- Enforce standfast orders or prohibit other unauthorized movements
- Screen and take custody of enemy agents or other hostile personnel discovered among the refugees
- Screen refugees for suspicious activities (including detection of infiltration by guerrillas, enemy agents, and escaping members of hostile forces)

.3

- Perform traffic control on designated refugee routes
- Escort refugee columns
- Direct or divert refugee columns to secondary roads and through areas not used by military elements
- Perform police duties such as traffic control or security at refugee assembly areas, collecting points, and centers or camps
- Enforce prescribed restrictive measures such as curfews, identification and pass systems, restricted area restraining lines, no-passage lines, checkpoints and roadblocks, travel permits or manifests, prohibition on the use of vehicles, priorities of movement, and limits on personal belongings.

Section II

6201. Commander's Legal Authority and Responsibilities

a. General. Leaders at all levels of command are responsible for protecting noncombatants and minimizing collateral damage to their property. Looting, vandalism, and brutal treatment of civilians are strictly prohibited by the Geneva Accords. Individuals who commit such acts may be severely punished. Civilians, and their religions and customs, must be treated with respect. Women are especially susceptible to many forms of abuse, and special care must be taken to protect them. Urban warfare offers many opportunities for abuses of the civilian population. This section focuses on those civilian-military confrontations that are most common in the urban environment.

b. Control Measures. Commanders may enforce control measures to conduct operations, maintain security, or ensure the safety and well-being of the civilians.

(1) Curfew. A commander with the mission of defending a town could establish a curfew to maintain security or to aid in control of military traffic. However, a curfew would not be legal if imposed strictly as punishment.

(2) Evacuation. A commander can require civilians to evacuate towns or buildings if the purpose of the evacuation is to use the town or building for essential military purposes, to enhance security, or to safeguard those civilians. The commander must specify and provide security for those being evacuated. Food, clothing, and sanitary facilities should be provided at the destination.

(3) Forced Labor. The Geneva Accords prohibit the use of civilians in combat operations. However, they may be required to perform some forms of forced labor prior to and after the conduct of combat operations. Guidelines for use of civilian labor should be published. The commander may force civilians over 18 years of age to work if the work does not oblige them to take part in military operations. An example of forced labor that is permitted would be maintenance of public utilities as long as those utilities are not used in the general conduct of the war. Jobs can also include services to the local population such as care of the wounded and burial. Civilians can also be forced to help evacuate and care for military wounded, as long as doing so does not involve any physical danger. Prohibited forced labor include digging entrenchments, constructing fortifications, transporting supplies or ammunition, or acting as guards. Only civilians who are volunteers can be employed in such work.

c. Civilian Resistance Groups. Another situation that commanders might encounter is combat with a civilian resistance group. The combatant status of these groups varies with their organization, methods, and adherance to the Geneva Accords. For example:

(1) Civilians accompanying their armed forces (with an identity card authorizing them to do so) are treated as EPWs when captured. Examples would include civilian members of military aircraft crews, war correspondents, supply contractors, and members of labor units or of service organizations responsible for the welfare of the armed forces.

(2) Captured civilians of a nonoccupying territory who have taken up arms against an invading enemy without time to form regular armed forces are considered EPWs if they wear a distinctive insignia that can be seen at a distance, carry their weapons openly, and operate according to the rules and customs of warfare. Other civilians who provide assistance to such groups may not be entitled to status as combatants, depending on whether they are actually members of the resistance group. They are normally best treated as combatants until a higher authority determines their status.

(3) Armed civilian groups that do not meet the criteria of a legal resistance group or individuals caught in the act of sabotage, terrorism, or espionage are not considered legal combatants. If captured, they will be considered criminals under the provisions of the law of land warfare. They should be detained in a facility separate from EPWs and should be quickly transferred to the MPs. Reprisals, mass punishments, hostage taking, corporal punishment, pillaging, or indiscriminate destruction of property are prohibited.

(4) The law of land warfare lets the commander control the civil population under the conditions already described using his own organic resources. However, language and cultural differences between U.S. and foreign personnel make it good practice to use local authorities such as the police for such purposes. Use of the police does not relieve a commander of his responsibility to safeguard civilians in his area.

d. Protection of Property. Like civilian personnel, civilian buildings and towns normally have a protected status. As such, they are not legal targets. They lose their protected status if authorities determine that the enemy is using them for military purposes. All doubt as to whether a town or building is a military target must be resolved before it can be targeted. For example:

(1) If the enemy is using a building or a portion of the town for military purposes—such as a supply point or a strongpoint—that building or that portion of the town is a legal target. However, engagement of the target should be limited to the buildings or portion of the town being occupied by the enemy.

(2) Normally, religious, historical, and cultural objects and buildings are not legal targets. They are sometimes marked with symbols to signify their status. Medical facilities are identified under the internationally recognized Red Cross, Red Crescent, Red Lion, or Red Star of David symbols. The fact that such symbols are absent does not relieve a commander of his responsibility to protect objects he recognizes as having religious, cultural, medical, or historical value.

(3) The misuse of such objects by the enemy is grounds to disregard their protected status. A demand should be made for the enemy to stop his misuse of the protected object before any action is taken. However, if an enemy FO or sniper uses a church for an OP, a commander would be justified in destroying it immediately because of the immediate danger posed to his troops. Another example is a religious shrine that is used as a telephone switchboard; in this case a warning would be appropriate because it would take some time to dismantle the system. Attacks on these types of targets should be limited to the least amount necessary to neutralize the enemy installations.

(4) The destruction, demolition, or military use of other buildings is permitted under the law of land warfare if required by clear military necessity. Destroying a house to obtain a better field of fire would be a legal act. But destroying it as a reprisal would not. However, firing on any houses that are occupied or defended by an enemy force is usually legal.

Chapter 7

The Urban Environment and Restrictions to Operations

"The responsibility of officers . . . and the training necessary are of a very different order from their responsibilities and training in ordinary military duties. Instead of striving to generate the maximum power with the forces available, the goal is to gain decisive results with the least application of force and the consequent minimum loss of life."

USMC, 1940 (FMFRP 12-15, *Small Wars Manual*)

Section I

Military Operations Other Than War

7101. Introduction. One of the most likely missions that U.S. Marines will undertake abroad will be military operations other than war (MOOTW). These missions typically will take place in the Third World and are normally associated with urban areas. MOOTW focuses on deterring war, resolving conflict, promoting peace, and supporting civil authorities in response to domestic crises. It is important to remember that political considerations permeate at all levels. Although some types of MOOTW, such as peace enforcement, may have the same characteristics of war, they normally have more restrictive rules of engagement (ROE).

Urban environments are inherently complex, and sometimes unpredictable. Local governments may have limited functions. Services may be overburdened, leading to allocation on the basis of personal connections and corruption. Services may be provided by a mixture of formal, high-technology and informal, labor-intensive systems. Large portions of the population may be poor, young, and underemployed. Nevertheless, there may be a high degree of social organization and interaction. Word of mouth and informal organizations may be as important as mass media and formal organizations to the communication structure of the area. Indigenous populations may have experienced centuries of foreign interventions and become quite adept at innovative ways of communicating. These factors and others often combine to make MOOTW a challenging and frustrating endeavor.

7102. Types of MOOTW. MOOTW encompasses a wide variety of military operations, that may take place in an urban environment. Forward deployed MAGTFs are ideal forces to respond to contingency operations involving MOOTW. These operations include:

- Combatting Terrorism
- DOD support to counterdrug operations
- Enforcement of Sanctions/MIO
- Enforcing Exclusion Zones

.3

- Humanitarian Assistance
- Military Support to Civil Authorities
- Nation Assistance/Support to Counterinsurgency
- NEO
- Peace Operations
- Recovery Operations
- Show of Force Operations
- Strikes and Raids.

7103. Recent Experience and Lessons Learned. Recent Marine Corps urban experiences—Lebanon (1982 - 1984); Grenada (1983); Panama (1989); and Somalia (1992 - 1994)— are examples of MOOTW. Although each operation was unique, the lessons learned are similar. For detailed information, commanders may obtain after action reports through the Marine Corps Lessons Learned System (MCLLS). Some of the lessons learned include:

(1) Demonstrate a powerful military presence. Where and when possible, use tanks, armored vehicles, heavy automatic weapons, fixed-wing aircraft, and attack helicopters.

(2) PSYOP, CI, HUMINT teams, and linguists, are essential to communications and intelligence gathering.

(3) Topographic detachment is useful to provide necessary sketches and maps.

(4) UAVs can be used for security, reconnaissance, and surveillance.

(5) Whenever entering an AO, isolate the area, patrol, and use intelligence to evaluate specific buildings or groups of buildings for search-and-clear operations. PSYOP, CI, HUMINT teams, and linguists can be used, as appropriate, to support patrols.

(6) Take advantage of the night by using night vision devices.

(7) The most effective means of dealing with snipers is with counter snipering operations.

(8) Snipers are also a tremendous asset to overwatch patrol movements, provide intelligence, and direct fire support. (See MCWP 3-15.3.)

(9) Random patrolling, traffic control points, random checkpoints/roadblocks, and ambushes disrupt enemy movements, reduce looting, seize curfew violators, and convince the population that order has been restored.

(10) Use armored vehicles to move quickly to new checkpoints and set up roadblocks or ambushes. They also increase protection for infantry. Their night vision sights also provide increased security and recognition during periods of limited visibility.

(11) Extensive patrolling provides current combat information used to maintain situational awareness. It helps to familiarize units with the AO and presents a strong military presence. (See Appendix A, Section VII)

(12) Tanks, armored vehicles, or attack helicopters increase military presence and can provide immediate, accurate gunfire.

(13) Before search and clear of a building, use loudspeakers to explain your intentions to noncombatants and enlist their cooperation.

(14) Use a minimum amount of force when dealing with noncombatants and their property. This will assist you in obtaining them to cooperate and may prevent turning them into active combatants. Assign engineers to search-and-clear forces, as necessary.

(15) MPs can instruct small unit leaders on the proper method of conducting vehicle, individual, and room search techniques. In some cases, it may be appropriate to attach MPs to search-and-clear forces. (See MCWP 3-34.1, *Military Police in Combat.*)

(16) Maintain a strong, mobile reserve.

(17) Avoid becoming predictable.

7104. Urban Terrorism. Terrorism is "the calculated use of violence or threat of violence to inculcate fear; intended to coerce or to intimidate governments or societies in the pursuit of goals that are generally political, religious, or ideological" (Joint Pub 1-02). Terrorism is a criminal act that is symbolic in nature. Its intent is to influence an audience other than the victim.

The likelihood of Marine involvement with terrorism increases in the urban environment. Terrorism in an urban environment can occur across the spectrum of conflict. Terrorists are an integral element in an insurgency and also play a major role in unconventional warfare. Terrorists can disrupt economic functions, expose a government's incompetence, eliminate opposition leaders, and elevate social anxiety. (See MCRP 3-02D, *Combating Terrorism*; and MCRP 3-02E, *The Individual's Guide for Understanding and Surviving Terrorism.*)

In countering this threat, Marines should determine whether it is internally or externally directed terrorism. Terrorism rooted externally must be severed from its roots. Against internal terrorism,

.3

Marines should attempt to penetrate the infrastructure and destroy the leadership of the terrorist groups.

The keys to defeating terrorists are awareness, education, and intelligence in order to deny, deter, delay, and detect terrorist acts. Rapid coordination among military units, military services, local police, and other government agencies is essential in denying the terrorists targets and refuge.

a. Nature of Urban Terrorism. Terrorism is generally an urban phenomenon. Incidents of terrorism are increasing in the international community, and in the United States, and pose a significant threat for U.S. military forces. Military personnel in noncombatant roles are considered terroist targets. For example, the attack on the Marine compound in Lebanon can be considered a terrorist attack instead of a military action because the Marines were officially in a noncombat activity (a multinational peacekeeping activity).

b. Terrorists in the Urban Area. Cities are target rich environments for terrorists. They offer ample cover and concealment. They contain large amounts of resources, transportation, communications, and potential recruits or sympathizers to their cause. Cities also provide the media attention essential for terrorists to accomplish their mission.

c. Combating Terrorism

(1) Counterterrorism. Counterterrorism includes the full range of offensive measures to prevent, deter, and respond to terrorism (Joint Pub 1-02). Population protection and control to combat the terrorist threat may be a major effort within the urban environment. The ability to protect the general population from attack and intimidation is essential.

(2) Antiterrorism. Antiterrorism includes defensive measures used to reduce the vulnerability of individuals and property to terrorist acts, to include limited response and containment by local military forces (Joint Pub 1-02). The extent of these measures varies based on assessments of the local threat. These include personnel awareness and knowledge of personal protection techniques. They also include crime prevention and physical security programs to "harden" the target, making installations and personnel less appealing as terrorist targets.

Section II

Restrictive Conditions

7201. Introduction. Commanders can expect to encounter restrictions on the use of force in MOOTW, especially in the urban environment. While we tend to focus on a traditional image of urban combat, MOUT extends beyond the house-to-house, close-combat effort to seize a city. Recent history provides many examples of the need for precise application of force to minimize collateral damage. MOOTW generally requires the application of force in varying degrees. Future missions in urban areas will likely follow this same pattern. While Marine forces must be prepared for more intensive urban combat, they must also train to apply only the level of force necessary to accomplish the mission. *Our tactics may have to rely more on innovation than firepower.*

Basic tenets of manuever warfare remain the same in MOOTW, but the tactics, techniques, and procedures may have to be modified to stay within established ROE and to avoid unnecessary collateral damage.

7202. Rules of Engagement. ROE are "directives issued by competent military authority which delineate the circumstances and limitations under which United States forces will initiate and/or continue combat engagement with other forces encountered" (Joint Pub 1-02). ROE for Marine forces are usually set by the JTF commander. They are based on National Command Authorities (NCA) guidance, the mission, the threat, the laws of war, and host nation constraints on force deployment.

MOOTW focuses on deterring war, resolving conflict, promoting peace, and supporting civil authorities in response to domestic crises. It is important to remember that political considerations permeate at all levels. Although some types of MOOTW, such as peace enforcement, may have the same characteristics of war, they normally have more restrictive rules of engagement (ROE). Restraint requires the careful balancing of the need for security, the conduct of operations, and the political objective. Excessive force antagonizes those parties involved, thereby damaging the legitimacy of the organization that uses it while possibly enhancing the legitimacy of the opposing party. Ensure all personnel understand the ROE and incorporate the practice into training exercises at all levels.

ROE must be fully integrated in the MAGTFs plans. As an example, table 7-1, presents the ROE used in Operation United Shield:

.3

ROE Used for Operation United Shield
Nothing in these Rules of Engagement limits your right to take appropriate action to defend yourself and your unit.
a. You have the right to use deadly force in response to a hostile act or when there is a clear indication of hostile intent.
b. Hostile fire may be returned effectively and promptly to stop a hostile act.
c. When US forces are attacked by unarmed hostile elements, mobs and/or rioters, US forces should use the minimum force necessary under the circumstances and proportional to the threat.
d. Inside designated security zones, once a hostile act or hostile intent is demonstrated, you have the right to use minimum force to prevent armed individuals/crew-served weapons from endangering US/UNOSOM II forces. This includes deadly force.
e. Detention of civilians is authorized for security reasons or in self-defense.
Remember: 1. The United States is not at war. 2. Treat all persons with dignity and respect. 3. Use minimum force to carry out mission. 4. Always be prepared to act in self-defense.

Table 7-1. ROE Used in Operation United Shield

7203. Planning. Commanders and planners thoroughly study each situation to identify the restrictive measures necessary to prevent needless civilian casualties and damage while also protecting the lives of the Marines who will execute the mission.

a. Mission/Command Guidance. The mission statement itself may determine much about the required level of force for a given operation. A humanitarian relief operation may require only individual weapons for security and personal protection. A raid on an insurgent communication center may need more. The requirements for an NEO conducted in a permissive environment would be different from those in a nonpermissive one. The key is to conduct a thorough analysis of the mission and determine the necessary combat power and how it must be applied to accomplish the mission. Many of these missions will be accompanied by a great deal of additional guidance concerning operational restrictions. These restrictions may come from the NCA, State Department, Department of Defense (DoD), or the joint task force (JTF) commander, and other sources. Such guidance must be carefully studied and incorporated into the plan. The commander can impose restrictions on his own forces that are more stringent then those dictated by higher authority. A careful analysis of the ROE will provide key insights into other restraints that the commander may wish to impose on his own forces.

b. Intelligence/Threat. Obtaining detailed and specific intelligence on the target and the area is essential to minimize collateral damage. Information on the physical layout of an area is often as important as information about the enemy. This is especially true for an NEO where avoiding conflict is a primary concern.

c. Location. The geographic area into which operations are to be conducted will normally dictate some restrictions on operations. In an urban environment, where the city is located in a desert, a common restriction is placed on the destruction of water producing facilities. If the operation is conducted where there is a large Hindu population, special protection may be given to cattle. If the operation is on an island nation, restrictions may be placed on destroying port or airfield facilities. In an urban environment with a large amount of shanty towns, restrictions may be placed on the use of incendiary devices.

d. Restrictions on Force Employment. The *right of self-defense is inherent in all military operations*. Marines should not be placed in a situation where they are not provided the means and authority to protect themselves or others. Beyond that, the restrictions placed on the forces conducting a particular operation depend on a variety of factors. These include:

- NCA guidance
- Type of mission
- Nature of the threat
- Political environment
- Host nation attitude.

e. Lethal Fires. Most restrictions in MOOTW are normally concerned with the application of fires. For instance, a NEO in a permissive environment may not require the landing of fire support assets However, a raid may require supporting arms to be landed with detailed instructions on their use. Aviation in MOOTW is commonly restricted in the amount and type of ordnance it can deliver because of the potential for collateral damage. However, in both a permissive and nonpermissive environment, supporting arms still should be planned. The most common types of fire support assets that will be restricted in the urban environment include artillery mortars, tanks, armored vehicles, and fixed-wing attack aircraft.

f. Search and Clear Buildings. Marine forces organize to search and clear a building under restrictive conditions in the same ways as discussed in Appendix C. However, because noncombatants may be occupying rooms within the building, clearing procedures must be modified as follows:

(1) Squad organization remains the same and is reinforced as appropriate (see Appendix A), except assault elements become search elements. Search elements enter and search rooms in the same manner as assault elements enter and clear rooms, as discussed in Appendix A.

(2) ROE must be identified and known by all personnel before entering or clearing a room or building.

(3) PSYOP or CA teams can help remove noncombatants before a search or battle starts.

(4) The building or area of search should be isolated with security forces and the intent to search the building should be communicated via loudspeakers.

(5) Assault forces move to the building and begin systematically to search room by room and floor by floor. Marines should carry ammunition and other equipment as in normal urban operations. However, Marines will be restricted in the use of their weapons and other equipment as dictated by the ROE. If permitted, a riot control agent (such as CS gas) can be used to assist in the clearing of a building. Additionally, shotguns and stun grenades are more appropriate than rifles and machine guns or fragmentation grenades for this type of restricted, close-combat operation.

g. Communications and Information Systems (CIS).

(1) CIS must be able to satisfy the C2 requirements of MOOTW. CIS systems must provide the commander with the ability to rapidly collect, process, analyze, and exchange information in order to keep the potential for violence to a minimum. These systems must make available the information needed, when it is needed, whenever it is needed. When supporting arms are on standby, fire support coordination and air traffic control nets are particularly critical.

(2) CIS officers (G/S-6) must carefully analyze the mission, situation, threat, and specific characteristics of the urban environment to design reliable, flexible, and responsive CIS. These systems should include single and multichannel radios, tactical telephones, and tactical satellite communications. A detailed backup CIS plan should also be prepared. The success of CIS in MOOTW depends on designing, planning, and employing CIS that satisfies the commander's C2 requirements.

Appendix A

Organization for Urban Combat and Fundamental Combat Skills

Successful combat operations in built-up areas depend on training in fundamental skills and proper employment techniques and procedures.

Before conducting urban operations, Marines must be trained in fundamental combat skills to conduct urban operations. Training for this unique environment will reduce casualties because properly prepared Marines will be able to fight more effectively. This appendix supersedes all other fundamental combat skills taught for fighting in urbanized terrain. Urban combat will require Marines to use their best initiative and judgment in applying the techniques and procedures described in this appendix.

Section I

Organization

1. Structure. Basic infantry organization does not change when the battlefield is moved into an urban area. However, infantry units may be task organized or reinforced down to squad level in order for small units to fight and win in this violent, three-dimensional battlespace. Several new terms that address organization and employment have been added to clearly define mission assignments. Figures A-1 through A-3 show platoon through fire team organization breakdown.

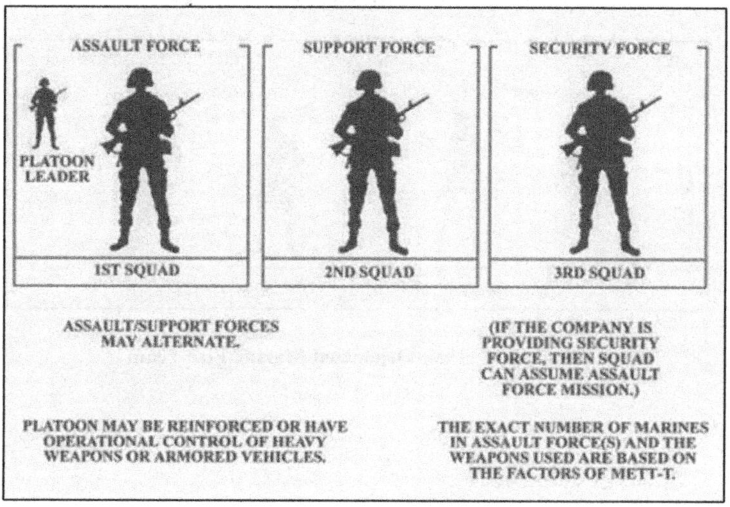

Figure A-1. Task-Organized Marine Platoon

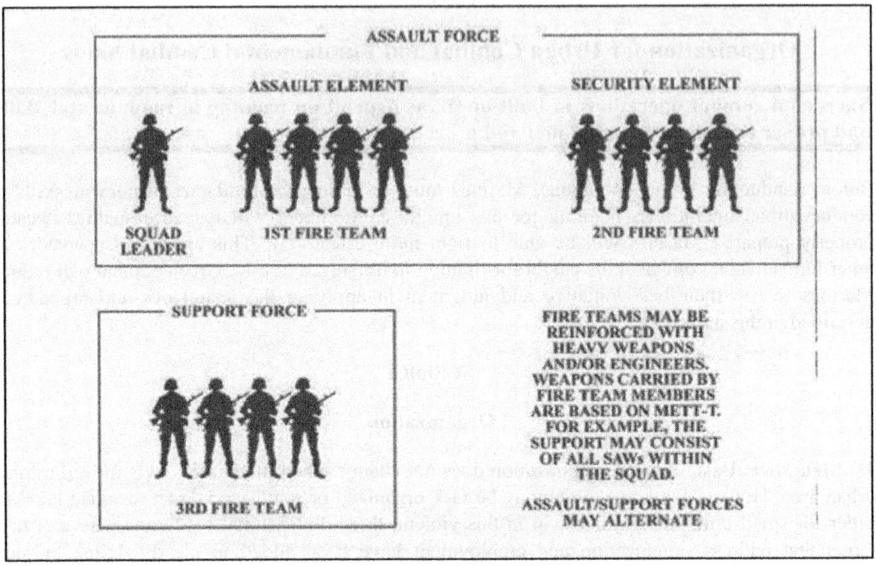

Figure A-2. Task-Organized Marine Squad

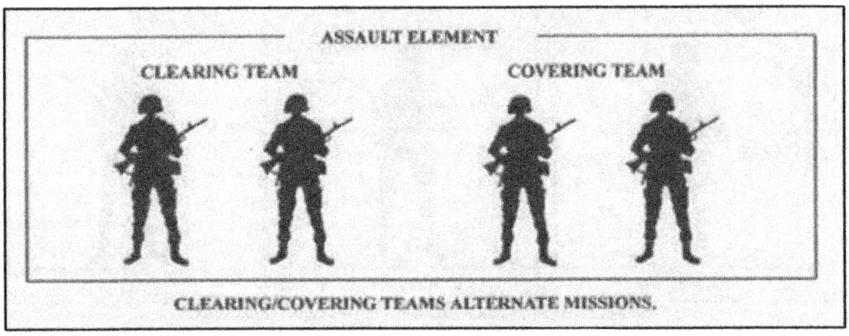

Figure A-3. Task-Organized Marine Fire Team

Section II

Weapons Handling and Firing Techniques

2. Weapons Carries. During MOUT, a Marine often finds himself very unexpectedly engaging targets at close ranges. Weapons carries provide a safe and effective method for handling the service rifle. For additional in-depth discussion of weapons handling and firing techniques, refer to MCWP 3-01X, *Field Firing of the M16A2 Rifle.*

a. Tactical Carry. The tactical carry is used when no immediate threat is present. It permits control of the rifle while moving and still allows quick engagement of the enemy. The buttstock of the rifle is placed alongside the body at approximately hip level, and the barrel is angled upward approximately 45 degrees in the general direction of the enemy (Figure A-4).

Figure A-4. Tactical Carry

b. Alert Carry. The alert carry is used when enemy contact is likely. Engagement of the enemy is faster from the alert carry than from the tactical carry. The buttstock of the rifle is placed in the shoulder with the muzzle angled down approximately 45 degrees and pointed in the likely direction of the enemy (Figure A-5).

In the alert carry, the Marine keeps both eyes open and scans for any threat with the weapon always in line with the Marine's LOS (this is called "guns 'n' eyeballs"). When a threat is spotted, the Marine quickly snaps his weapon into a firing position, looks over the rear sight aperture, and aligns the tip of the front sight post center mass on the target. Once the threat has been eliminated, the Marine returns to the alert carry and continues to scan for targets. Accuracy is the primary goal.

Figure A-5. Alert Carry

c. Ready Carry. The ready carry is employed when contact with the enemy is imminent. It allows for immediate target engagement. The buttstock of the rifle is in the shoulder with the muzzle of the rifle pointed in the direction of the enemy (Figure A-6). As in the alert carry, sight alignment and sight picture are achieved as the shot is fired. Marines should always strive for a clear tip of the front sight post and center mass hold.

Figure A-6. Ready Carry

d. Short Stocking. Rifles may be "short stocked" to increase their maneuverability in an enclosed area and to reduce possible "target indicators" for the enemy (muzzles sticking around corners, for example). The weapon is held in such a manner as to reduce the length of the weapon without sacrificing too much accuracy (Figure A-7). The stock is positioned so that the pistol grip is behind the Marine's head. The Marine may use his index finger or thumb to manipulate the trigger. The handguard of the weapon is placed against the Marine's cheek, and a firm stockweld is attained. The Marine should use the tip of the front sight post. Short stocking may be conducted from either side. However, care should be taken not to mask the ejection port cover. The short stocking technique may be used in the tactical, alert, or ready carry.

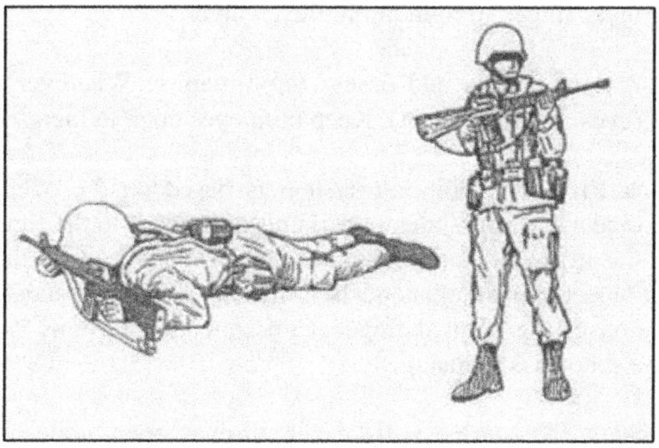

Figure A-7. Short Stocking

e. Weak-Hand Carries. Marines may carry their weapon on the weak-hand side to obtain maximum cover and fire from the weak-hand position to avoid exposing themselves. However, firing from the weak-hand position may reduce a Marine's ability to fire accurately. This may be acceptable when Marines are providing supporting fires, but in close kill-or-be-killed engagements, rapid, accurate firing from the strong-hand position is preferred.

3. Firing Techniques. The speed and uncertainty of combat require Marines to act without hesitation while achieving accurate target engagement. Several firing techniques are listed below. For an in-depth discussion of firing techniques that can be applied to MOUT operations, see MCWP 3-01X.

a. Pieing. Pieing is an effective technique for clearing dead space inside rooms and buildings to gain security of hallways, stairwells, mouseholes, and so on. It is conducted by using the ready carry position or short stocking technique. The weapon is aimed at a sector of a window, doorway, corner, or hallway and slowly moved at different angles, sectoring off the window, doorway, corner, or hallway until each sector is cleared of any threat. As soon as a hostile threat is seen, immediate, accurate fire can be placed on the threat. Figures A-11, A-34, A-35, A-38, A-47, and A-49 show examples of the pieing technique.

b. Aimed Quick Fire. The Marine's initial focus is on the target. As the rifle is brought up, the firing eye looks through or just over the rear sight aperture, and the front sight post is used to aim at the target. The Marine fires two quick shots. Focus remains on the front sight post throughout the aiming process.

Immediately after target engagement, the Marine scans for additional targets. To scan, perform the following steps:

(1) Lower the weapon to look over the sights.

(2) Place the trigger finger straight along the receiver.

(3) Scan the area for targets and assess the situation. Wherever the head moves, the muzzle moves (eyes, muzzle, target). Keep both eyes open to increase the field of view.

c. Pointing Quick Fire. The pointing system is based on the phenomenon that when a person looks at an object and simultaneously points a finger at it, the finger aligns itself on the point of focus of the eyes with no conscious effort on the part of the individual. When a Marine looks at an object and simultaneously brings his rifle to his shoulder, the rifle in effect becomes an extension of the pointed finger. Consequently, it aligns itself naturally with the object on which the shooter is focusing.

When a target appears, the Marine will keep both eyes open, concentrating intensely on a small, specific focal point near the base of the target mass. The rifle is brought simultaneously to the hollow of the shoulder. The head is held high, stock welded to the jaw. The eyes are 2 - 3 inches over the top of the sights, staring intently at the target. As soon as the rifle is brought to the shoulder, two quick shots are fired. Focus on the target is not broken during the interval between initially seeing the target and discharging the rifle.

d. Instinctive Shooting. There may be situations in which a Marine is surprised and may need to react immediately. If possible, the Marine should engage the threat by using the tip of the front sight post. However, *speed* may be more important. The Marine's weapon and body are quickly "pointed," and the target is engaged. It is important that the body be turned with the weapon in order to achieve a natural point of aim. Simply pointing the weapon will usually result in a miss. Once the first two shots have been fired and the Marine regains the initiative, the weapon should be quickly moved to the Marine's shoulder and the tip of the front post used for sighting subsequent shots.

Section III

Movement

Movement techniques within built-up areas are the first fundamental skills that Marines should master. These movement techniques should be practiced until they become habitual. To reduce exposure to enemy fire, Marines should avoid silhouetting themselves, avoid open areas, and select their next covered position before moving.

4. Crossing a Wall. Each Marine should learn the correct method of crossing a wall (Figure A-8). After reconnoitering the other side, the Marine quickly rolls over the wall, keeping a low silhouette. The speed of his movement and a low silhouette deny the enemy a good target.

Figure A-8. Crossing a Wall

5. Observation Around Corners. The area around a corner should be observed before the Marine moves beyond it. The most common mistake the Marine makes at a corner is allowing his weapon to extend beyond the corner before observing, thereby exposing his position. Using the short stocking technique (Paragraph 2.d.) reduces exposure. Additionally, Marines can use a technique called "popping the corner" to reduce exposure time (Figure A-9). The Marine will get into a prone position near the corner of a building or obstacle around which he needs to observe. The weapon is short stocked, and the muzzle is pointed in the direction the Marine is looking. This allows the Marine to engage a target, if necessary, when he observes around a corner. The Marine will crawl to the corner but not expose himself. He will raise his upper body onto his elbows. The Marine will then push his body forward with his feet and legs without moving his elbows. His upper body, with the weapon ready, will move forward. The final position will expose the weapon, the Marine's helmet, and a minimal amount of the Marine's face. The Marine's forearms will come to rest on the deck giving him a low profile, the ability to observe around the corner, and the immediate capability to engage targets with his weapon.

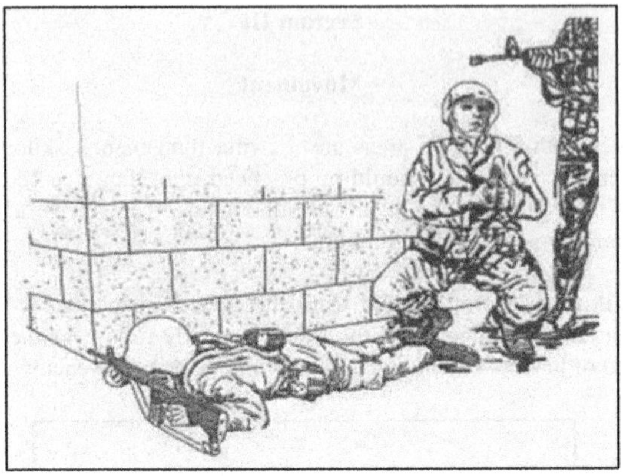

Figure A-9. Correct Technique for Observing Around a Corner

6. Movement Past Windows. Windows present another hazard to Marines. The most common mistake made when passing a window is exposing the head to enemy observation from inside the room. If a Marine shows his head, an enemy gunner inside the building could engage him through the window without exposing himself to return fires.

a. The correct technique for passing a window is for the Marine to stay below the window level. He makes sure he does not silhouette himself in the window; he "hugs" the side of the building. An enemy gunner inside the building would have to expose himself to covering fires to engage the Marine (Figure A-10A).

Figure A-10A. Moving Past Windows

b. The most common mistake Marines make when passing a basement window is not being aware of it. A Marine should not walk or run past a basement window because he presents a good target to an enemy gunner inside the building. When using the correct procedure for negotiating a basement window, the Marine stays close to the wall of the building and steps or jumps past the window without exposing his legs (Figure A10-B).

c. An alternate method is for the pointman to pie off the window (or other opening) as he approaches it. As the pointman moves toward the window, the second Marine (behind the pointman) steps out to provide security to the direct front. The pointman continues to cover the opening by pieing, all the while using the near edge for cover. The pointman covers and sectors off that part of the interior that can be viewed from the outside. The second Marine moves with the pointman and maintains security to the front, eventually moving past the pointman as the pointman clears the inside corners of the opening. Once the opening is cleared, the second man may assume the duties of point, or when subsequent Marines move up to secure the opening, the original pointman may reassume those duties. Figure A-11 illustrates this alternate method.

Figure A-10B. Marine Passing Basement Windows

7. Use of Doorways. Doorways should not be used as entrances or exits because they are normally covered by enemy fire. If a Marine must use a doorway as an exit, he should move quickly through it to his next position, staying as low as possible to avoid silhouetting himself (Figure A-12). Preselection of positions, speed, a low silhouette, and the use of covering fires should be emphasized when exiting doorways.

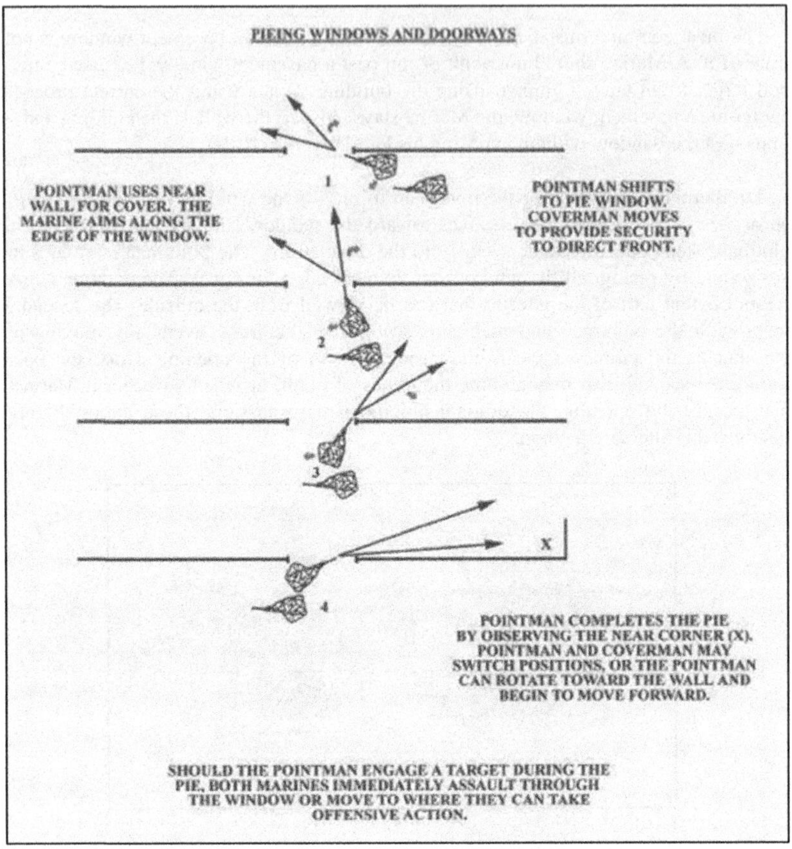

Figure A-11. Alternate Technique for Moving Past Windows

8. Movement Parallel to Buildings. Marines and small units may not always be able to use the inside of buildings as a route of advance. In such cases, they should move on the outside of the buildings (Figure A-13). Smoke, covering fires, and cover and concealment should be used to hide movement. When correctly moving along the outside of a building, the Marine hugs the side of the building, stays in the shadow, presents a low silhouette, and moves rapidly to his next position (Figure A-14). If an enemy gunner inside the building fires on a Marine, he exposes himself to fire from other squad/fire team members who are covering him. Furthermore, an enemy gunner farther down the street may have difficulty detecting and engaging the Marine.

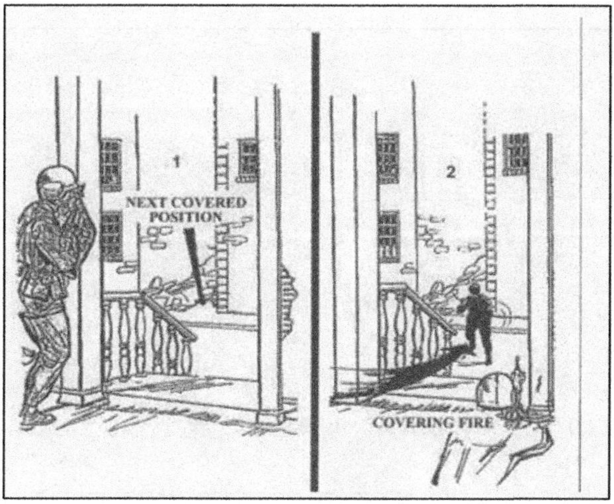

Figure A-12. Exiting a Doorway

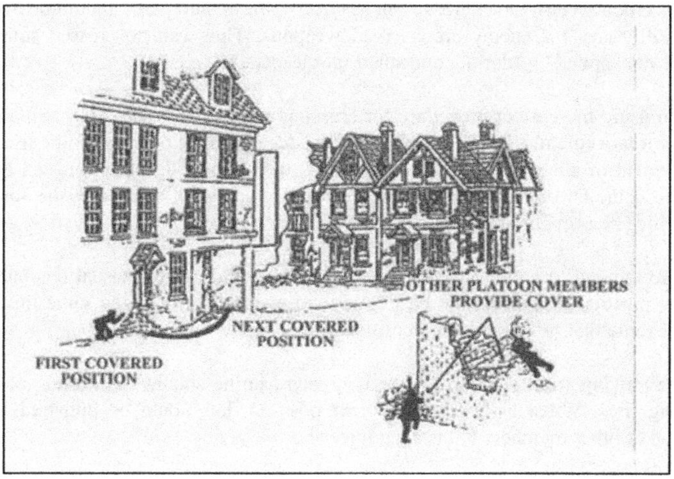

Figure A-13. Moving Outside a Building

Figure A-14. Selection of the Next Position

9. Crossing Open Areas. Open areas such as streets, alleys, and parks should be avoided. They are natural kill zones for enemy crew-served weapons. They can be crossed safely if certain fundamentals are applied by Marines and small-unit leaders.

a. When using the correct procedure for crossing an open area, the Marine develops a plan for his own movement. (Smoke from hand grenades or smoke pots should be used to conceal the movement of all Marines.) The Marine runs the shortest distance between buildings and moves along the far building to the next position. By doing so, he reduces the amount of time during which he is exposed to enemy fire.

b. Before moving to another position, the Marine should make a visual reconnaissance and select the position that offers the best cover and concealment. At the same time, he should select the route that he will take to get to that position.

c. When moving from position to position, each Marine should be careful not to mask his supporting fires. When he reaches his next position, he should be prepared to cover the movement of other members of his assault force or element.

10. Assault Element Employment. Moving as an assault element from building to building or between buildings presents a problem because an assault element presents a large target to enemy fire. When moving from the corner of one building to another, the assault element should move across the open area in groups of two, maintaining their integrity within their respective clearing or covering team (Figure A-15). The assault element leader (fire team leader) is responsible for the secure movement of his unit. Smoke should be used to screen the assault element's movement. The covering team will move first to establish security on the far side. Their movement is protected by the fire of the clearing team. Once the covering team is in position, they provide cover as the clearing team moves across and prepares, then enters, the structure. Moving from the side of one building to the side of another presents a similar problem, and the technique of movement employed is the same (Figure A-16).

There may be situations in which it is best to move in larger groups. An example is that Marines crossing in pairs may key enemy snipers in to the movement, setting them up for good shots at the next pair to cross. Leaders should remain aware of the situation and threat and plan movements accordingly.

Figure A-15. Assault Element Movement

11. Movement Inside a Building. When moving within a building that is under attack (Figure A-17), Marines should avoid silhouetting themselves in doors and windows. If forced to use a hallway (Figure A-18), Marines should move by hugging the wall to avoid presenting targets to the enemy.

Figure A-16. Movement to an Adjacent Building

Figure A-17. Movement Within a Building Under Attack

Figure A-18. Moving Down a Hallway

Section IV

Entry Techniques

Marines should enter a building with minimum exposure. They should select an entry point before moving toward the building; avoid windows and doors; use smoke to conceal their advance to the building; use demolitions, tank rounds, and so on to make new entrances; precede the entry of the clearing team with a grenade; enter immediately after the grenade explodes; and be covered by the covering team.

12. Upper Building Levels. Clearing a building from the top down is the preferred method. Clearing or defending a building is easier from an upper story. Gravity and the building's floor plan become assets when throwing hand grenades and moving from floor to floor.

a. An enemy who is forced to the top of a building may be cornered and fight desperately or escape over the roof, but an enemy who is forced down to the ground level may withdraw from the building, thus exposing himself to friendly fires from the outside.

b. Various means such as ladders, drainpipes, vines, helicopters, or the roofs and windows of adjoining buildings may be used to reach the top floor or roof of a building. In some cases, one Marine can climb onto the shoulders of another and reach high enough to pull himself up. Another method is to attach a grappling hook to the end of a scaling rope so that a Marine can scale a wall, spring from one building to another, or gain entrance through an upstairs window.

13. Use of Ladders. Ladders offer the quickest method to gain access to the upper levels of a building (Figure A-19). Units can get ladders from local civilians or stores or obtain material to build ladders through supply channels. If required, ladders can be built with resources that are available throughout the urban area; for example, lumber can be taken from inside the walls of buildings (Figure A-20). Although ladders may not permit access to the top of some buildings, they may offer security and safety through speed.

14. Use of Grappling Hook. A suitable grappling hook and rope are selected to meet the need. The grappling hook should be sturdy, portable, easily thrown, and equipped with hooks that can hold inside a window. The scaling rope should be 5/8 of an inch to 1 inch in diameter and long enough to reach the objective window. Knots are tied in the rope at 1-foot intervals to make climbing and holding easier. Marines should follow the procedures outlined below.

a. When throwing the grappling hook, the Marine should stand as close to the building as possible (Figure A-21). If the Marine stands close to the building, he is less exposed to enemy fires and he does not have to throw the hook as far.

Figure A-19. Using Ladders To Get to Upper Levels

Figure A-20. Getting Lumber From Inside the Walls

Figure A-21. Grappling Hook Thrown at Close Range

b. Making sure there is enough rope to reach the target, the Marine should hold the hook and a few coils of rope in the throwing hand. The remainder of the rope, in loose coils, should be in the other hand. The rope should be allowed to play out freely. The throw should be a gentle, even, upward lob of the hook, with the other hand releasing the rope as it plays out.

c. Once the grappling hook is inside the window (or on the roof), the Marine should pull on the rope to obtain a good hold before beginning to climb. When using a window, the hook should be pulled to one corner to ensure a good "bite" and to reduce the Marine's exposure to lower windows during the climb.

d. The use of grappling hooks is the least preferred method for gaining entry to upper levels of buildings. Grappling hooks should be used only as a last resort and away from potential enemy positions. This method may potentially be used on adjacent buildings that offer concealed locations and a roof that connects to enemy positions.

15. Scaling Walls. When forced to scale a wall during exposure to enemy fire, all available means of concealment should be used. Smoke and diversionary measures improve the chances of a successful exposed movement. When using smoke for concealment, Marines should plan for wind direction. They should use fire, shouts, and fake movement to distract the enemy.

a. A Marine scaling an outside wall is vulnerable to enemy sniper fire. Marines moving from building to building and climbing buildings should be covered with friendly fire. Areas between buildings offer good fields of fire to the enemy. Properly positioned friendly weapons can suppress and eliminate enemy fire. The M203 grenade launcher is effective in clearing the enemy from rooms inside buildings (Figure A-22).

Figure A-22. Employment of the M203 Grenade Launcher

b. A Marine scaling a wall with a rope should avoid silhouetting himself in windows of uncleared rooms or exposing himself to enemy fires from lower windows. He should climb with his weapon slung over the firing shoulder so that he may quickly bring it to a firing position. He should clear the lower room with a hand grenade before going outside the window. Before doing this, the Marine first loosens the safety pin so that he needs only one hand to throw the grenade. The objective upper story window should not be entered before a hand grenade has been thrown in.

c. The Marine enters the objective window with a low silhouette (Figure A-23). Entry can be head first; however, a preferred method is to hook a leg over the windowsill and enter sideways, straddling the ledge.

16. Rappelling. Rappelling (Figure A-24) is an entry technique that Marines can use to descend from the rooftop of a tall building into a window. (See Army manual TC 21-24 for more information on rappelling.)

17. Entry at Lower Levels. Buildings should be cleared from the top down. However, it may be impossible to enter a building at the top; therefore, entry at the bottom or lower level may be the only course of action. When entering a building at the lower level, Marines avoid entry through windows and doors because both can be easily boobytrapped and are usually covered by enemy fire.

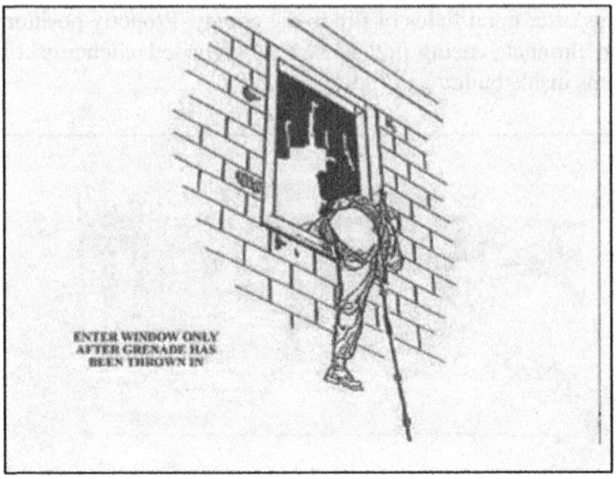

Figure A-23. Entering the Objective Window

Figure A-24. Rappelling

a. Ideally, when entering at lower levels, demolitions, artillery, tank fire, antiarmor weapons fire, or similar means are used to create a new entrance to avoid boobytraps. Quick entry is required to follow through on the effects of the blast and concussion.

b. When the only entry to a building is through a window or door, supporting fire should be directed at that location. (Specific lower level entry techniques are shown in Figure A-25.) If no supporting fire is available, AT4s can be employed instead.

c. Before entering, Marines throw a hand grenade into the new entrance to reinforce the effects of the original blast. When making a new entrance in a building, they consider the effects of the blast on the building and adjacent buildings. If there is the possibility of a fire in an adjacent building, Marines coordinate with adjacent units and obtain permission before starting the operation. In wooden-framed buildings, the blast may cause the building to collapse. In stone, brick, or cement buildings, supporting fires are aimed at the corner of the building or at weak points in the building construction.

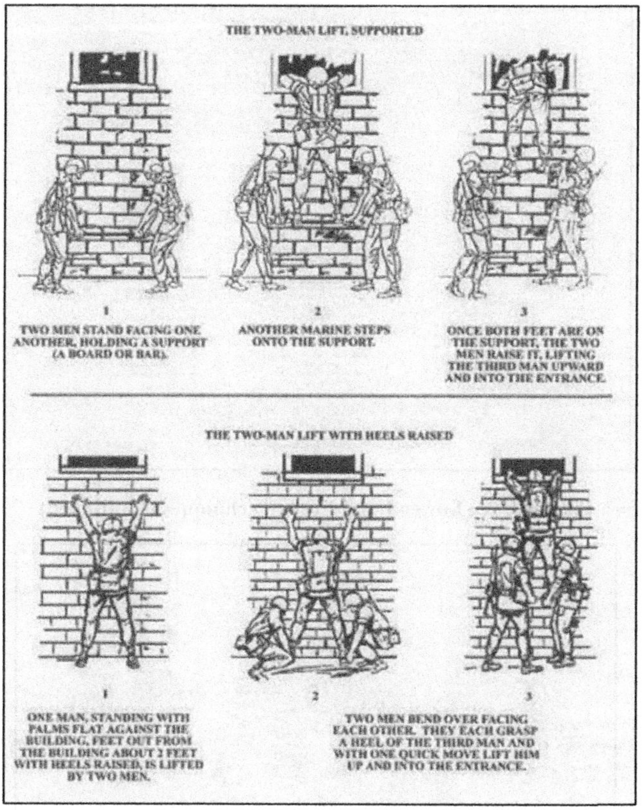

Figure A-25. Lower Level Entry Techniques

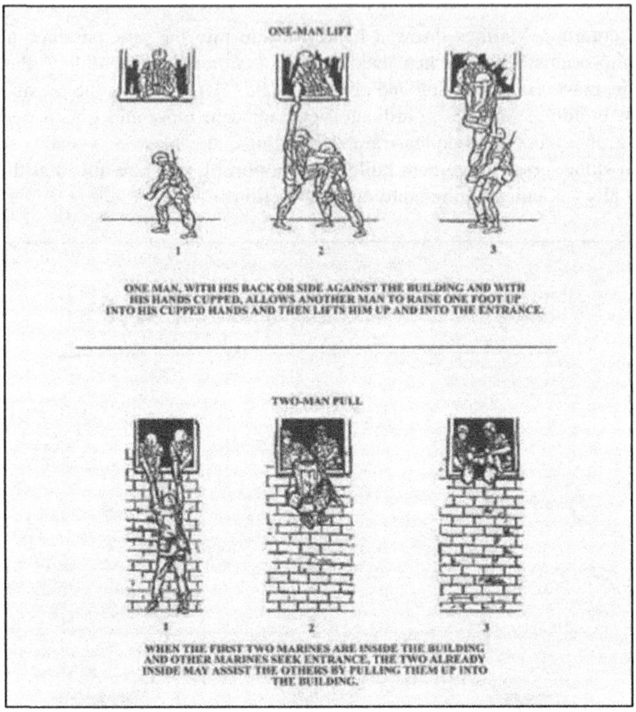

Figure A-25. Lower Level Entry Techniques (Continued)

Figure A-25. Lower Level Entry Techniques (Continued)

18. Hand Grenades. Combat in built-up areas (mainly during the attack) requires extensive use of hand grenades. The Marine should throw a grenade before entering rooms or negotiating staircases, mouscholes, and so on. This usually requires the use of both hands and employs both the overhand and underhand methods of throwing.

a. Two techniques may be used in the throwing of a grenade. The *preferred technique* involves throwing the grenade hard enough that it bounces or skips around, making it difficult to pick up. The hard-throw, skip/bounce technique may be used by Marines in training and combat. The *least preferred technique*, which involves cooking-off the grenade for two seconds then throwing it to prevent the enemy from grabbing the grenade and tossing it back, may be used by Marines, as appropriate, during actual combat but not in training.

b. Nonverbal or verbal alerts are used, as appropriate, before throwing a grenade. A visual showing (preferred) of the grenade to be thrown is made to assault element members, and a visual acknowledgment from them is received. A nonverbal alert may ensure that the enemy is surprised when the grenade is thrown. If the situation demands, a voice alert can be used, but the element of surprise may be lost. When or if a voice alert is used, the voice alert is "**FRAG OUT**;" when an enemy grenade has been identified, friendly forces shout, "**GRENADE**."

c. The construction material used in the building being cleared influences the use of grenades. In some situations, concussion grenades may be preferred over fragmentary grenades during offensive operations or when defending from hasty defensive positions. If the walls of a building are made of thin material, such as sheetrock or thin plyboard, the Marine should either lie flat on the floor with his helmet pointing toward the area of detonation or move away from any wall that might be penetrated by grenade fragments.

d. Marines should throw grenades into an opening before entering a building to eliminate any enemy that might be near the entrance (Figure A-26). Using the M203 grenade launcher is the best method for putting a grenade in an upper story window.

e. When a hand grenade must be used, the Marine throwing the grenade should stand close to the building, using it for cover. At the same time, the individual and the rest of the element should have a planned area to which they can move for safety if the grenade does not go through the window, but falls back to the ground.

f. The Marine throwing the grenade should step out far enough to lob the grenade into the upper story opening. The Marine's weapon should be kept in the nonthrowing hand so that it can be used if needed. The weapon should never be laid down outside or inside the building. Once the grenade has been thrown into the opening, assaulting forces should move swiftly to enter the building. This technique should be employed only when the window has already been broken. Otherwise, the chances are high that the thrown grenade will bounce off of the window and fall back onto the ground without going into the room.

Figure A-26. Hand Grenade Thrown Through Window

WARNING

After throwing the grenade, the Marine should immediately announce, "FRAG OUT," to indicate that a grenade has been thrown. He then takes cover because the grenade may bounce back or be thrown back, or the enemy may fire at him.

g. If Marines must enter the building by using the stairs, they first look for boobytraps. Then they throw a grenade through the stairwell door, let it detonate, and move quickly inside. They can use the staircase for cover.

h. The best way to enter a building is to breach the exterior wall. Again, a grenade should be thrown through the hole while using all available cover, such as the lower corner of the building (Figure A-27).

i. Another way to enter a room is to blast mouseholes with demolitions. In moving from room to room through mouseholes, Marines should use grenades to clear the room as they do in moving through open doorways.

j. Although buildings are best cleared from the top down, this is not always possible. While clearing the bottom floor of a building, Marines may encounter stairs that must also be cleared. Once again, grenades play an important role. Before climbing the stairs from the bottom floor, Marines should first inspect for boobytraps then toss a grenade to the head of the stairs, if appropriate (Figure A-28). The use of a grenade in these situations is based on stair/building construction and building layout. *Throwing grenades up staircases is extremely* *dangerous, and extreme caution and common sense should be used.* In most situations, throwing a grenade up a staircase is not recommended. (Again, the grenade is being used in the initial building entry or clearing of the bottom floor of a building having a staircase.) If a grenade is thrown too hard or the staircase is too steep, the grenade can roll back down the staircase. For this reason, if a grenade is thrown, it should be thrown underhand to reduce the risk of it bouncing back and rolling down the stairs. A nonverbal or, as appropriate, voice alert is used when throwing the grenade. Once the first grenade has detonated, another grenade should be thrown over and behind the staircase banister and into the hallway, destroying any enemy hiding to the rear.

Figure A-27. Entering Through a Breached Hole

Figure A-28. Tossing a Grenade Up a Stairway

k. After the stairs have been cleared, assaulting forces move to the top floor and clear it. Upon clearing the top floor, forces move downstairs to clear the center and bottom floors and to continue with the mission.

Note: Because large quantities of hand grenades are used when clearing buildings, a continuous supply must be available to forces having this mission within a built-up area.

Section V

Clearing Techniques

This section will discuss basic clearing techniques; however, the fundamentals may have to be modified depending on the various situations and building/room layouts that Marines may encounter. Caution should be taken when using a stacked type of position as personnel may be concentrating in a danger area where they could be exposed to fires penetrating walls or explosive effects from boobytraps or grenades thrown by the enemy. *Marines should always remain aware of the protective, or lack of protective, capabilities of the materials comprising the walls, floors, ceilings, and doors within the areas they are clearing.*

19. Clearing a Room, Door Closed or Open, From Stacked Positions. The following discussion and series of figures will focus on the two-man clearing team in the conduct of clearing a room. The covering team is positioned by the assault element leader to provide security as the clearing team begins its procedures to enter and clear a room.

a. Shooter Number Two is behind Shooter Number One (stacked position). This positioning is the same if the door is already open or nonexistent (Figure A-29).

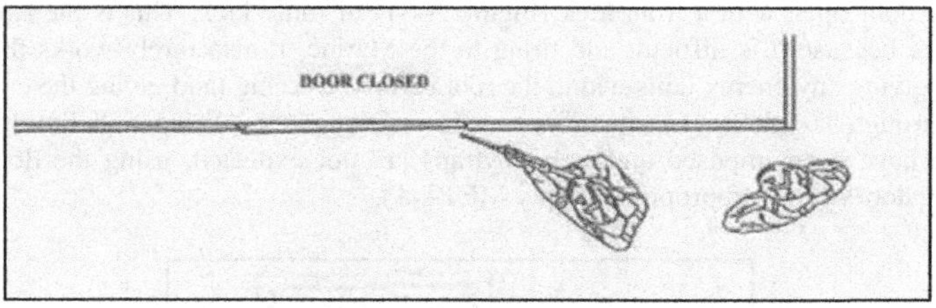

Figure A-29. Clearing a Room, Door Closed

b. Shooter Number One determines the direction in which the door opens then quickly informs Shooter Number Two and forces the door open.

When a door is the only means of entering a room, Marines should beware of fire from enemy soldiers within the room and of boobytraps. When opening a door, Marines should not expose themselves to fires through the door. The two-man clearing team forces the door open by directing a short burst of automatic fire through the door around the latch (Figure A-30) and then kicking it open. Another method of opening the door is to use a shotgun with 12-gauge slugs to shoot the hinges and/or the latch. When the doorknob is locked and/or deadbolt locks are present, shooting the hinges of the door is recommended. The door may then be opened by kicking or using a battering ram.

Figure A-30. Shooting the Door Open

Marines can also use an ax or demolitions if they are available. As a last resort, Marines can kick the door open with a front kick (Figure A-31) or mule kick. This is the least preferred technique because it is difficult and tiring to the Marine. It also rarely works the first time, thereby giving any enemy units within the room ample warning (and giving the enemy time to shoot through the door). In situations where restrictions on firing and limiting structural damage have been imposed and/or boobytraps are not expected, using the door handle to open the door may be appropriate (apply METT-T).

Figure A-31. Opening a Door With a Front Kick

c. Once the door is open, Shooter Number Two immediately throws a hand grenade into the room. Shooter Number One maintains cover for Shooter Number Two as he prepares to throw the grenade into the room. Shooter Number Two then returns to the original stacked position behind Shooter Number One (Figure A-32). If Marines have no grenades, the procedures remain the same except for throwing the grenade. *Note: When the door is already open or nonexistent, the stacked position is always used.*

WARNING

Because fragments from M67 fragmentation grenades may injure Marines outside the room, they should not be used. Marines should use MK3A2 offensive hand grenades instead. Also, the technique of cooking-off hand grenades can be deadly unless properly performed.

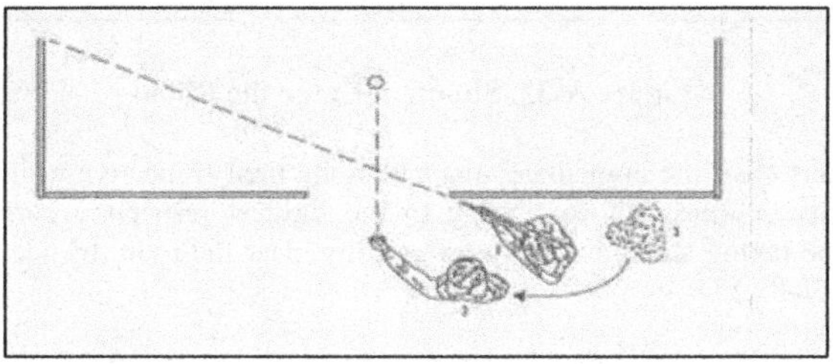

Figure A-32. Throwing a Grenade Into a Room

(1) Throwing Grenades. Two techniques are available for using grenades in room clearing. The *preferred technique* for Marines is to throw a concussion hand grenade into the room so hard that it skips and bounces, making it difficult for the enemy to pick up and throw back. The skip/bounce technique should be used by Marines during training and combat. The *least preferred technique* is to cook-off a concussion hand grenade by removing the grenade's safety pin, releasing the safety lever, counting off two seconds by thousands (one thousand and one, one thousand and two), and then throwing the grenade into the room. Cooking-off the grenade will be used only as appropriate during combat.

(2) Nonverbal and Verbal Alerts. To alert all that a grenade will be thrown, a visual showing of the grenade is made to assault element members, and a visual acknowledgment from them is received. A nonverbal alert may ensure that the enemy is surprised when the grenade is thrown. If the situation demands, a voice alert can be used, but the element of surprise may be lost. When or if a voice alert is used, the voice alert is "**FRAG OUT**;" when an enemy grenade has been identified, friendly forces shout, "**GRENADE**." This allows Marines to distinguish between warnings for outgoing and incoming grenades.

d. After the grenade explodes, Shooter Number One steps across the threshold and clears his immediate area (Figure A-33). He engages targets from the ready carry position. Shooter Number Two follows immediately behind Shooter Number One, buttonhooks, and clears his area.

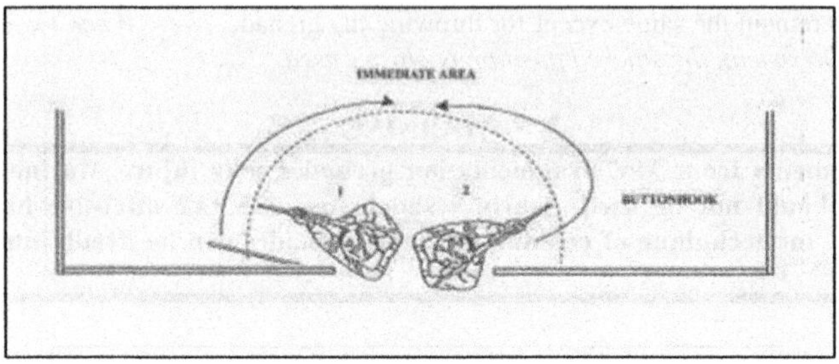

Figure A-33. Shooters Enter the Room

e. Both shooters clear the immediate area and along their respective walls, starting from the nearest respective corner and continuing to the farthest respective corner (Figure A-34). Shooters use the pieing technique to systematically clear the room by sector. (See Paragraph 3.a. of this appendix.)

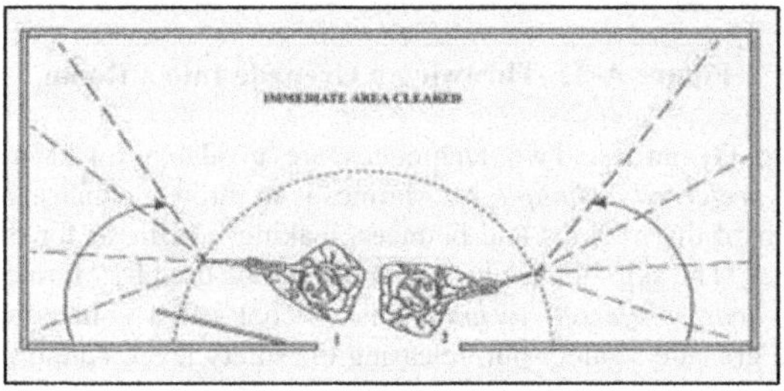

Figure A-34. Clearing a Room

f. Both shooters then establish a dominant position in the room one step away from the wall and two steps into the room and clear the room by sector, pieing to the opposite side of the room (Figure A-35). Meanwhile, the covering team (two-man team) in position outside the room being cleared provides security.

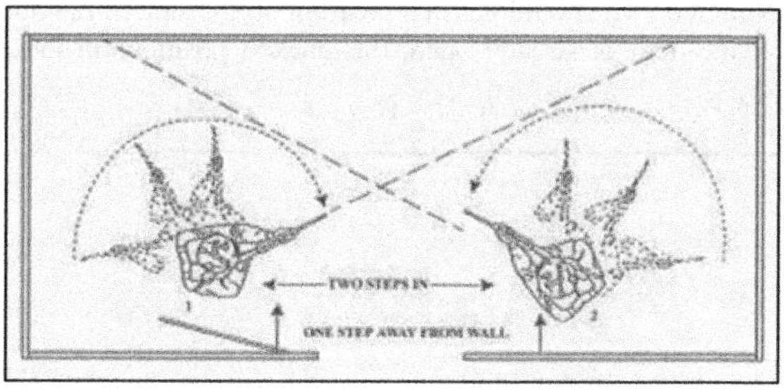

Figure A-35. Clearing a Room (Continued)

g. Another option for entering a room (situation dependent): If Shooter Number One of the clearing team enters a room without Shooter Number Two immediately following him (this could be due to room size, entering the room through a narrow opening, or a number of factors caused by the situation), then Shooter Number One will position himself inside the room to the left or right of the door and quickly scan the entire room. He then gives the command to Shooter Number Two, "**NEXT MAN IN, LEFT (RIGHT)**." Shooter Number Two shouts, "**COMING IN, LEFT (RIGHT)**," enters the room, positions himself up against the wall to the left (right) of the entrance as designated, and scans the room. Once in position, the clearing team leader can call in additional members with the "**NEXT MAN IN**" command, as the situation dictates. It is critical that all assault element members tell each other where they are to avoid fratricide.

h. When the clearing team has cleared the room and is ready to exit the room, it uses the following voice alerts. Once a room has been cleared, the clearing team yells, "**CLEAR**," to inform the covering team. Before leaving the room and rejoining the covering team, the clearing team yells, "**COMING OUT**" and waits to hear the covering team yell, "**ALL CLEAR**." The clearing team then executes its movement out of the room. The assault element then marks the room according to unit SOP.

When moving up or down a staircase, the appropriate team yells, "**COMING UP**" or "**COMING DOWN**." As before, it waits for the response "**ALL CLEAR**" before executing movement and rejoining its element. When exiting a building, the clearing team or assault element yells, "**COMING OUT**" and waits for the support force or covering team to respond with "**ALL CLEAR**."

20. Clearing a Room, Door Closed, Split Positions

a. Before opening a door, Shooter Number Two of the clearing team positions himself opposite Shooter Number One on the other side of the door and away from the wall in a safe position that allows Shooter Number One to shoot the door-opening mechanism (Figure A-36). This positioning allows Shooter Number Two to be in position to immediately move to a wall position opposite Shooter Number One. No matter what method is used to open the door, Shooter Number Two should get in a position on the side of the door opposite Shooter Number One. If the door is already open, the stacked positions in Paragraph 19.a. of this section are used.

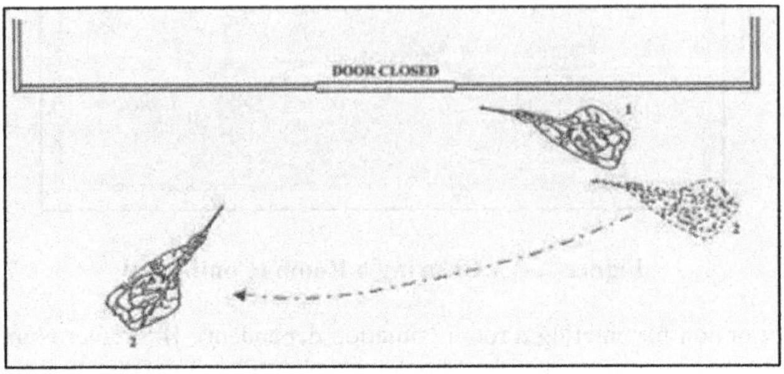

Figure A-36. Positioning To Open a Door

b. After Shooter Number One shoots the door open, Shooter Number Two immediately moves to a kneeling position against the wall (Figure A-37). Shooter Number One kicks the door open from a standing position. If the Marines have grenades, then Shooter Number Two immediately throws a grenade into the room as discussed in Paragraph 19.c. of this section. If the Marines do not have any grenades, the stand/kneel positioning of shooters places the shooters' weapons at different levels, thus allowing each shooter to engage targets inside the room while staying out of the other shooter's line of fire. The shooters scan the room for targets.

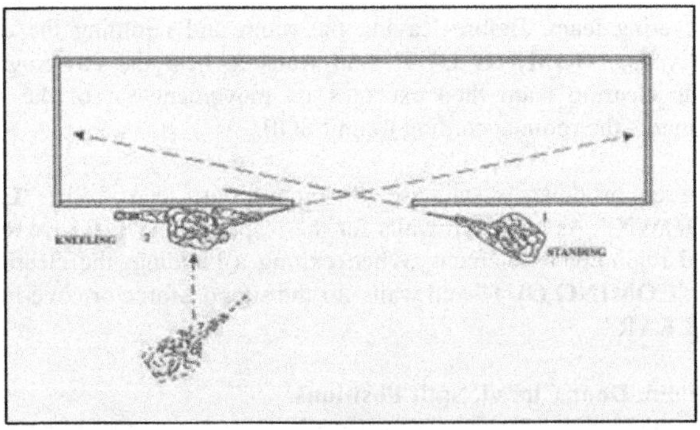

Figure A-37. Positioning To Enter a Room

c. If no grenade was thrown: Once the initial scan of the room is complete, Shooter Number One yells, "**MOVE,**" and both shooters pivot from their respective positions toward each other, pieing (see Paragraph 3.a. of this appendix) while maintaining eye-muzzle-target contact. Shooter Number Two is kneeling, and Shooter Number One is still standing (Figure A-38). Both shooters clear their respective sector of fire.

If a grenade was thrown: After the explosion, Shooter Number One yells, "**MOVE**," and both shooters pivot from their respective positions toward each other, immediately enter the room, and buttonhook (Figures A-38, A-39A, and A-39B). Shooter Number Two goes from kneeling to standing as he pivots. Both shooters clear their respective sector of fire.

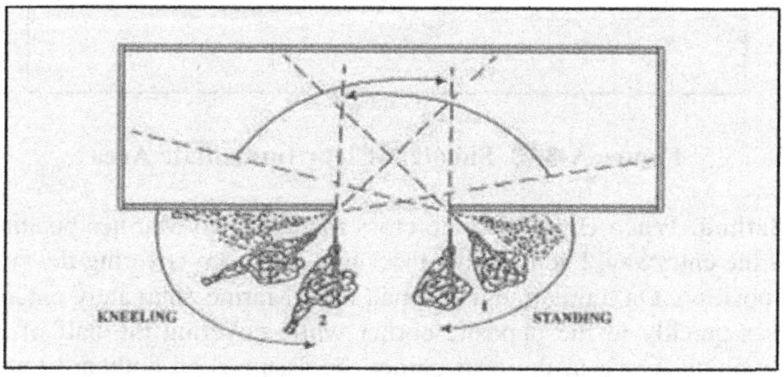

Figure A-38. Shooters Clear Sectors of Fire From Doorway

d. If no grenade was thrown: After scanning is completed, Shooter Number Two stands up and yells, "**READY**," and Shooter Number One Yells, "**MOVE**." Both shooters enter the room together (Figure A-39A).

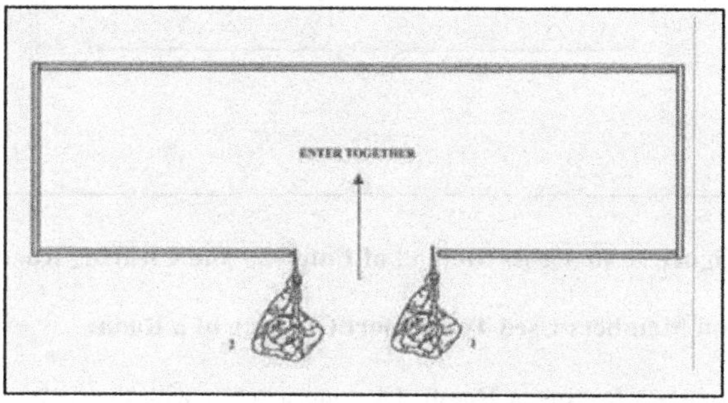

Figure A-39A. Shooters Enter the Room Together

e. Both shooters pass through the doorway together and immediately face respective corners, using the adjacent wall for cover. Both shooters buttonhook their respective areas (Figure A-39B). At this point, clearing the room is the same as discussed in Paragraphs 19.e. - 19.h.

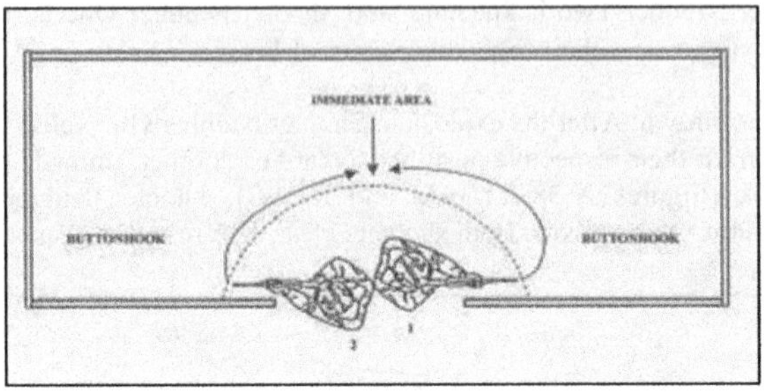

Figure A-39B. Shooters Clear Immediate Area

f. Cross Method. When employing the cross method, two Marines position themselves on either side of the entryway. Each Marine faces into the room covering the corner of the room opposite his position. On a prearranged signal, each Marine alternately enters the room. Each Marine crosses quickly to the opposite corner while covering the half of the room toward which he is moving. Once in the near corner, he assumes an outboard kneeling position to reduce his silhouette and continues to maintain coverage of his half of the room. He may change the position of his weapon to best cover the room. (Figure A-40)

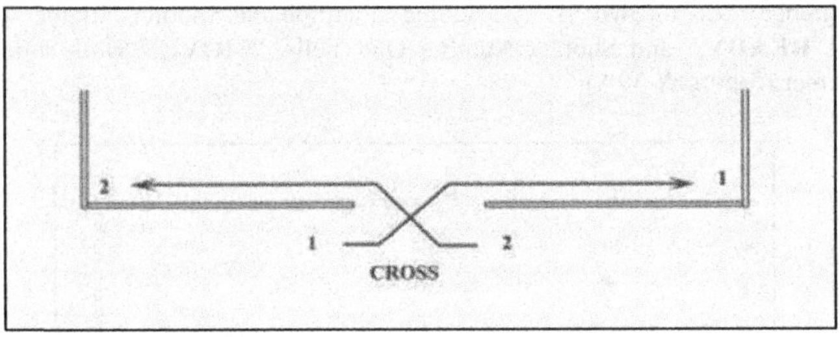

Figure A-40. Cross Method of Entering and Clearing Rooms

21. Covering Team Members Used To Support Clearing of a Room

a. Three Marines Clearing a Room. Clearing a room with three men involves the same procedures used by the aforementioned two-man clearing team, with a third Marine (from the covering team) following immediately (stacked) behind Shooters Number One and Two. Shooter Number Three moves to one side of the door and establishes a center sector of fire coverage (Figure A-41). The remaining covering team Marine provides outside security. If the split position is used, Shooters Number Two and Three are placed on the opposite side of the door.

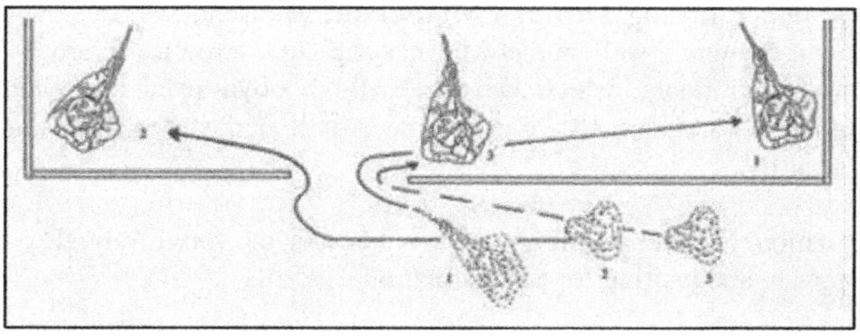

Figure A-41. Three-Marine Clearing Team (Stacked)

b. Four Marines Clearing a Room. When four Marines (entire assault element) are used to clear a room, outside security should be established by another assault element. (If outside security cannot be established, then the two- or three-Marine clearing team should be used.) The two-Marine clearing team procedures are used for initial entry into the room. Shooters Number Three and Four (split position example) stack up opposite Shooters Number One and Two. Shooters Number Three and Four follow Shooter Number Two into the room. Shooter Number Three crosses the threshold and establishes a center sector of fire coverage. Shooter Number Four follows Shooter Number Three and establishes a center sector of fire coverage opposite Shooter Number Three (Figure A-42). If the stacked position is used to enter a room, all shooters stack up on the same side of the door and follow each other in with the same positioning.

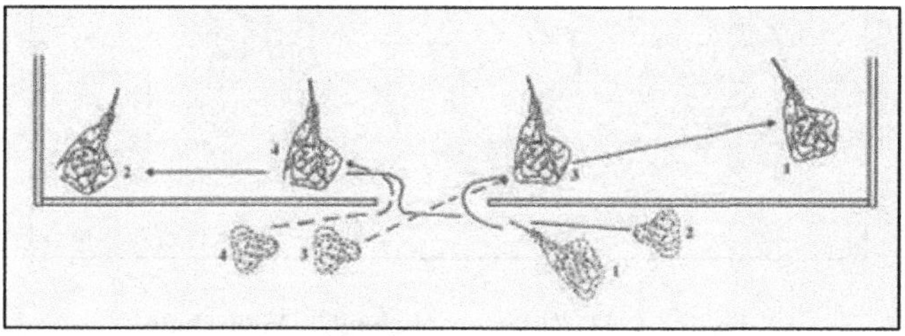

Figure A-42. Four-Marine Clearing Team (Split)

c. Single Marine Clearing a Room. This is the *least preferred* method of clearing a room. However, there may be times when this is the only method available, for example, when gaining access to an upper story window. When this is the case, the Marine gains entry to the room by shooting, kicking, ramming, or breaking a window. He then throws a grenade into the room if one is available and no restrictions apply and takes cover. Immediately after the grenade explodes, he crosses the threshold using the buttonhook technique depicted in earlier figures, scans the room for targets, and engages targets with two shots as they appear.

22. Clearing a Room, Entering Through Mousehole. Mouseholes measure about 2 feet wide and are blown or cut through a wall so that Marines can enter a room (Figure A-43). Mouseholes are safer entrances than doors, which can be easily boobytrapped and should be avoided. Procedures discussed in Paragraph 19 are used. The assault element leader positions the covering team to provide security.

a. In this situation, Shooter Number One would enter the room through a mousehole after throwing a grenade and waiting for the detonation.

b. Shooter Number One will position himself inside the room to the left or right of the mousehole and quickly scan the entire room. He then gives the command to Shooter Number Two, "**NEXT MAN IN, LEFT (RIGHT)**." Shooter Number Two shouts, "**COMING IN, LEFT (RIGHT)**," enters the room, positions himself against the wall to the left (right) of the entrance, and scans the room.

c. The clearing team continues clearing the room as discussed in Paragraph 19.

Figure A-43. Entering Through a Mousehole

23. Clearing an L-Shaped Hallway. After entering a room or building, the clearing team may encounter an L-shaped hallway that should be cleared. The following procedures are used:

a. The clearing team finds an L-shaped hallway.

b. Each shooter takes a dominant position.

c. Shooter Number Two aims his weapon to cover the dead space (Figure A-44).

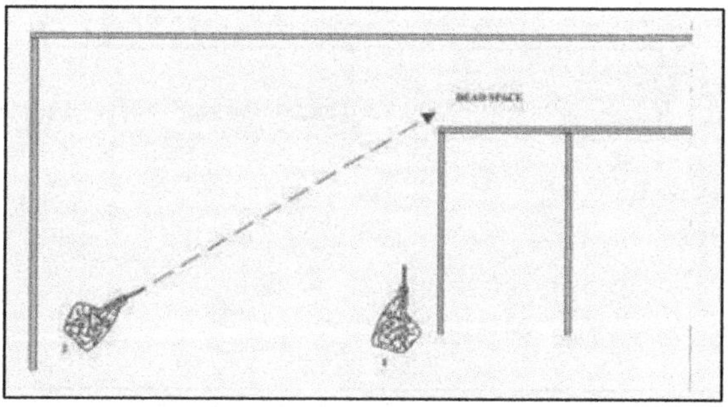

Figure A-44. Positioning To Clear an L-Shaped Hallway

d. Shooter Number Two, with his weapon covering the dead space, pies as far as possible and moves forward to increase his angle of fire further into the dead space. Shooter Number One moves accordingly with Shooter Number Two (Figure A-45).

e. Shooter Number Two stops pieing just before he reaches the position where he cannot cover the dead space. Shooter Number One stops short of the hallway and goes to a kneeling position.

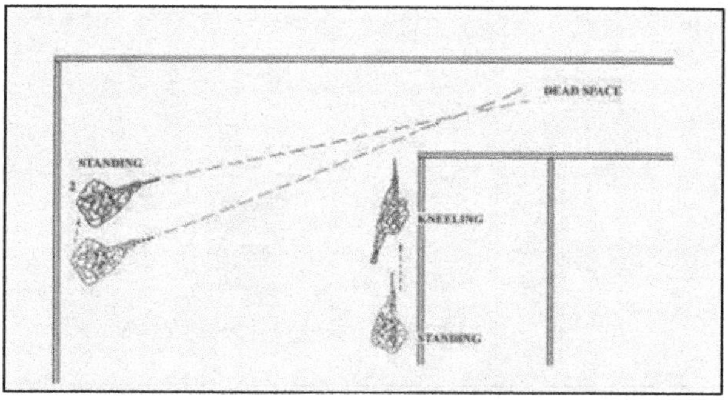

Figure A-45. Clearing an L-Shaped Hallway

f. Shooter Number One says, "**READY!**"

g. Shooter Number Two then says, "**MOVE!**" and both shooters move simultaneously to clear the hallway (Figure A-46). Shooter Number Two remains standing, and Shooter Number One remains kneeling as he pivots around the corner of the wall into the hallway.

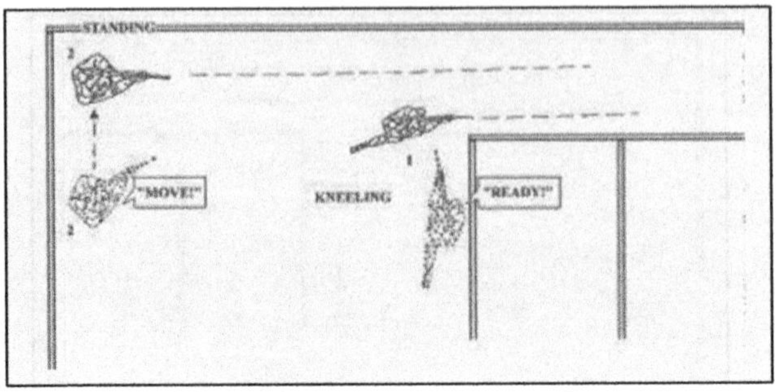

Figure A-46. Clearing an L-Shaped Hallway (Continued)

24. Clearing a T-Shaped Hallway. After entering a room or building, the clearing team may find a T-shaped hallway that should be cleared. The following procedures are used:

a. Both shooters approach the T-shaped intersection together, pieing their individual sectors while maintaining eye-muzzle-target contact (Figure A-47).

b. Both shooters move to dominant positions without entering the intersection.

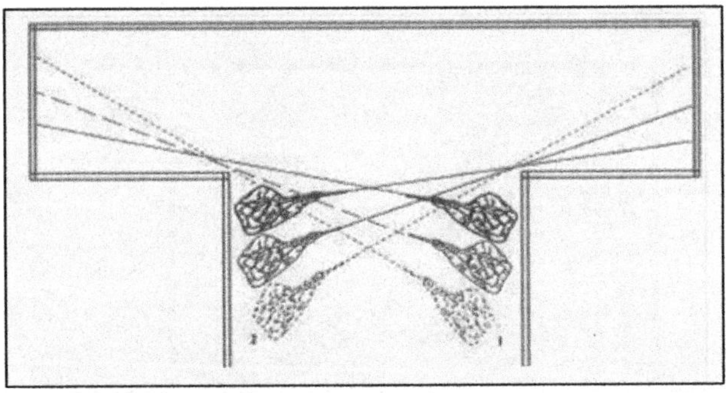

Figure A-47. Positioning To Clear a T-Shaped Hallway

c. Shooter Number One clears his sector of fire and then says, "**READY!**" (Figure A-48).

d. Shooter Number Two clears his sector of fire and then says, "**MOVE!**"

e. Both shooters buttonhook into their next positions.

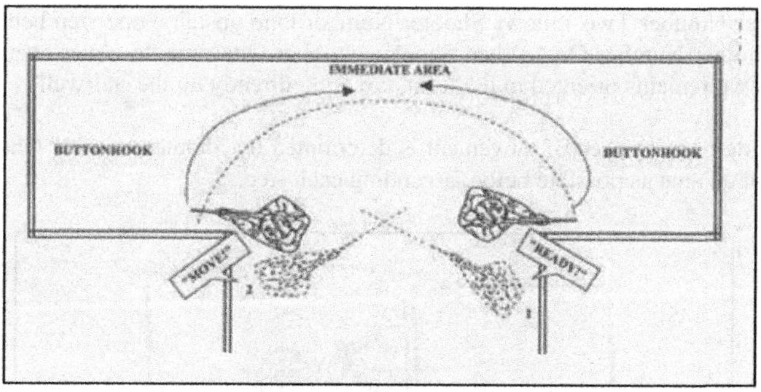

Figure A-48. Clearing a T-Shaped Hallway

f. Both shooters clear their immediate areas along their respective walls, starting from the nearest respective corner and continuing to the farthest respective corner (Figure A-49).

g. Both shooters establish dominant positions that give them control of the hallway and the doorways leading into hallways. The covering team can then be called forward to provide security while the clearing team clears one end of the hallway and associated rooms. The clearing team then clears the other end of the hallway.

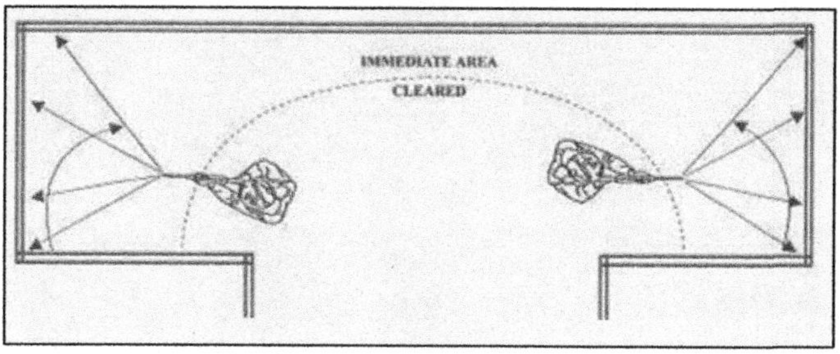

Figure A-49. Clearing a T-Shaped Hallway (Continued)

25. Clearing a Stairwell. After entering a building, the assault element may encounter stairwells. The following procedures are used:

a. Shooter Number One leads upstairs, one step ahead of Shooter Number Two (Figure A-50). Upon reaching the point just before he can be engaged from above, Shooter Number One turns around and covers overhead. From this point, Shooter Number One ascends the stairs, moving backward while covering behind and above.

b. Shooter Number Two follows Shooter Number One upstairs, one step behind and to the side of Shooter Number One. When Shooter Number One turns to cover overhead, Shooter Number Two remains oriented to the front, covering directly up the stairwell.

The clearing team's speed of movement is determined by Shooter Number One. Marines pie as much of an area as possible before ascending each step.

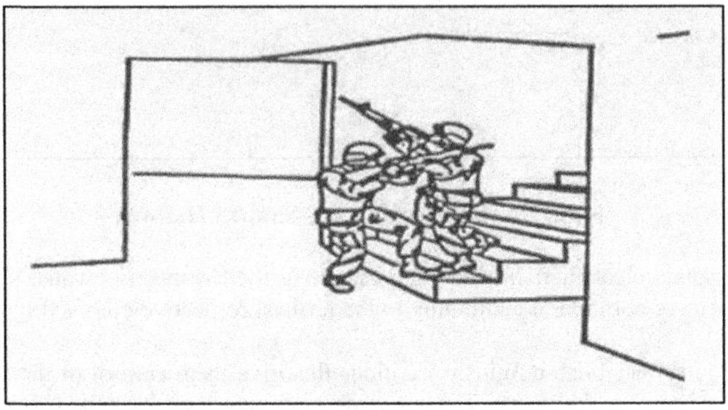

Figure A-50. Clearing a Stairwell

Section VI

Firing Positions

Whether a unit is attacking, defending, or conducting retrograde operations, its success or failure depends on the ability of the individual Marine to place accurate fire on the enemy while providing the least exposure to return fire. Consequently, Marines should immediately seek and properly use firing positions.

26. Hasty Firing Position. A hasty firing position is one that is normally occupied in the attack or the early stages of the defense. It is a position from which a Marine can place fire on the enemy while using available cover for protection from return fire. The Marine may occupy this position voluntarily, or he may be forced to occupy it because of enemy fire. In either case, the position lacks preparation before occupation. Some of the more common hasty firing positions in a built-up area involve firing around corners of buildings, firing from behind walls, firing from windows, firing from unprepared loopholes, and firing from the peak of a roof.

a. Corners of Buildings. The corner of a building provides cover for a hasty firing position if used properly.

(1) The shooter should be capable of both right-handed and left-handed firing of his weapon (using the "short stocking" technique) to be effective around corners. Short stocking the weapon will prevent the muzzle from protruding and keep the weapon ready to fire the instant visual contact with the enemy is made. Furthermore, it reduces the Marine's exposure as a target. (Figure A-51)

(2) A common mistake made when firing around corners is firing from the standing position. The shooter exposes himself at the height the enemy would expect a target to appear and risks exposing the entire length of his body as a target for the enemy.

Figure A-51. Firing, Using Either Hand, Around the Corner of a Building

b. Walls. When firing from behind walls, a Marine should fire around cover—not over it (Figure A-52). Marines in this situation should be able to fire from the right or left shoulder. When firing from behind cover, a Marine should remain far enough back from that cover so that the weapon does not extend beyond it. This aids in weapon retention by ensuring that no enemy on the other side of the wall can get control of the barrel.

Figure A-52. Firing Around Cover

c. Windows. In a built-up area, windows provide convenient firing ports. A Marine should avoid firing from the standing position because it exposes most of his body to return fire from the enemy and could silhouette him against a light-colored interior beyond the window. This is an obvious indicator of the firer's position, especially at night when the muzzle flash can easily be observed. When using the proper method of firing from a window (Figure A-53), the Marine is well back into the room to prevent the muzzle flash from being seen, and he is kneeling to limit exposure and avoid silhouetting himself.

Figure A-53. Firing From a Window

d. Loopholes. A Marine may fire through a hole torn in the wall and avoid windows (Figure A-54). He stays well back from the loophole so the muzzle of the weapon does not protrude beyond the wall and the muzzle flash is concealed.

Figure A-54. Firing From a Loophole

e. Roof. The peak of a roof provides a vantage point for snipers that increases their field of vision and the ranges at which they can engage targets (Figure A-55). A chimney, a smokestack, or any other object protruding from the roof of a building can reduce the size of the exposed target and should be used.

Figure A-55. Firing From the Peak of a Roof

f. No Position Available. When a Marine is subjected to enemy fire and none of the positions mentioned are available, he should try to expose as little of himself as possible. When a Marine is in an open area between buildings (a street or alley) with no cover available and he is fired upon by an enemy from one of the buildings to his front, he should lie prone as close as possible to a building on the same side of the open area as the enemy. To engage the Marine, the enemy must then lean out of the window and expose himself to return fire.

g. No Cover Available. When no cover is available, target exposure can be reduced by firing from the prone position, by firing from shadows, and by presenting no silhouette against buildings.

27. Prepared Firing Position. A prepared firing position is one built or improved to allow the shooter to engage a particular area, avenue of approach, or enemy position, while reducing his exposure to return fire. Examples of prepared positions include: barricaded windows, fortified loopholes, sniper positions, antiarmor positions, and machine gun positions.

a. The natural firing port provided by windows can be improved by barricading the window and leaving a small hole for the firer's use (Figure A-56). The barricading may be accomplished by using materials torn from the interior walls of the building or any other available material. When barricading windows, a Marine should avoid:

(1) Barricading only the windows that will be used as firing ports. The enemy will soon determine that the barricaded windows are firing positions.

(2) Neat square or rectangular holes that are easily identified by the enemy. A barricaded window should not have a neat, regular firing port. The window should keep its original shape so that the position of the firer is hard to detect. Firing from the bottom of the window gives the firer the cover advantage of the wall because the firing port is less obvious to the enemy. Sandbags are used to reinforce the wall below the window and to increase protection for the firer. All glass should be removed from the window to prevent injury to the firer. Lace curtains permit the firer to see out and prevent the enemy from seeing in. Wet blankets should be placed under weapons to reduce dust that may create a signature. Wire mesh placed over the window keeps the enemy from throwing in hand grenades.

b. Although windows are usually good firing positions, they do not always allow the firer to engage targets throughout his sector.

(1) To avoid establishing a pattern of always firing from windows, an alternate position is required such as the prepared loophole (Figure A-57). This involves cutting or blowing a small hole in the wall to allow the shooter to observe and engage targets in his sector.

(2) Sandbags are used to reinforce the walls below, around, and above the loophole. If the position is on the second floor or higher, two layers of sandbags are placed on the floor under the firer to protect him from an explosion on a lower floor. A wall of sandbags, rubble, furniture, and so on should be constructed to the rear of the position to protect the firer from explosions in the room.

Figure A-56. Window Firing Position

(3) A table, a bedstead, or other available material provides overhead cover for the position. This prevents injury to the shooter from falling debris or explosions above his position.

(4) The position should be camouflaged by knocking other holes in the wall. This makes it difficult for the enemy to determine which hole the fire is coming from. Siding material should be removed from the building in several places to make loopholes less noticeable.

Figure A-57. Prepared Loopholes

c. A chimney or other protruding structure provides a base from which a sniper position can be prepared. Part of the roofing material is removed to allow the sniper to fire around the chimney. He should stand inside the building on the beams or on a platform with only his head and shoulders above the roof (behind the chimney). Sandbags placed on the sides of the position protect the sniper's flanks.

d. When the roof has no protruding structure to provide protection, the sniper position should be prepared from underneath on the enemy side of the roof (Figure A-58). The position is reinforced with sandbags, and a small piece of roofing material should be removed to allow the sniper to engage targets in his sector. The missing piece of roofing material should be the only sign that a position exists. Other pieces of roofing should be removed to deceive the enemy as to the true sniper position. The sniper should be invisible from outside the building, and the muzzle flash should be hidden from view.

Figure A-58. Sniper Position

e. Some rules and considerations for selecting and occupying individual firing positions are:

(1) Make maximum use of available cover and concealment.

(2) Avoid firing over cover; when possible, fire around it.

(3) Avoid silhouetting against light-colored buildings, the skyline, and so on.

(4) Carefully select a new firing position before leaving an old one.

(5) Avoid setting a pattern; fire from both barricaded and unbarricaded windows.

(6) Keep exposure time to a minimum.

(7) Begin improving a hasty position immediately after occupation.

(8) Use construction material that is readily available in the built-up area for prepared positions.

(9) Remember that positions that provide cover at ground level may not provide cover on higher floors.

f. In attacking a built-up area, recoilless weapon and ATGM crews may be hampered in choosing firing positions by the backblast of their weapons. They may not have enough time to knock out walls in buildings and clear backblast areas. They should select positions that allow the backblast to escape, such as corner windows where the round fired goes out one window and the backblast escapes from another. The corner of a building can be improved with sandbags to create a firing position (Figure A-59).

Figure A-59. Corner Firing Position

g. The rifle squad is often reinforced with attached antitank weapons during an attack on and in defense of a built-up area. Therefore, the rifle squad leader should be able to choose good firing positions for the antitank weapons under his control.

h. Various principles for employing antitank weapons have universal applications, such as making maximum use of available cover, trying to achieve mutual support, and allowing for the backblast when positioning recoilless weapons, TOWs, Dragons, and AT4s.

i. Operating in a built-up area presents new considerations. Marines should select numerous alternate positions, particularly when the structure does not provide cover from small-arms fire. They should position their weapons in the shadows and within the building.

j. Recoilless weapons and ATGMs firing from the top of a building can use the chimney for cover (Figure A-60). The rear of this position should be reinforced with sandbags.

Figure A-60. Antitank Weapon Firing From a Rooftop

k. When selecting firing positions for recoilless weapons and ATGMs, Marines should make maximum use of rubble, corners of buildings, and destroyed vehicles to provide cover for the crew. Recoilless weapons and ATGMs can also be moved along rooftops to obtain a better firing angle from which to engage enemy armor. When buildings are elevated off the ground, positions can be prepared using a building for overhead cover (Figure A-61). The backblast under the building must not damage or collapse the building or injure the crew.

Note: When firing from a slope, ensure that the angle of the launcher relative to the ground or firing platform is not greater than 20 degrees. When firing within a building, ensure that the enclosure is at least 10 feet by 15 feet, is clear of debris and loose objects, and has windows, doors, or holes in the walls to allow the backblast to escape.

l. The machine gun can be emplaced almost anywhere. In the attack, windows and doors offer ready-made firing ports (Figure A-62). For this reason, the enemy normally has windows and doors under observation and fire; they should therefore be avoided. Any openings in walls that were created during the fighting may be used. When other holes are not present, small explosive charges can create loopholes (Figure A-63). Regardless of what openings are used, machine guns should be sited within the building and in the shadows.

Figure A-61. Prepared Positions Using a Building for Overhead Cover

Figure A-62. Emplacement of a Machine Gun in a Doorway

m. Upon occupying a building, Marines board up all windows and doors. By leaving small gaps between the slots, Marines can use windows and doors as good alternate firing positions.

n. Loopholes should be used extensively in the defense. They should not be constructed in any logical pattern, nor should they all be at floor or tabletop level. Varying their height and location makes them hard to pinpoint and identify. Dummy loopholes, knocked-off shingles, or holes that are cut but not intended to be used as firing positions aid in deception. Loopholes located behind shrubbery, under doorjambs, and under the eaves of a building are hard to detect.

Figure A-63. Use of a Loophole With a Machine Gun

o. Increased fields of fire can be obtained by locating the machine gun in the corner of the building or sandbagging it under a building (Figure A-64). Available materials, such as desks, overstuffed chairs, couches, and other items of furniture, should be integrated into the construction of bunkers to add both cover and concealment (Figure A-65).

Figure A-64. Sandbagged Machine Gun Emplacement Under a Building

p. Although grazing fire is desirable when employing the machine gun, it may not always be practical or possible. Where destroyed vehicles, rubble, and other obstructions restrict the fields of grazing fire, the gun can be elevated to a position from which it can deliver plunging fire over obstacles. Firing from loopholes on the second or third story may be necessary. A firing platform can be built under the roof (Figure A-66) and a loophole constructed. Again, the exact location of the position should be concealed by knocking off shingles in isolated patches over the entire roof.

Figure A-65. Corner Machine Gun Bunker

Figure A-66. Firing Platform Built Under a Roof

28. Target Acquisition. Built-up areas provide unique challenges to acquiring targets. Buildings mask movement and the effects of direct and indirect fires. Also, the rubble from destroyed buildings provides concealment and protection for attackers and defenders alike, making target acquisition difficult.

a. The techniques of patrolling and using OPs apply in the city as well as in wooded terrain. These techniques enable units to locate the enemy, to develop targets for direct and indirect fires in the defense, and to find uncovered avenues of approach in the offense.

b. Most weapons and vehicles produce recognizable signatures. These come from design features or from the environment in which the equipment is used. For example, firing a tank main gun in dry, dusty, and debris-covered streets raises a dust cloud; a tank being driven in built-up areas produces more noise than one moving through an open field; people moving through rubble on a street or in the halls of a damaged building create more noise than in a wooded area. Marines should learn to recognize signatures so they can locate and identify targets. Seeing and hearing assist in detecting and identifying signatures that lead to target location, identification, and rapid engagement. Marines should look for targets in areas where they are most likely to be employed.

c. Target acquisition should be continuous, whether halted or moving. Built-up areas provide both the attacker and defender with good cover and concealment but usually favor the defender. This makes rapid and accurate target acquisition extremely important.

d. When a unit is moving and enemy contact is likely, the unit should employ an overwatching element. This principle applies in built-up areas as it does in other kinds of terrain except that in built-up areas the overwatching element should observe both the upper floors of buildings and the street level.

e. Stealth should be used when moving in built-up areas because little distance separates attackers from defenders. Only hand and arm signals should be used until contact is made. The unit should stop periodically to listen and observe to ensure that it is not being followed or that the enemy is not moving parallel to the unit's flank for an ambush. Routes should be carefully chosen so that buildings and piles of rubble can be used to mask the unit's movement.

f. Observation duties should be clearly issued to squad members to ensure all-around security as they move. This security continues at the halt. All of the senses should be used to acquire targets, especially hearing and smelling. Marines soon recognize the sounds of vehicles and people moving through streets that are littered with rubble. The smell of fuel, cologne, or food cooking can disclose enemy positions.

g. OPs are "positions from which military observations are made, or fire directed and adjusted, and which possess appropriate communications" (Joint Pub 1-02). They are positions from which Marines can watch and listen to enemy activity in a specific sector. They warn the unit of enemy approach and should be positioned in the upper floors of buildings to give Marines a better vantage point than they would have at street level.

h. In the defense, a platoon leader positions OPs for local security as ordered by the company commander. The platoon leader selects the general location, but the squad leader establishes the OP (Figure A-67). Normally, there is at least one OP per platoon. An OP consists of two to four men and is within small-arms supporting range of the platoon. Leaders look for positions that allow good observation of the target sector. Ideally, an OP has a field of observation that overlaps those of adjacent OPs. The position selected for the OP should provide cover and concealment for units moving to and from the OP. The upper floors of houses or other buildings should be used. The squad leader should not routinely select obvious positions such as water towers or church steeples because those positions naturally attract the enemy's attention.

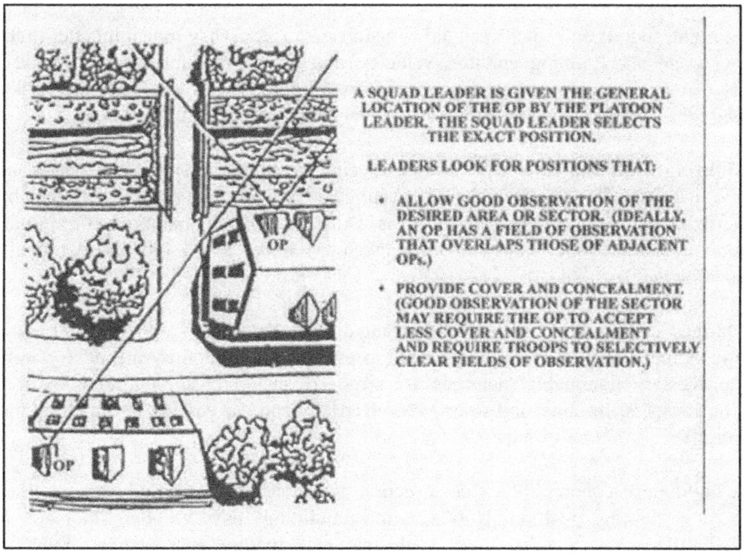

Figure A-67. Selection of an OP Location

i. The Marine should be taught how to scan a target area from OPs or from fighting positions. Use of proper scanning techniques enables observers to quickly locate and identify targets. The Marine searches quickly without optics for obvious targets, using all of his senses to detect target signatures. If no targets are found and time permits, he makes a more detailed search (using binoculars, if available) of the terrain in the assigned sector using the 50-meter method. First, he searches a strip 50 meters deep from right to left. He then searches a strip from left to right that is farther out, overlapping the first strip. This process is continued until the entire sector has been searched. In the city core or core periphery where the observer is faced with multistory buildings, the overlapping sectors may extend up rather than out.

j. Marines who man OPs and other positions should employ target acquisition devices. These devices include binoculars, image intensification devices, and thermal sights. All of these devices can enhance the unit's ability to detect and engage targets, especially at night or during periods of reduced visibility.

k. Target acquisition techniques used at night are similar to those used during the day. At night, whether using daylight optics or the unaided eye, the observer should not look directly at an object but a few degrees off to the side. The side of the eye is more sensitive to dim light. When scanning with off-center vision, the Marine moves his eyes in short, abrupt, irregular movements. At each likely target area, he pauses for a few seconds to detect any motion.

l. At night, sounds and smells can aid in acquiring targets. They transmit better in the cooler, damper night air. Running engines, vehicles, and people moving through rubble- covered streets can be heard for great distances. Odors from diesel fuel, gasoline, cooking food, burning tobacco and aftershave lotion can reveal enemy and friendly locations.

29. Firefighting Planning and Operations. Incendiary ammunition, special weapons, and the ease with which incendiary devices can be constructed from gasoline and other flammables make fire a true threat in operations in built-up areas. During defensive operations, firefighting should be a primary concern. Proper steps should be taken to reduce the risk of a fire that could make a chosen position indefensible.

a. Marines choose or create positions that do not have large openings. These positions provide as much built-in cover as possible to prevent penetration by incendiary ammunition. All unnecessary flammable materials are removed, including ammunition boxes, furniture, rugs, newspapers, curtains, and so on. The electricity and gas coming into the building should be shut off.

b. A building of concrete block construction with concrete floors and a tin roof is an ideal place for a fighting position. However, most buildings have wooden floors or subfloors, wooden rafters, and wooden inner walls that may require improvement. Inner walls are removed and replaced with blankets to resemble walls from the outside. Sand is spread 2 inches deep on floors and in attics to retard fire.

c. All available firefighting gear is prepositioned so that it can be used during combat. For the individual Marine, such gear includes entrenching tools, helmets, sand, and blankets. These items are supplemented with fire extinguishers.

d. Fire is so destructive that it can easily overwhelm personnel regardless of extraordinary precautions. Marines should plan routes of withdrawal so that evacuation can be accomplished from their fighting positions. This allows Marines to exit through areas that are free of combustible material but that still provide cover from enemy direct fire.

e. The confined space and large amounts of combustible material found in built-up areas may influence the enemy to use incendiary devices. First-aid problems that are expected include burns and smoke/flame inhalation. These can easily occur in buildings, thereby rendering the victims combat ineffective. Although there is little defense against flame inhalation and lack of oxygen, smoke inhalation may be greatly reduced by wearing the individual protective mask. Regardless of the fire hazard, defensive planning for combat in built-up areas should include location of corpsmen. Corpsmen must reach victims and their equipment and should stock extra supplies for the treatment of burns and inhalation injuries.

f. Offensive operations also require plans for firefighting because the success of the mission can easily be threatened by fire. Poorly planned use of incendiary munitions can make fires so extensive that they become obstacles to offensive operations. The enemy may use fire to cover his withdrawal and to create obstacles and barriers to the attacker.

g. Every Marine participating in an attack should be ready to deal with the problems of fire. The normal firefighting equipment available includes the entrenching tool, helmet (for carrying sand or water), and blankets (for snuffing out small fires). Fire extinguishers are normally available on vehicles.

30. Employment of Snipers. The value of the sniper to a unit operating in a built-up area depends on several factors. These factors include the type of operation, the level of conflict, and the ROE. Where ROE allow destruction, snipers may not be needed because other weapons systems available to the infantry commander have a greater destructive effect. However, snipers can contribute to the fight. Where the ROE prohibit collateral damage, snipers may be the most valuable tool the commander has. (See MCWP 3-15.3 (under development).)

a. Sniper effectiveness depends in part on the terrain. Control is decentralized by the characteristics of an urban area. To provide timely and effective support, the sniper should have a clear picture of the commander's mission and intent.

b. Where possible, snipers should be positioned in masonry buildings. These buildings should also offer long-range fields of fire and all-around observation. The sniper has an advantage because he does not have to move with or be positioned with lead elements. He may occupy a higher position to the rear or flanks of, and some distance away from, the element that he is supporting. By operating far from the other elements, a sniper avoids decisive engagement but remains close enough to kill distant targets that threaten the unit. Snipers should not be placed in obvious positions such as church steeples and rooftops because the enemy often watches these and targets them for destruction. Also, snipers should not be positioned where there is heavy traffic; these areas also invite enemy observation.

c. Snipers should operate throughout the AO, moving with and supporting the companies as necessary. Some teams may be designated to operate independently of other forces. They may be tasked to search for targets of opportunity, especially for enemy snipers. The sniper team may occupy multiple positions. A single position may not afford adequate observation for the entire team without increasing the risk of detection by the enemy. Separate positions should maintain mutual support. Alternate and supplementary positions should also be established in urban areas.

d. Snipers may be assigned the following tasks:

- Killing enemy snipers (countersniper fire)

- Killing targets of opportunity (These targets should be prioritized by the commander. Types of targets might include enemy snipers, leaders, vehicle commanders, radio men, sappers, and machine gun crews.)
- Denying enemy access to certain areas or avenues of approach (controlling key terrain)
- Providing fire support for covering barricades and other obstacles
- Maintaining surveillance of flank and rear avenues of approach (screening)
- Supporting local counterattacks with precision fire.

Section VII

Advancing/Patrolling Along City Streets

31. Advancing/Patrolling Along City Streets. During MOUT operations, skills for advancing/ patrolling along city streets will be required. The platoon, squad, or fire team mission may be such that building-to-building clearing is neither required nor desired until contact with the enemy has been made. This section will discuss advancing/patrolling along urban streets. The standard tactics, techniques, and procedures for the conduct of a patrol, as outlined in MCWP 3-11.6 *Scouting and Patrolling* (draft), apply except as otherwise discussed within this section.

a. Organization. The platoon's three-squad organization or the squad's three-fire team organization is already structured to conduct an advance/patrol down an urban street. The typical patrol is organized into assault, support, security, and command and control elements. Security is divided into front, flank, and rear security responsibilities. The platoon or squad can move down the street in a double or single column.

b. Formation and Movement

(1) Double Column. The double column is the *preferred method* for moving along an urban street (Figures A-68 and A-69). The double column provides 360 degrees of security and allows for mutual support and interlocking sectors of fire. The double column is flexible but can make control more difficult. There should be at least two squads or fire teams to conduct a double column. The two squads or fire teams move down the sides of the street using the walls of the buildings for cover and concealment. The pointmen for the two columns should always stay abreast of each other and not move in front of or behind the opposite column. The Marines following in trace of the pointmen stagger themselves accordingly. The platoon or squad leaders move where they can best control their units. If a squad is conducting the advance/patrol along the street, the trail fire team is never split.

(a) Sectors of Responsibility. Every Marine moving in a column has a particular sector for whose security he is responsible. The pointmen are responsible for security to the direct front and for pieing off windows and doorways. They are also responsible for staying in line with one another.

The Marines immediately behind the pointmen are designated as the covermen. They are responsible for providing guidance to their respective pointmen from the leader of the unit and for covering the 45-degree oblique angle across from one another. The second coverman is also responsible for covering the direct front when the pointman is pieing and checking for the enemy. The third and fourth covermen in the column are responsible for the far-side flank security; one covers the upper level and one covers the lower level building windows, doors, and stairwells. Marines will alternate providing security on upper level and lower level windows, doors, and stairwells. The fourth Marine also provides rear security if he is the last man in the column or interlocks his sector of fire with that of the lead Marine in the next fire team in the column. Should the second coverman have to take the point while the pointman pies a window or door, the third coverman would automatically shift his sector of fire to the 45-degree oblique angle. If the second coverman has to assist the pointman in pieing a large opening, the third coverman would provide security to the direct front while the fourth coverman would take the 45-degree oblique angle. Marines following in the column would adjust their sectors of fire accordingly.

The two columns provide mutual support for each other and the greatest all-around security. Teamwork and coordination between the two columns are vital.

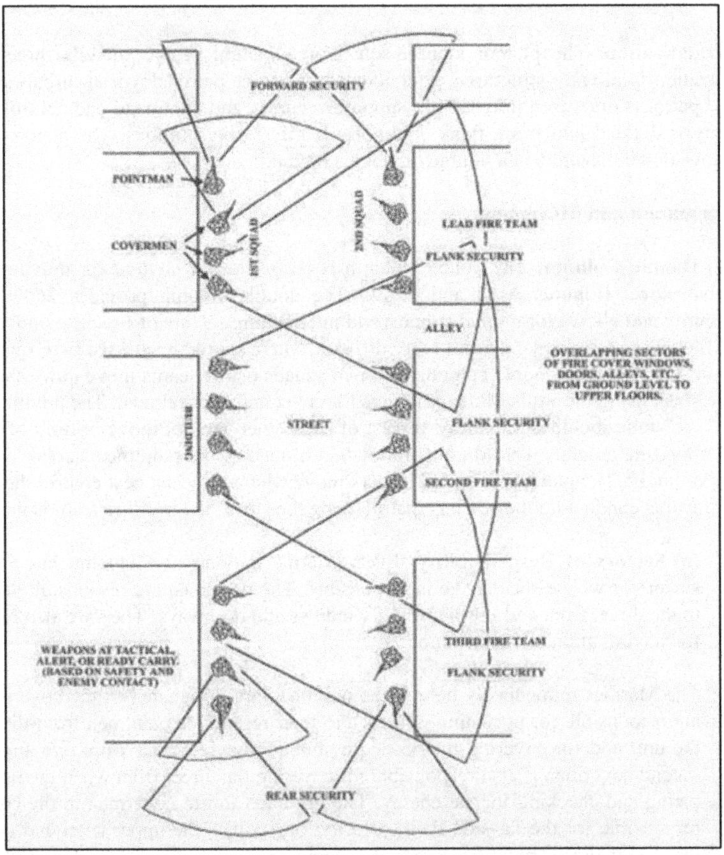

Figure A-68. Advancing/Patrolling Along an Urban Street (Squad, Double Column)

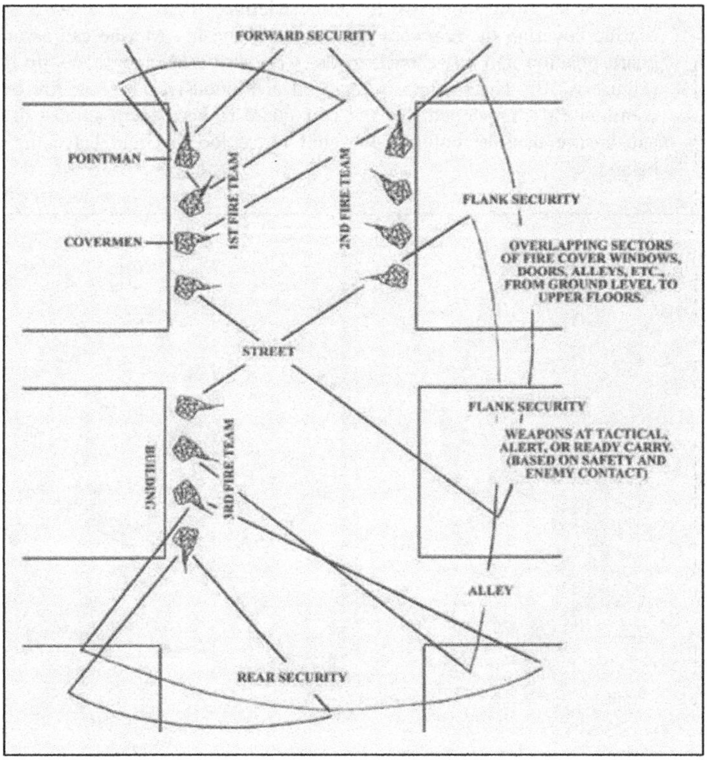

Figure A-69. Advancing/Patrolling Along an Urban Street (Fire Team, Double Column)

(b) Rear Security. The last two Marines in the squad of each column are responsible for the rear security of each of their formations. They will move in bounds in order to allow a Marine to always provide active security to the squad's rear (Figures A-70 and A-71). Two methods of bounding may be used: *alternate* or *successive bounding.*

1 **Alternate Bounding.** In this method, the two rear-security Marines alternate being the last Marine in the squad. The last Marine in the formation will have his weapon at the ready and will be in the kneeling position along the wall of the building. The second to last Marine will move in trace of the unit until he has moved approximately 10 to 15 meters. At this point, he will step out one pace from the wall. He will then assume a good kneeling firing position and announce, "**SET**." The last Marine will rise from his position, turn around, and move next to the wall to catch up with the column. It is important that the Marine moving back

does not step out and mask the other Marine's field of fire. At this point, the Marine covering the rearward movement of the last Marine can assume the rear guard position and move back to the wall as the Marine passes by his position (Figure A-70). The process is repeated continuously. The rear fire team leaders coordinate the movements of the rear guard to keep them abreast of each other and ensure that the column does not move too fast and leave the rear guard behind.

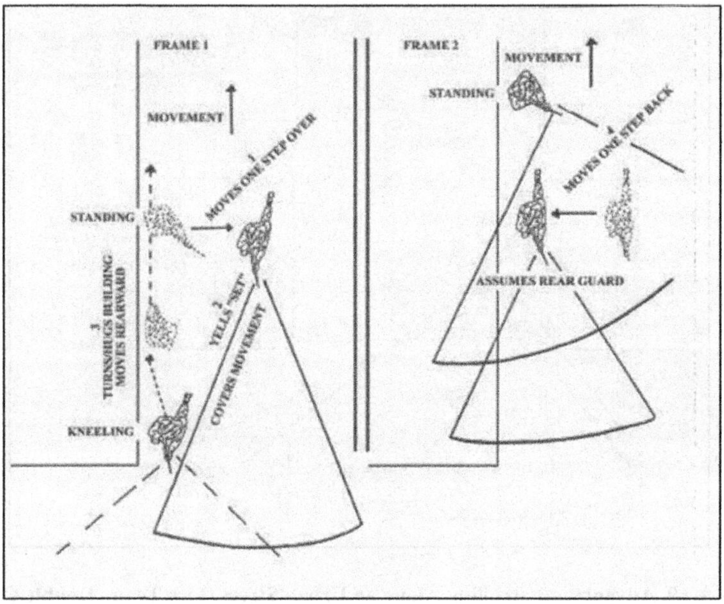

Figure A-70. Rear Security (Alternate Bounds)

2 **Successive Bounding.** In this method, the last Marine in the squad continues to be the rear guard after bounding. The same procedures are employed as in alternate bounding; however, the rear-guard Marine bounds abreast of the Marine providing cover, turns around, drops to a good kneeling position, provides rear-guard security, and announces, "**SET**." The covering Marine moves behind the rear-guard Marine and continues to move 10 to 15 meters before repeating the process (Figure A-71). The rear fire team leaders coordinate the movements of the rear guard as in alternate bounding.

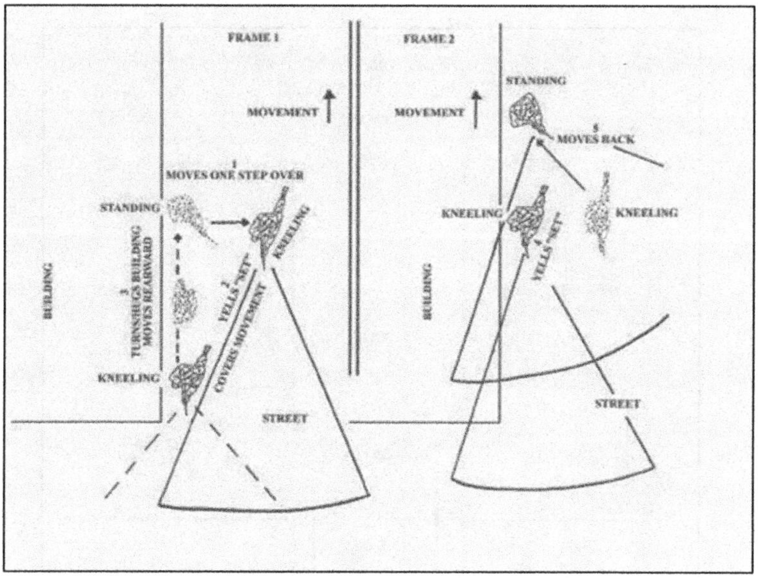

Figure A-71. Rear Security (Successive Bounds)

(2) Single Column. The single column is used to move rapidly along a street or when streets are restricted to a single column. The single column is the least preferred advance/patrol method of moving along an urban street. The same principles apply as in the double column, except the single column does not provide 360-degree security. The single column is normally used when speed is desired and contact with the enemy is considered to be unlikely.

32. Crossing a Street Intersection. Four-way intersections are very dangerous areas that should be crossed quickly. Security at an intersection is extremely important for the protection of crossing units. When moving, the pointmen notify the unit leaders, through the covermen, of an approaching intersection. Crossing the intersection will force the three elements of a unit moving along a city street to rotate positions and assume appropriate responsibilities.

a. First Fire Team To Cross. The lead (1st) fire teams in the two columns will simultaneously set security around the corners of the building. They will provide mutually supporting security to the direct front, security to the building on the far side of the intersection, and intersection security (Figure A-72). The middle (2nd) fire teams provide mutually supporting flank and rear security. On order, the rear (3rd) fire teams move forward, cross the intersection, and assume the lead (1st) fire teams' forward security mission.

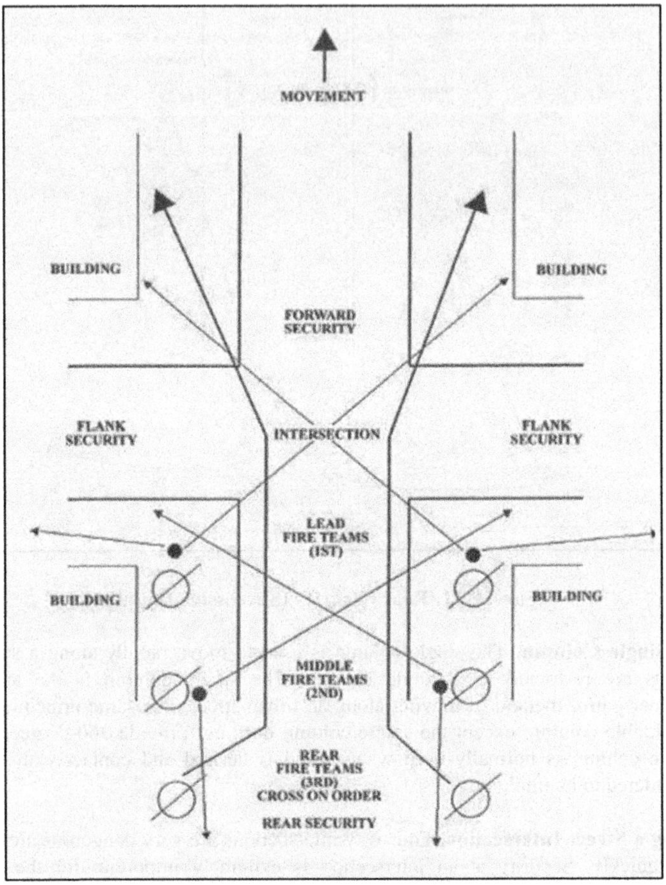

Figure A-72. Crossing an Intersection (Frame 1)

b. Second Fire Team to Cross. The 1st fire teams (now the middle fire teams) continue to provide intersection security and assume rear area security. The 2nd fire teams (now rear fire teams) cross the intersection on order. The second fire teams position themselves to provide rear and intersection security to support the covering of the last fire teams (Figure A-73).

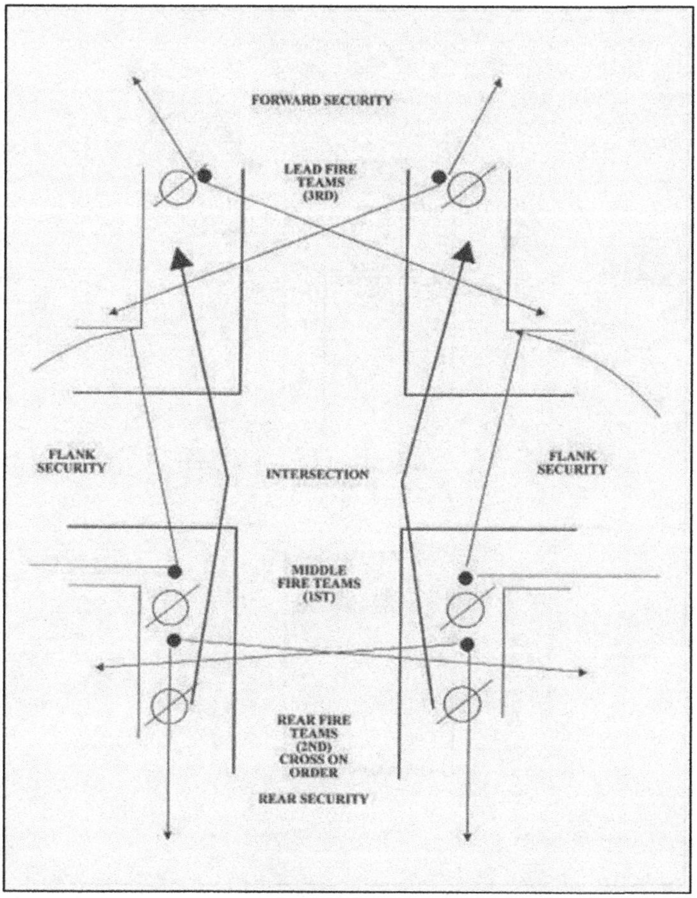

Figure A-73. Crossing an Intersection (Frame 2)

c. **Third Fire Team To Cross.** The 1st fire teams (now the rear fire teams) provide rear security and cross the intersection on order. The 1st fire teams bypass the 2nd fire teams to become the middle fire teams and provide mutually supporting forward and flank security. The 2nd fire teams become the rear security (Figure A-74).

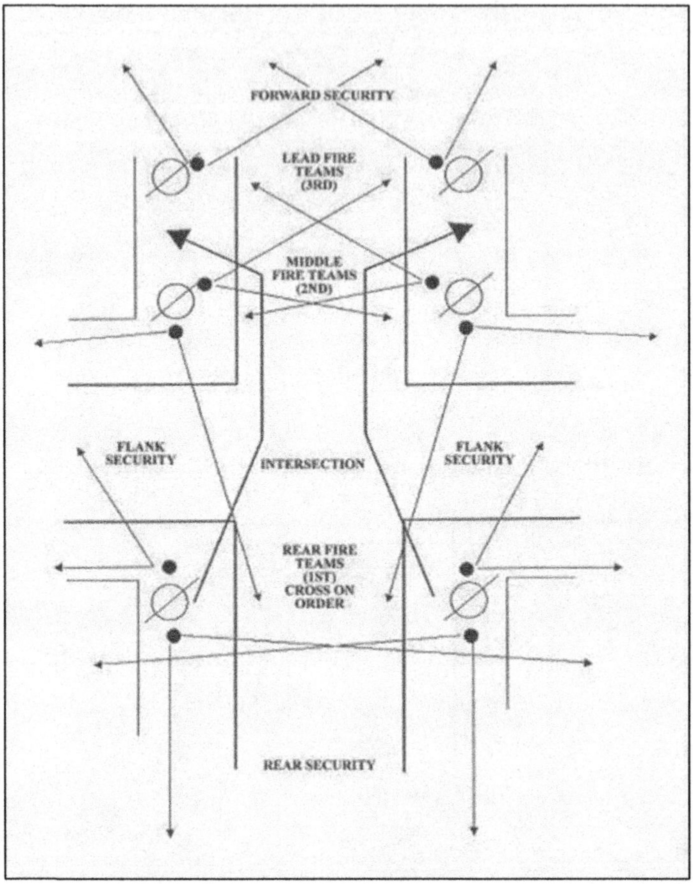

Figure A-74. Crossing an Intersection (Frame 3)

d. The squads and three fire teams have now crossed the intersection, and the fire teams have rotated (Figure A-75). This rotation is convenient because it alternates the lead units, who may quickly become fatigued during MOUT operations. The same procedures are used to cross an intersection with a single column.

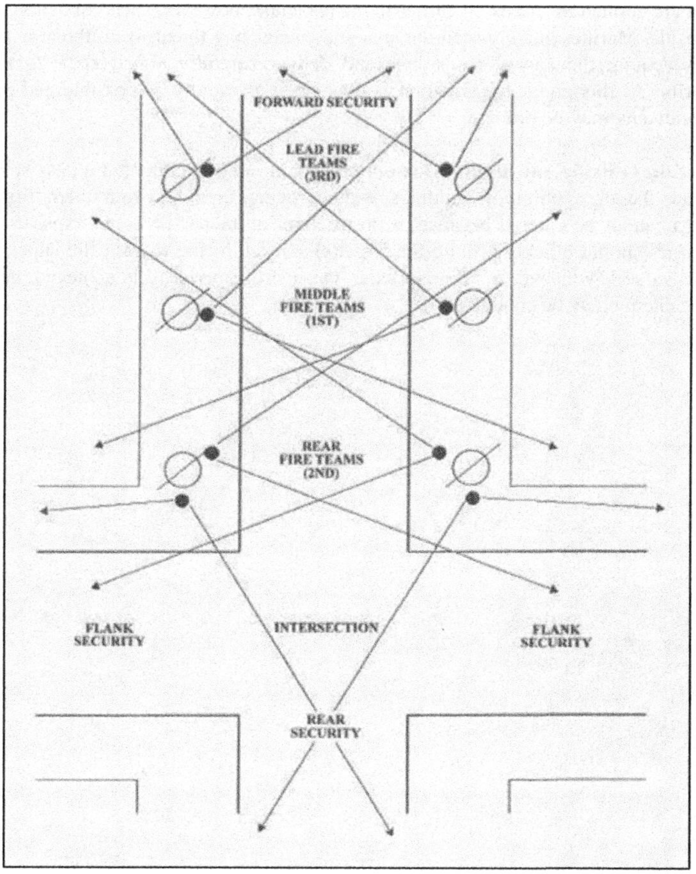

Figure A-75. Crossing an Intersection (Frame 4)

33. Reacting to Enemy Contact. When a platoon or squad is moving along a city street, contact with the enemy could happen at any time, even if the area has been declared secure. If a unit engages or is engaged by the enemy, there are two basic options that Marines may follow. The first and most preferred option is to immediately return fire and conduct a hasty clearing of a structure to seek cover inside of the building to which the unit was adjacent. The second and least preferred method is to remain outside of the building(s) and fight from the street.

a. Hasty Clearing. To conduct a hasty clearing of a structure in order to seek cover, Marines immediately return fire and enter buildings as quickly as possible. The basic principles of entry are applied as discussed earlier in the appendix; however, speed of action is essential. Once inside, Marines quickly scan the area and engage any threat. If no threat is present, the Marines acquire the enemy's positions and deliver carefully aimed shots to achieve fire superiority. At this point, the situation is developed as rapidly as possible and an attack to clear the enemy may be ordered.

b. Remain Outside and Fight. This option should not be conducted unless strict ROE do not allow the occupation of buildings without being fired on from them first or unless buildings cannot be entered because of obstruction, obstacles, or boobytraps. In such cases, the Marines should quickly acquire the target(s) and return fire to gain fire superiority. They should also seek whatever cover may exist. Once fire superiority is achieved, an assault to clear the enemy may be conducted.

Section VIII

Navigation in Built-Up Areas

Navigation in built-up areas present a unique set of challenges. Deep in the city core, the normal terrain features depicted on maps may not apply—buildings become the major terrain features, and units become tied to streets. Fighting in the city destroys buildings, whose rubble then blocks streets. Street and road signs are destroyed during the fighting if they are not switched or removed by the defenders. Operations in subways and sewers present other unique challenges. However, maps and photographs are available to help units overcome these problems. Additionally, the global positioning system (GPS) can help supplement navigational abilities in built-up areas.

34. Military Maps. The military city map is a topographical map of a city that delineates streets and shows street names, important buildings, and other urban elements. The scale of a city map can vary from 1:25,000 to 1:50,000 to 1:100,000 depending on the importance and size of the city, the density of detail, and intelligence information available.

a. Special maps prepared by supporting topographical engineers can assist units in navigating in built-up areas. These maps are designed or modified to provide information that is not covered on a standard military map. This may include maps of road and bridge networks, railroads, built-up areas, and electric power fields. Special maps can be used to supplement military city maps and topographical maps.

b. Once in the built-up area, Marines may use street intersections as reference points as they would use hills and streams in rural terrain. City maps supplement or replace topographical maps as the basis of navigation. These maps enable units moving in the built-up area to know where they are and to move to new locations even though streets have been blocked or a key building has been destroyed.

c. The techniques of compass reading and pace counting can still be used, especially in a blacked-out city where street signs and buildings are not visible. The presence of steel and iron in the MOUT environment may cause inaccurate compass readings. Sewers may be navigated in much the same way. Maps providing the basic layout of the sewer system are maintained by city sewer departments. This information includes directions in which the sewer lines run and distances between manhole covers. Along with basic compass and pace-count techniques, such information enables a unit to move through the city sewers with accuracy.

d. Operations in a built-up area adversely affect the performance of some types of communications-electronic devices such as the GPS, the Position Location Reporting System (PLRS), or other data-distribution systems. These systems function the same as some communications equipment—by LOS. They cannot determine underground locations or positions within a building. These systems should be employed on the tops of buildings, in open areas, and down streets where obstacles will not affect LOS readings.

e. City utility workers are assets to units fighting in built-up areas. They can provide maps of sewers and electrical fields and information about the city. This is especially important with regard to the use of the sewers. Sewers can contain pockets of methane gas that are highly toxic to humans. City sewer workers know the locations of these danger areas and can advise a unit on how to avoid them.

35. Aerial Photographs. Current aerial photographs are also excellent supplements to military city maps and can be substituted for a map. A topographical map or military city map could be obsolete if compiled many years ago. A recent aerial photograph shows changes that have taken place since the map was made. This could include destroyed buildings and streets that have been blocked by rubble as well as enemy defensive preparations. More information can be gained by using aerial photographs and maps together than by using either one alone.

Section IX

Camouflage

To survive and win in combat in built-up areas, a unit should supplement cover and concealment with camouflage. To properly camouflage men, carriers, and equipment, Marines should study the surrounding area and make their fighting positions blend with the local terrain.

36. Application. Only the proper amount of material needed for camouflaging a position should be used. Excess material could reveal the position. Material should be obtained from a wide area around the position. For example, Marines defending a cinder-block building do not strip the front, sides, or rear of the building to camouflage a position.

a. Buildings provide numerous opportunities for establishing concealed positions. Armored vehicles can often find prospective positions under archways or inside small industrial or commercial structures. Thick masonry, stone, or brick walls offer excellent protection from direct fire and may provide concealed routes into or out of the position.

b. After camouflage is completed, the Marine inspects his position from the enemy's viewpoint. He makes routine checks to see whether the camouflage remains natural looking and actually conceals the position. If it does not look natural, the Marine should rearrange or replace it until a natural look is achieved.

c. Positions should be progressively camouflaged as they are prepared. Work should continue until all camouflage is complete. If the enemy has air superiority, work may be possible only at night. Shiny or light-colored objects that attract attention from the air should be hidden.

d. Camouflage face paint is issued in three standard, two-tone sticks. When issue-type face-paint sticks are not available, burnt cork, charcoal, or lampblack can be used to tone down exposed skin. Mud may be used as a last resort because it dries and may peel off, leaving the skin exposed, and may also contain harmful bacteria.

37. Use of Shadows. Buildings throw sharp shadows that can be used to conceal vehicles and equipment (Figure A-76). Marines should avoid movement or positions in areas that are not in shadows. Vehicles may have to be moved periodically as shadows shift during the day. Emplacements inside buildings provide better concealment and cover as well.

38. Color and Texture. Standard camouflage-pattern painting of equipment is not as effective in built-up areas as a solid, dull, dark color hidden in shadows. Because repainting vehicles before entering a built-up area is not always practical, the lighter sand-colored patterns should be subdued with mud or dirt.

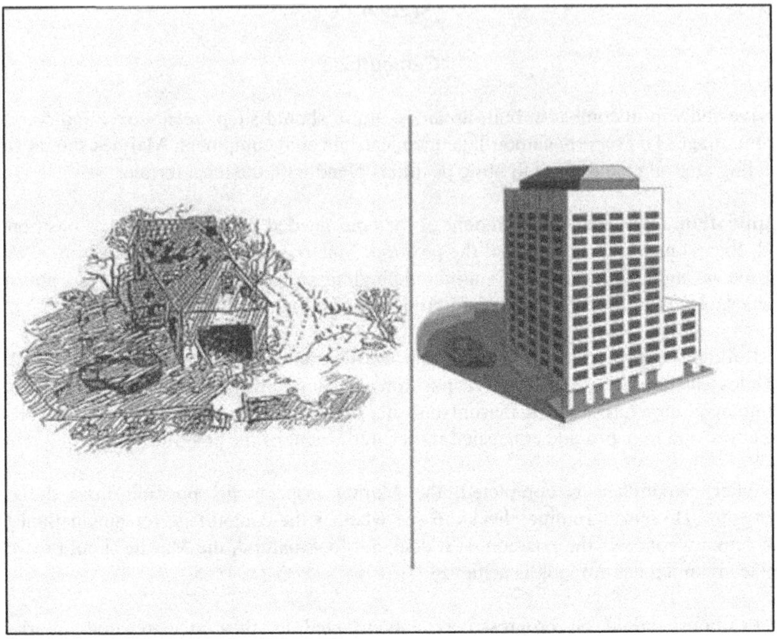

Figure A-76. Use of Shadows for Concealment

a. The need to break up the silhouette of helmets and individual equipment exists in built-up areas the same as it does elsewhere. However, burlap or canvas strips are a more effective camouflage than foliage (Figure A-77). Predominant colors are normally browns, tans, and sometimes grays rather than greens, but each camouflage location should be evaluated to determine ideal colors and patterns.

Figure A-77. Helmet Camouflaged With Burlap Strips

b. CPs and logistics emplacements are easier to camouflage and better protected if located underground. Antennas can be remoted to upper stories or to higher buildings based on remote capabilities. Field telephone wire should be laid in conduits, in sewers, or through buildings.

c. Marines should consider the background to ensure that they are not silhouetted or skylined, but rather blend into their surroundings. To defeat enemy urban camouflage, Marines should be alert for common camouflage errors such as:

- Tracks or other evidence of activity
- Shine or shadows
- Unnatural color or texture
- Muzzle flash, smoke, or dust
- Unnatural sounds and smells
- Movement.

d. Dummy positions can be used effectively to distract the enemy and make him reveal his position by firing on them.

e. Built-up areas afford cover, resources for camouflage, and locations for concealment. Following are the basic rules of cover, camouflage, and concealment to which Marines should adhere:

(1) Use the terrain and alter camouflage habits to suit your surroundings.

(2) Employ deceptive camouflage of buildings.

(3) Continue to improve positions. Reinforce fighting positions with sandbags or other fragment- and blast-absorbent material.

(4) Maintain the natural look of the area.

(5) Keep positions hidden by clearing away minimal debris for fields of fire.

(6) Choose firing ports in inconspicuous spots when available.

Note: Remember that a force that *covers* and *conceals* itself has a significant advantage over a force that does not.

Appendix B

Employment and Effects of Weapons

Success against the enemy in urban operations to a large extent depends on the effective employment of weapons. Marines at all levels must understand the effects that their weapons will produce in urban combat.

1. Effectiveness of Weapons and Demolitions. The characteristics and nature of combat in built-up areas affect the results and employment of weapons. Leaders at all levels must consider the following factors in various combinations when choosing weapons to employ.

a. Hard, smooth, flat surfaces are characteristic of urban targets. Rounds rarely impact perpendicular to these flat surfaces, but rather tend to impact at some angle of obliquity. This reduces the effect of a round and increases the threat of ricochets. The tendency of rounds to strike glancing blows against hard surfaces means that up to 25 percent of impact-fuzed explosive rounds may not detonate when fired into rubbled areas.

b. Engagement ranges in MOUT are close. Studies and historical analyses have shown that only 5 percent of all targets are more than 100 meters away. About 90 percent of all targets are located 50 meters or less from the identifying Marine. Few personnel targets will be visible beyond 50 meters, and most occur at 35 meters or less. Minimum arming ranges and unit safety from backblast or fragmentation effects must be considered.

c. Engagement times are short because of the close fight. Enemy personnel present only fleeting targets. Enemy-held buildings or structures are normally covered by mutually supporting fires and often cannot be engaged with deliberate, well-aimed shots without prolonging exposure to their observation and weapons.

d. Depression and elevation limits for some weapons create dead space. Tall buildings form deep canyons that are often safe from indirect fires. Some weapons, such as the M203, can fire rounds to ricochet behind cover and inflict casualties. Target engagement from oblique angles, both horizontal and vertical, demands superior marksmanship skills.

e. Smoke from burning buildings, dust from explosions, shadows from tall buildings, and the lack of light penetrating inner rooms all combine to reduce visibility and to increase a sense of isolation. Added to this is the masking of fires caused by rubble and manmade structures. Targets, even at close range, tend to be indistinct.

f. Urban fighting often leads to confused melees with several small units attacking on converging axes. The risks from friendly fires, ricochets, and fratricide must be considered during the planning phase of operations, and control measures must be continually adjusted to lower these risks. Marines and leaders must maintain a sense of situational awareness and clearly mark their progress in accordance with unit SOPs to avoid fratricide.

g. The firer and target may each be inside or outside the same or separate buildings. The enclosed nature of combat in built-up areas means that a weapon's signature effect, such as a muzzle blast or backblast, must be considered, as well as the round's impact on the target.

h. Usually the manmade structure that is protecting the enemy must be attacked before enemy personnel inside can be engaged. Therefore, weapons and demolitions can be chosen for employment based on their effects against masonry and concrete rather than against enemy personnel.

i. Modern engineering and design improvements mean that most large buildings constructed since World War II are resilient to the blast effects of bomb and artillery attack. Even though modern buildings may burn easily, they often retain their structural integrity and remain standing. Once high-rise buildings burn out, they are still useful to the military and are almost impossible to damage further. A large structure can take 24 to 48 hours to burn out and become cool enough for people to enter.

j. The most common worldwide building type is the 12- to 24-inch brick building. Table B-1 lists the frequency of occurrence of building types worldwide.

Type of Building	Frequency of Occurrence (Percentage)
30-inch stone	1
8- to 10-inch reinforced concrete	6
12- to 24-inch brick	63
6-inch wood	16
14-inch steel and concrete (heavy clad)	2
7-inch steel and concrete (light clad)	12

Table B-1. Types of Buildings and Frequency of Occurrence

2. M16 Rifle and M249 Squad Automatic Weapon. The M16A2 rifle and the M249 squad automatic weapon are the most common weapons used in combat in built-up areas. The M16A2 rifle and the M249 are used to kill enemy personnel, to suppress enemy fire and observation, and to penetrate light cover. Leaders can use 5.56-mm tracer fire to designate targets for other weapons.

a. Employment. Close combat is the predominant characteristic of urban engagements. Marine riflemen must be able to hit small, fleeting targets from bunker apertures, windows, and loopholes. This requires pinpoint accuracy with weapons fired in the semiautomatic mode. Killing an enemy through an 8-inch loophole at a range of 50 meters is a challenge, but one that may be common in combat in built-up areas.

(1) When fighting and engaging targets inside buildings, the weapons handling and firing techniques discussed in Appendix A are used.

(2) Within built-up areas, burning debris, reduced ambient light, strong shadow patterns of varying density, and smoke all limit the effectiveness of night vision and sighting devices. Aiming stakes and noise/light boobytraps coupled with night vision goggles and firing techniques can be used in the defense. During the offense, night vision goggles and illumination munitions can be used. Any Marine using night vision goggles should be teamed with at least one Marine not wearing them.

b. Weapon Penetration. The penetration that can be achieved with a 5.56-mm round depends on the range to the target and the type of material being fired against. Single 5.56-mm rounds are not effective against structural materials (as opposed to partitions) when fired at close range—the closer the range, the less the penetration.

(1) For the 5.56-mm round, maximum penetration occurs at 200 meters. At ranges of less than 25 meters, penetration is greatly reduced. At 10 meters, penetration by the M16 round is poor as a result of the tremendous stress placed on this high-speed round, which causes it to yaw upon striking a target. Stress causes the projectile to break up, and the resulting fragments are often too small to penetrate.

(2) Even with reduced penetration at short ranges, interior walls made of thin wood paneling, sheetrock, or plaster offer no protection against 5.56-mm rounds. Common office furniture such as desks and chairs cannot stop these rounds, but a layer of books 18 to 24 inches thick can.

(3) Wooden-framed buildings and single cinder-block walls offer little protection from 5.56-mm rounds. When clearing such structures, Marines must ensure that friendly casualties do not result from rounds passing through walls, floors, or ceilings.

(4) Armor-piercing rounds are slightly more effective than ball ammunition in penetrating urban targets at all ranges. They are also more likely to ricochet than ball ammunition, especially when the target presents a high degree of obliquity.

c. Protection. The following common barriers in built-up areas stop a 5.56-mm round fired at less than 50 meters:

- One thickness of sandbags
- A 2-inch concrete wall (unreinforced)
- A 55-gallon drum filled with water or sand
- A small ammunition can filled with sand

- A cinder block filled with sand (block will probably shatter)
- A plate-glass windowpane at a 45-degree angle (glass fragments will be thrown behind the glass)
- A brick veneer
- A car body (5.56-mm round will penetrate but normally not exit).

d. Wall Penetration. Although most structural materials repel single 5.56-mm rounds, continued and concentrated firing can breach some typical urban structures (See Table B-2).

Type	Penetration	Rounds Required
8-inch reinforced concrete	Initial Loophole	35 250
14-inch triple brick	Initial Loophole	90 160
12-inch cinder block with single-brick veneer	Loophole Breach hole	60 250
9-inch double brick	Initial Loophole	70 120
16-inch tree trunk or log wall	Initial[1]	1 to 3
12-inch cinder block (filled w/sand)	Loophole	35
24-inch double sandbag wall	Initial[1]	220
3/8-inch mild steel door	Initial[1]	1

[1]Penetration only, no loophole

Table B-2. Structure Penetration Capabilities of the 5.56-mm Round Against Typical Urban Targets (Range 25 to 100 Meters)

(1) The best method for breaching a masonry wall is by firing short bursts (3 - 5 rounds) in a U-shaped pattern. The distance from the gunner to the wall should be minimized for best results—ranges as close as 25 meters are relatively safe from ricochet. Ballistic eye protection, a protective vest, and a helmet should be worn.

(2) Ball ammunition and armor-piercing rounds produce almost the same results, but armor-piercing rounds are more likely to fly back at the firer. The 5.56-mm round can be used to create either a loophole (about 7 inches in diameter) or a breach hole (large enough for a man to enter). When used against reinforced concrete, the M16 rifle and M249 cannot cut the reinforcing bars.

3. Medium/Heavy Machine Guns and Sniper Rifles (7.62 mm and .50 caliber)

a. Machine Guns. In the urban environment, the .50-caliber machine gun and the 7.62-mm M240G machine gun provide high-volume, long-range, automatic fires for the suppression or destruction of targets. They provide final protective fires along fixed lines and can be used to penetrate light structures; the .50-caliber machine gun is most effective in this role. Tracers from both machine guns are likely to start fires, but the .50-caliber tracer is more apt to do so.

(1) Employment. The primary consideration affecting the employment of machine guns within built-up areas is the limited availability of long-range fields of fire. Although machine guns should be emplaced at the lowest level possible, grazing fire at ground level is often obstructed by rubble.

(a) The .50-caliber machine gun is often employed on its vehicular mount during both offensive and defensive operations. If necessary, it can be mounted on the M3 tripod for use in the ground role or in the upper level of buildings. When mounted on a tripod, the .50-caliber machine gun can be used as an accurate, long-range weapon and can supplement sniper fires.

(b) The M240G machine gun is cumbersome, making it difficult to use inside while clearing a building. However, it is useful outside to suppress and isolate enemy defenders. The M240G can be fired from either the shoulder or the hip to provide a high volume of assault and suppressive fires. The use of the long sling to support the weapon and ammunition is preferred.

(c) Because of their reduced penetration power, M240G machine guns are less effective against masonry targets than .50-caliber machine guns. However, their availability and light weight make them well suited to augment heavy machine gun fire, to be used in areas where .50-caliber machine guns cannot be positioned, or as a substitute when heavy machine guns are not available. The M240G machine gun can be employed on its tripod to deliver accurate fire along fixed lines and can then quickly be converted to bipod fire to cover alternate fields of fire.

(2) Penetration. The ability of the 7.62-mm and .50-caliber rounds to penetrate is also affected by the range to the target and type of material against which the rounds are fired. Close ranges affect penetration of the 7.62-mm round less than the 5.56-mm round; the .50-caliber round's penetration is reduced the least.

(a) At 50 meters, the 7.62-mm ball round cannot penetrate a single layer of sandbags. It can penetrate a single layer at 200 meters, but not a double layer. The armor-piercing round does only slightly better against sandbags. It cannot penetrate a double layer but can penetrate up to 10 inches of sand at 600 meters.

(b) The penetration of the 7.62-mm round is best at 600 meters, but most urban targets are closer. The longest engagement range is usually 200 meters or less. Table B-3 lists the penetration capabilities of a single 7.62-mm (ball) round at closer ranges.

Range (meters)	Pine Board (inches)	Dry, Loose Sand (inches)	Cinder Block (inches)	Concrete (inches)
25	13	5	8	2
100	18	4.5	10	2
200	41	7	8	2

Table B-3. Penetration Capabilities of a Single 7.62-mm (Ball) Round

(c) The .50-caliber round is also optimized for penetration at long ranges (about 800 meters). For hard targets, .50-caliber penetration is affected by obliquity and range. Both armor-piercing and ball ammunition penetrate 14 inches of sand or 28 inches of packed earth at 200 meters if the rounds impact perpendicular to the flat face of the target. Table B-4 depicts the effect of a 25-degree obliquity on a .50-caliber penetration.

Thickness (feet)	100 Meters (rounds)	200 Meters (rounds)
2	300	1,200
3	450	1,800
4	600	2,400

Table B-4. Number of Rounds Needed To Penetrate a Reinforced Concrete Wall at a 25-Degree Obliquity

(3) Protection. Barriers that offer protection against 5.56-mm rounds are also effective against 7.62-mm rounds with some exceptions. The 7.62-mm round can penetrate a windowpane at a 45-degree obliquity, a hollow cinder block, or both sides of a car body. It can also easily penetrate wooden-framed buildings. The .50-caliber round can penetrate all of the commonly found urban barriers except a sand-filled 55-gallon drum.

(4) Wall Penetration. Continued and concentrated machine gun fire can breach most typical urban walls. Such fire cannot breach thick reinforced concrete structures or dense natural stone walls. Internal walls, partitions, plaster, floors, ceilings, common office furniture, home appliances, and bedding can be easily penetrated by both 7.62-mm and .50-caliber rounds (Tables B-5 and B-6).

(a) The M240G machine gun can be hard to hold steady enough to repeatedly hit the same point on a wall. The dust created by the bullet strikes also makes precise aiming difficult. Firing with a tripod is usually more effective than without, especially if sandbags are used to steady the weapon. Short bursts of three to five rounds fired in a U-type pattern are best.

(b) Breaching a brick veneer over cinder block presents a special problem for the M240G machine gun. Rounds penetrate the cinder block but leave a net-like structure of unbroken block. Excessive ammunition is required to destroy the net because most rounds only pass through a previously eroded hole. One or two minutes of work with an entrenching tool, crowbar, or ax can remove this web and allow entry through the breach hole.

(c) The .50-caliber machine gun can be fired accurately from the tripod using the single-shot mode. This is the most efficient method for producing a loophole. Automatic fire in three- to five-round bursts in a U-type pattern is more effective in producing a breach.

Type	Thickness (Inches)	Hole Diameter (Inches)	Rounds Required
Reinforced concrete	8	7	100
Triple brick wall	14	7	170
Concrete block with single brick veneer	12	6 and 24	30 and 200
Cinder block (filled)	12	N/A[1]	18
Double brick wall	9	N/A[1]	45
Double sandbag wall	24	N/A[1]	110
Log wall	16	N/A[1]	1
Mild steel door	3/8	N/A[1]	1

[1]Penetration only, no loophole

Table B-5. Structure Penetrating Capabilities of 7.62-mm Round (NATO) Against Typical Urban Targets (Range 25 Meters)

Type	Thickness (Inches)	Hole Diameter (Inches)	Rounds Required
Reinforced concrete	10	12	50
		24	100
	18	7	140
Triple brick wall	12	8 and 26	15 and 50
Concrete block with single brick veneer	12	10 and 33	25 and 45
Armor plate	1	N/A[1]	1
Double sandbag wall	24	N/A[1]	5
Log wall	16	N/A[1]	1

[1]Penetration only, no loophole

Table B-6. Structure Penetrating Capabilities of .50-caliber Ball Against Typical Urban Targets (Range 35 Meters)

b. Sniper Rifles. The 7.62-mm match ball ammunition fired from the M-40A1 sniper rifle will perform similarly to the 7.62-mm ball round. The M-82A1A special application sniper rifle (SASR) can fire the .50-caliber armor-piercing incendiary (API) ammunition, which will have the same performance as when fired from the M-2 HB. The M-82A1A also fires the .50-caliber RAUFOSS ammunition, which contains a tungsten penetrator and a more powerful explosive charge than the API ammunition. When the RAUFOSS penetrator punches through the metal, the explosive detonates inside, acting as an incendiary round. Although the SASR's exact penetration capabilities have yet to be defined, it has penetrated an inch of steel at 2000 yards. Both sniper rifles are organic to the Marine infantry battalion.

4. Grenade Launchers, 40-mm (M203 and MK19). Both the M203 dual-purpose weapon and the MK19 grenade machine gun fire 40-mm HE and HE dual-purpose (HEDP) ammunition. Ammunition for these weapons is not interchangeable, but the grenade and fuze assembly that actually hits the target is identical. Both weapons provide point and area destructive fires as well as suppression. The MK19 has a much higher rate of fire and a longer range. The M203 is much lighter and more maneuverable.

a. Employment. The main consideration affecting the employment of 40-mm grenades within built-up areas is the typically short engagement range. The 40-mm grenade has a *minimum arming range of 14 to 28 meters.* If the round strikes an object before it is armed, it will not detonate. Both the HE and HEDP rounds have 5-meter burst radii against exposed forces, which means that the *minimum safe firing range for combat is 33 meters*. The 40-mm grenades can be used to suppress the enemy in a building or inflict casualties by firing through apertures or windows. The MK19 can use its high rate of fire to concentrate rounds against light structures. This concentrated fire can create extensive damage. The 40-mm HEDP round can penetrate the armor on the flank, rear, and top of Soviet-made BMPs and BTRs. Marines can use the M203 from upper stories to deliver accurate fire against the top decks of armored vehicles. Multiple hits are normally required to achieve a kill.

b. Weapon Penetration. The 40-mm HEDP grenade has a small shaped charge that penetrates better than the HE round. It also has a thin wire wrapping that bursts into a dense fragmentation pattern, creating casualties out to 5 meters. Because they explode on contact, 40-mm rounds achieve the same penetration regardless of range. Table B-7 lists the penetration capabilities of the HEDP round.

Target	Penetration (Inches)
Sandbags	20
Sand-filled cinder block	16
Pine logs	12
Armor plate	2

Table B-7. Penetration Capabilities of the HEDP Round

(1) If projected into an interior room, the 40-mm HEDP can penetrate all interior partition-type walls. It splinters plywood and plaster walls, making a hole large enough to fire a rifle through. It is better to have HEDP rounds pass into a room and explode on a far wall, even though much of the round's energy is wasted penetrating the back wall (Figure B-1). The fragmentation produced in the room causes more casualties than the high-explosive jet formed by the shaped charge.

(2) The fragments from the HEDP round do not reliably penetrate interior walls. They are also stopped by office furniture, sandbags, helmets, and protective vests (flak jackets). The M203 dual-purpose weapon has the inherent accuracy to place grenades into windows at 125 meters and bunker apertures at 50 meters. These ranges are significantly reduced as the angle of obliquity increases. Combat experience shows that M203 gunners cannot consistently hit windows at 50 meters when forced to aim and fire quickly.

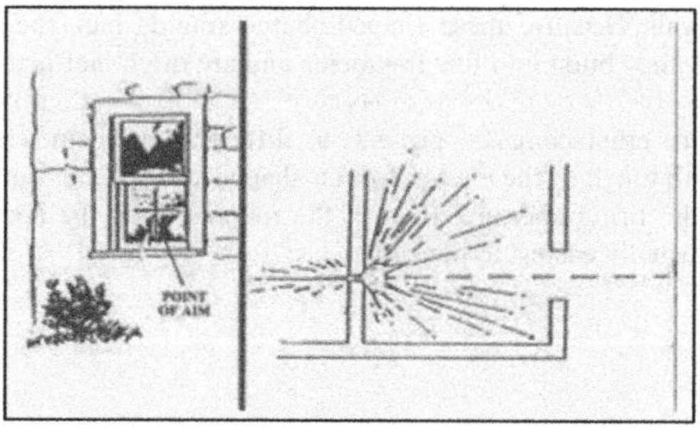

Figure B-1. Aim Point for 40-mm HEDP

c. Wall Penetration. The M203 cannot reasonably deliver the rounds needed to breach a typical exterior wall. The MK19 can concentrate its fire and achieve wall penetration. Firing from a tripod using a locked-down traversing and elevating mechanism is best for this role. Brick, cinder block, and concrete can be breached by using the MK19; individual HEDP rounds can penetrate 6 to 8 inches of brick. The only material that has proven resistant to concentrated 40-mm fire is dense stone such as that used in some European building construction. No precise data exist as to the number of rounds required to produce loopholes or breach holes with the MK19. However, the rounds' explosive effects are dramatic and should exceed the performance of the .50-caliber machine gun.

5. Light and Medium Recoilless Weapons. Light and medium recoilless weapons are used to attack enemy personnel, field fortifications, and LAVs. They have limited capability against main battle tanks, especially those equipped with reactive armor, except when attacking from the top, flanks, or rear. This category of weapons includes the AT4, the M47 Dragon, and the SMAW.

a. Employment. Other than defeating LAVs, the most common task for which light recoilless weapons are used is to neutralize fortified firing positions. Because of the design of the warhead and the narrow blast effect, these weapons are not as effective in this role as heavier weapons such as a tank main gun round. Their light weight allows Marines to carry several AT4s. Light recoilless weapons can be fired from the tops of buildings or from areas with extensive ventilation.

(1) Light and medium recoilless weapons with the exception of the SMAW employ shaped-charge warheads. As a result, the hole they punch in walls is often too small to use as a loophole. The fragmentation and spall that these weapons produce are limited. Normally, shaped-charge warheads do not neutralize enemy forces behind walls unless they are located directly in line with the point of impact.

(2) Against structures, shaped-charge weapons should be aimed about 6 inches below or to the side of a firing aperture (Figure B-2). This enhances the probability of killing the enemy behind the wall. A round that passes through a window wastes much of its energy on the back wall. Because these shaped-charge rounds lack the wire wrapping of the 40-mm HEDP, they burst into few fragments and are often ineffective casualty producers.

(3) Sandbagged emplacements present a different problem (Figure B-3). Because sandbags absorb much of the energy from a shaped charge, the rounds should be aimed at the center of the firing aperture. Even if the round misses the aperture, the bunker wall area near it is usually easiest to penetrate.

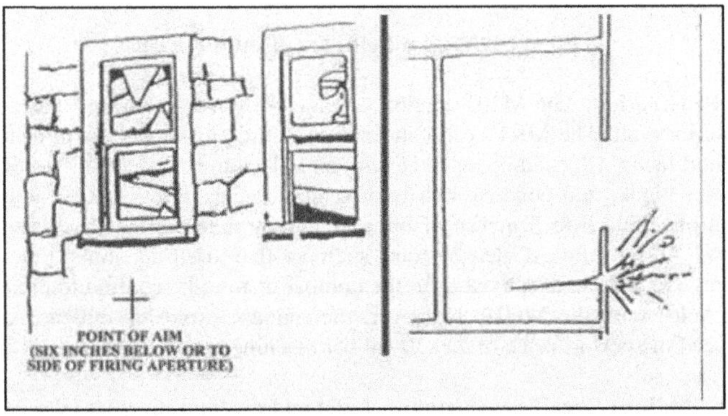

Figure B-2. Point of Aim for a Shaped-Charge Weapon Against a Masonry Structure

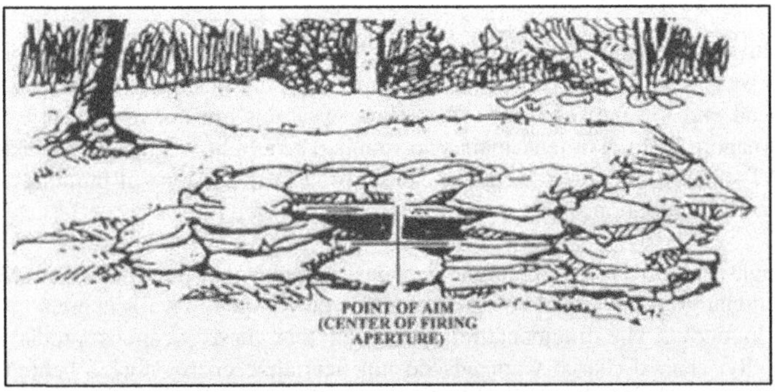

Figure B-3. Point of Aim for Sandbagged Emplacement

(4) Light and medium recoilless weapons obtain their most effective short-range antiarmor shots by firing from upper stories or from the flanks and rear. When firing at main battle tanks, these weapons should always be employed against these weaker areas in volley or paired firing. They normally require multiple hits to achieve a kill on a tank. Flank, top, and rear shots hit the most vulnerable parts of armored vehicles. Firing from upper stories protects the firer from tank main gun and coaxial machine gun fire because tanks cannot sharply elevate their cannons. The BMP-2 *can* elevate its 30-mm cannon to engage targets in upper stories. The BTR-series armored vehicles *can* also fire into upper stories with their heavy machine gun.

(5) Modern IFVs such as the BMP-2 and the BTR-80 have significantly improved frontal protection against shaped-charge weapons. Many main battle tanks have some form of reactive armor in addition to their thick armor plate. Head-on, ground-level shots against these vehicles have little probability of obtaining a kill. Even without reactive armor, modern main battle tanks are hard to destroy with a light antiarmor weapon.

(6) The easiest technique to improve the probability of hitting and killing an armored vehicle is to increase the firing depression angle. A 45-degree downward firing angle doubles the probability of a first-round hit as compared to a ground-level shot.

b. Backblast. Backblast characteristics must be considered when employing all recoilless weapons. During combat in built-up areas, the backblast area in the open is more hazardous because of the loose rubble and the channeling effect of the narrow streets and alleys. Figure B-4 shows the backblast areas of Marine light recoilless weapons in the open.

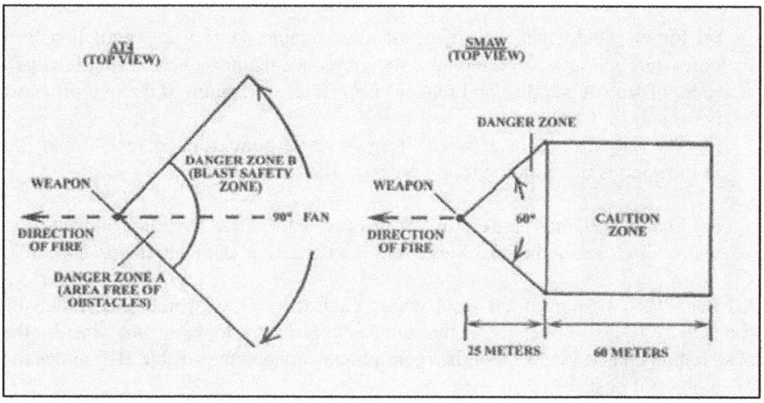

Figure B-4. Backblast Areas of Light Recoilless Weapons in the Open

(1) When firing recoilless weapons in the open, Marines should protect themselves from blast and burn injuries caused by the backblast. All personnel should be out of the danger zone. Anyone not able to vacate the caution zone should be behind cover. Marines in the caution zone should wear helmets, protective vests, and eye protection. The firer and all Marines in the area should wear earplugs.

(2) Extensive testing on the effects of firing recoilless weapons from within enclosures has been conducted since 1948. These tests showed that firing these weapons from enclosures presented no serious hazards, even when the overpressure was enough to produce structural damage to the building. The following were other findings of these tests.

(a) Little hazard exists to the gunnery or crew from any type of flying debris. Loose items were not hurled around the room.

(b) No substantial degradation occurs to the operator's tracking performance as a result of obscuration or blast overpressure.

(c) The most serious hazard that can be expected is hearing loss. This must be evaluated against the advantage gained in combat from firing from cover. To place this hazard in perspective, a gunner wearing earplugs and firing the loudest combination (the Dragon from within a masonry building) is exposed to less noise hazard than if he fired a LAW or AT4 in the open without earplugs.

(d) The safest place for other Marines in the room with the firer is against the wall from which the weapon is fired. Plastic ignition plugs are a hazard to anyone standing directly behind a LAW or TOW when it is fired.

(e) Firers should take advantage of all available sources of ventilation by opening doors and windows. Ventilation does not reduce the noise hazard, but it helps clear the room of smoke and dust and reduces the effective duration of the overpressure.

(f) The only difference between firing these weapons from enclosures and firing them in the open is the duration of the pressure fluctuation.

(g) Frame buildings, especially small ones, can suffer structural damage to the rear walls, windows, and doors. Large rooms suffer slight damage, if any.

(3) Recoilless weapons fired from within enclosures create some obscuration inside the room but almost none from the gunner's position looking out. Inside the room, obscuration can be intense, but the room remains inhabitable. Table B-8 shows the effects of smoke and obscuration.

Building	Weapon	From Gunner's Position Looking Out	Inside the Room	From Outside at a Distance
Masonry	LAW	None	Moderate	Slight smoke
	Dragon	Slight	Moderate	Small flash

Bunker	Dragon TOW	None None	Slight Slight	Moderate flash Moderate smoke
Small frame	LAW Dragon	None None	Moderate Severe	Moderate Moderate
Medium frame	LAW Dragon	None None	Slight Severe	Moderate Slight flash
Large frame	LAW Dragon TOW	None Slight None	Slight Severe Severe	None Slight flash Slight smoke

Table B-8. Smoke and Obscuration

(4) The Dragon causes the most structural damage, but only in frame buildings. There does not seem to be any threat of injury to the gunner because the damage is usually to the walls away from the gunner. The most damage and debris is from flying plaster chips and pieces of wood trim. Large chunks of plasterboard can be dislodged from ceilings. The backblast from a LAW, AT4, Dragon, or TOW rarely displaces furniture. Table B-9 shows the test results of structural damage and debris.

		Damage		
Building	**Weapon**	**Structure**	**Wall Covering**	**Debris Movement**
Masonry	LAW Dragon	None None	Slight Slight	Slight Slight
Bunker	Dragon TOW	None None	None None	None Leaves and dust disturbed
Small frame	LAW Dragon	None Severe	Slight Severe	None None
Medium frame	LAW Dragon	None Slight	None Slight	Slight Lamp and chair overturned
Large frame	LAW Dragon TOW	None Slight Slight	Slight Moderate Severe	Slight None None

Table B-9. Structural Damage and Debris Movement

(5) To fire a LAW, AT4, or SMAW from inside a room, the following safety precautions must be taken (Figure B-5).

(a) The building should be of a sturdy construction.

(b) The ceiling should be at least 7 feet high with loose plaster or ceiling boards removed.

(c) The floor size should be at least 15 feet by 12 feet. (The larger the room, the better.)

(d) At least 20 square feet of ventilation (room openings) should exist to the rear or side of the weapon. An open 7- by 3-foot door would provide minimum ventilation.

(e) All glass should be removed from windows, and small, loose objects should be removed from the room.

(f) Floors should be wet to prevent dust and dirt from blowing around and obscuring the gunner's vision.

(g) All personnel in the room should be forward of the rear of the weapon.

(h) All personnel in the room should wear helmets, protective vests, ballistic eye protection, and earplugs.

(i) If the gunner is firing from the prone position, his lower body must be perpendicular to the bore of the weapon, or the blast could cause injury to his legs.

Figure B-5. Firing an AT4 From Inside a Building

c. Weapon Penetration. The most important tasks to be performed against structures are the neutralization of fortified firing positions, personnel, and weapons behind barriers. Recoilless weapons can be used in this role; none, however, are as effective as heavy, direct-fire weapons or standard demolitions. Each recoilless weapon has different penetrating ability against various targets. Penetration does not always mean the destruction of the integrity of a position. Usually only those enemy soldiers directly in the path of the spall from a high-explosive antitank (HEAT) round become casualties. Other soldiers inside a fortification could be deafened, dazed, or shocked but could eventually return to action.

(1) AT4. The AT4 is heavier than the LAW and has a diameter of 84 millimeters, which gives the warhead much greater penetration. The AT4 can penetrate more than 17.5 inches (450 mm) of armor plate. Its warhead produces highly destructive results behind the armor. Tests against typical urban targets are still ongoing, but the AT4 should penetrate at least as well as the 90-mm recoilless rifle, if not better. The AT4 has a *minimum arming distance of 10 meters*, which allows it to be fired successfully against close targets. Firers should be well covered and protected when firing at close targets.

(2) MK153 83-mm SMAW. The SMAW is a lightweight assault weapon that is easily carried and placed into action by one man. It is used against fortified positions, but it is also effective against LAVs. The SMAW has a 9-mm spotting rifle and a 3.8-power telescope that ensure accuracy over ranges common to combat in built-up areas. The SMAW has excellent incapacitating effects for enemies behind walls and inside bunkers and can arm within 10 meters. Its fuze has the ability to distinguish between armor and soft earth, maximizing its capabilities against buildings, bunkers, or light armor. The warhead detonates immediately against hard targets but delays detonation against soft targets and burrows in to explode inside. The SMAW can destroy most bunkers with a single hit. The 83-mm HE warhead can create an 8-inch mousehole in reinforced concrete but will not cut the steel reinforcing bar. The SMAW is an outstanding urban support weapon that can be decisive during an urban assault. The SMAW allows for quick reduction of obstacles and bunkers as well as creation of entry points. SMAWs fired in a volley can be devastating to a building.

d. Wall Breaching. Wall breaching is a common combat task in built-up areas for which light recoilless weapons can be used. Breaching operations improve mobility by providing access to building interiors without using existing doors or windows. Breaching techniques can also be used to create loopholes for weapons positions or to allow hand grenades to be thrown into defended structures. Breaching holes for unit mobility should be about 24 inches (60 centimeters) in diameter. Loopholes should be about 8 inches (20 centimeters) in diameter (Figure B-6). Only the SMAW provides a one-shot wall breaching ability. However, in breaching walls a number of shots should be planned.

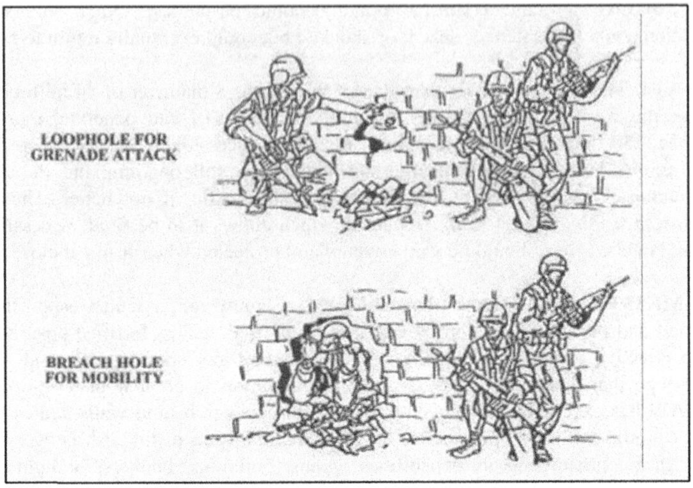

Figure B-6. Tactical Use of Holes in Masonry Walls

(1) Of all the common building materials, heavy stone is the most difficult to penetrate. The AT4 normally will not penetrate a heavy European-style stone wall. Surface cratering is usually the only effect it will have.

(2) Layered brick walls are also difficult to breach with light recoilless weapons. Some brick walls can be penetrated by multiple firings, especially if they are less than three bricks thick. Multiple AT4 rounds fired at the same spot on an 8-inch (double-brick) wall normally produce a loophole. The SMAW produces a hole in brick walls that is often large enough to be a breach hole.

(3) Wooden structural walls offer little resistance to light recoilless weapons. Even heavy timbered walls are penetrated and splintered. Multiple AT4 rounds fired at the same area of a wood-framed wall usually produce a man-sized hole. The SMAW has a devastating effect against wood-framed walls. A single round produces a breach hole as well as significant spall.

(4) Because of its high velocity, the AT4 may penetrate a soft target such as a car body or frame building before exploding.

(5) None of the light recoilless weapons are as effective against structural walls as demolitions or heavier weapons such as tank main guns or field artillery. Of all the light recoilless weapons, the SMAW is the most effective.

6. Antitank Guided Missiles. ATGMs are used mainly to defeat main battle tanks and other armored combat vehicles. They have a moderate capability against bunkers, buildings, and other fortified targets commonly found during combat in built-up areas. This category of weapons includes the TOW and Dragon missiles.

a. Employment. TOWs and Dragons provide overwatch antitank fires during the attack of a built-up area and an extended range capability for the engagement of armor during the defense. Within built-up areas, they are best employed along major thoroughfares and from the upper stories of buildings to attain long-range fields of fire. Their *minimum firing range of 65 meters* could limit firing opportunities in the confines of densely built-up areas.

(1) Obstacles. Rubble or other obstacles could interfere with the flight of missiles when fired from street level. At least 3.5 feet (1 meter) of vertical clearance over such obstacles must be maintained. Figure B-7 shows the most common obstacles to ATGM flight that are found in built-up areas. *Power lines are a special obstacle and present a unique threat to ATGM gunners. If the power in the lines has not been interrupted, the ATGM guidance wires could create a short circuit. This would allow extremely high voltage to pass to the gunner in the brief period before the guidance wires melted. This voltage could either damage the sight and guidance system or injure the gunner. Before any ATGM is fired over a power line, an attempt must be made to determine whether or not the power to that line has been interrupted.*

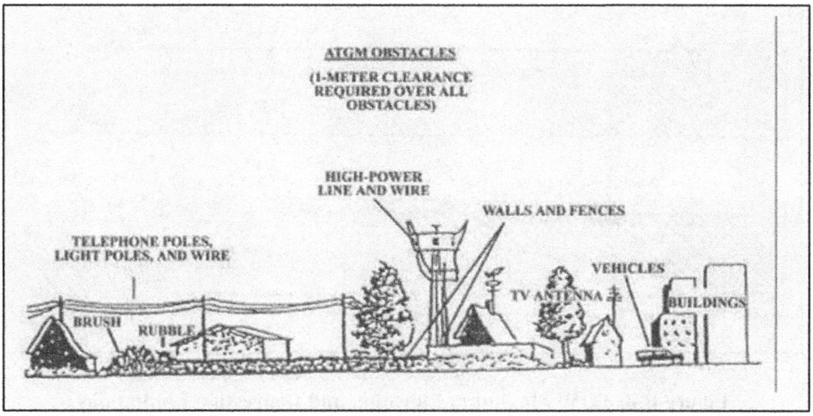

Figure B-7. Common Obstacles to ATGM Flights

(2) Dead Space. Three aspects of dead space that affect ATGM fires are arming distance, maximum depression, and maximum elevation.

(a) Both the Dragon and TOW missiles have a minimum arming distance of 65 meters, which severely limits their use in built-up areas. Few areas in the inner city permit fires much beyond the minimum arming distance. However, ground-level, long-range fires down streets or rail lines and across parks or plazas are possible. ATGMs may be used effectively from upper stories or roofs of buildings to fire into other buildings.

(b) The TOW is limited much more than the Dragon by its maximum depression and elevation. The maximum depression and elevation limits of the TOW mount could result in dead space and preclude the engagement of close targets (Figure B-8). A target located at the minimum arming range (65 meters) cannot be engaged by a TOW crew located any higher than the sixth floor of a building due to maximum depression limits. At 100 meters, the TOW crew can be located as high as the ninth floor and still engage the target.

(3) Backblast. Backblast for ATGMs is of more concern during combat in built-up areas than in open country. Any loose rubble in the caution zone could be picked up and thrown by the backblast, and the channeling effect of walls and narrow streets is even more pronounced due to the greater backblast. If the ATGM backblast strikes a wall at an angle, it can pick up debris or be deflected and cause injury to unprotected personnel (Figure B-9). Both types of ATGM can be fired from inside some buildings. In addition to the helmet and protective vest, eye protection and earplugs should be worn by all personnel in the room.

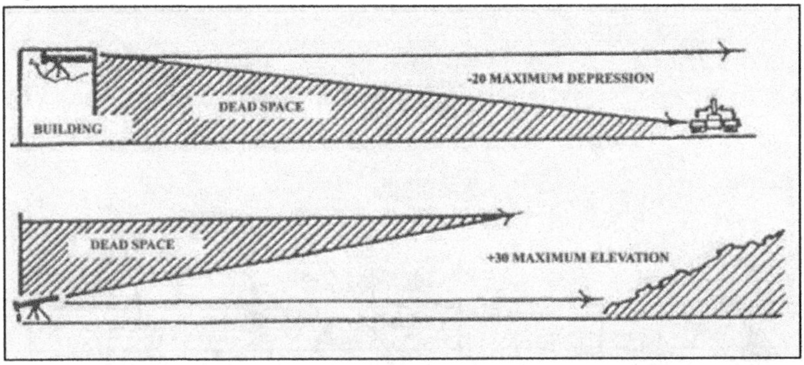

Figure B-8. TOW Maximum Elevation and Depression Limitations

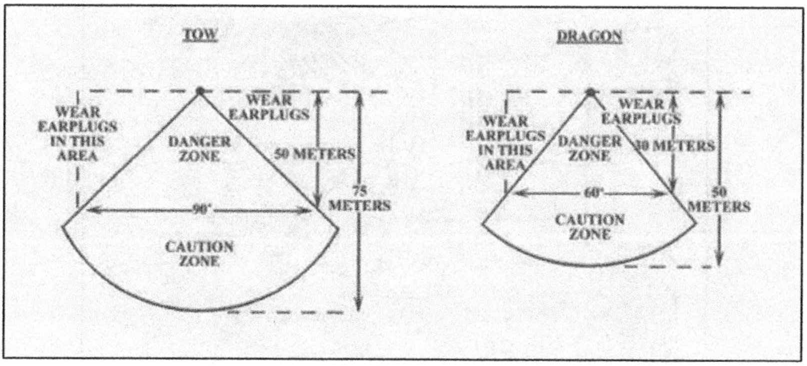

Figure B-9. ATGM Backblast in an Open Street

(a) To fire a TOW from inside a room, the following safety precautions must be taken (Figure B-10):

- The building must be of sturdy construction.
- The ceiling should be at least 7 feet high.
- The floor size of the room should be at least 15 by 15 feet, or larger if possible.
- At least 20 square feet of room ventilation should exist, preferably to the rear of the weapon. An open 7- by 3-foot door is sufficient. Additional ventilation can be created by removing sections of interior partitions.
- All glass must be removed from the windows and all small, loose objects removed from the room.
- All personnel in the room should be forward of the rear of the TOW.
- All personnel in the room should wear ballistic eye protection and earplugs.
- A clearance of 9 inches (23 centimeters) must exist between the launch tube and the aperture from which it is fired. (See FM 23-34, *TOW Weapon System*, for more detailed safety information.)

Figure B-10. TOW Fired From Inside a Room

(b) To fire a Dragon from inside a room, the following safety precautions must be taken.

- The building must be of sturdy construction.
- The ceiling should be at least 7 feet high.
- The floor size should be at least 15 by 15 feet, or larger if possible.
- At least 20 square feet of ventilation should exist (room openings), preferably to the rear of the weapon. An open 7- by 3-foot door would provide minimum ventilation.
- All glass should be removed from windows, and small, loose objects should be removed from the room.
- The room should be clean or the floors must be wet to prevent dust and dirt (kicked up by the backblast) from obscuring the vision of other soldiers in the room.
- All personnel in the room must be forward of the rear of the weapon.
- All personnel in the room must wear ballistic eye protection and earplugs.
- At least a 6-inch clearance must exist between the launch tube and the aperture from which it is fired.

b. Weapon Penetration. ATGMs can penetrate and destroy heavily armored tanks. They have large warheads that contain shaped charges. Because of their size, these warheads can achieve significant penetration against typical urban targets. Penetration, however, does not mean a concurrent destruction of the structural integrity of a position. The shaped-charge warhead produces relatively little spall. Enemy personnel not standing directly behind or near the point of impact of an ATGM may escape injury.

(1) Standard TOW Missiles. The basic TOW missile can penetrate 8 feet of packed earth, 4 feet of reinforced concrete, or 16 inches of steel plate. The improved TOW (ITOW), the TOW 2, and the TOW 2A have all been modified to improve their penetration. They all penetrate better than the basic TOW. All TOW missiles can defeat triple sandbag walls, double layers of earth-filled 55-gallon drums, and 18-inch log walls.

(2) TOW 2B. The TOW 2B uses a different method of defeating enemy armor. It flies over the target and fires an explosively formed penetrator down onto the top armor, which is thinner. *Because of this design feature, the TOW 2B missile cannot be used to attack nonmetallic structural targets. When using the TOW 2B missile against enemy armor, gunners must avoid firing directly over other friendly vehicles, disabled vehicles, or large metal objects such as water or oil tanks.*

(3) Dragon Missile. The Dragon missile can penetrate 8 feet of packed earth, 4 feet of concrete, or 13 inches of steel plate. It can attain effective short-range fire from upper stories or from the rear or flanks of a vehicle. These engagements are targeted against the most vulnerable parts of tanks and can entrap tanks in situations where they are unable to counterfire. Elevated firing positions increase the first-round hit probability. Firing down at an angle of 20 degrees increases the chance of a hit by 67 percent at 200 meters. A 45-degree downward angle doubles the first-round hit probability, compared to a ground-level shot.

c. Breaching Structural Walls. Firing ATGMs is the least efficient means to defeat structures. Because of their small basic load and high cost, ATGMs are better used against tanks or enemy-fortified firing positions. They can be effective against bunkers or other identified enemy firing positions.

7. Flame Weapons. Flame weapons are characterized by both physical and psychological casualty-producing abilities. Flame does not need to be applied with pinpoint accuracy, but it also must not spread to structures needed by friendly forces. Large fires in built-up areas are catastrophic. If they burn out of control, fires can create an impenetrable barrier for hours. The Marine Corps does not possess flame-generating equipment. However, the M34 WP grenade can be used to ignite and destroy flammable objects, especially wooden structures. The Army's M2A1-7 portable flamethrower is stored in war reserve status as a standard "C" item. Its availability is limited; however, Marines could be issued flame-support weapons for urban operations depending on the status of international agreements on the use of flame weapons that are in effect at the time.

a. Employment. Flame weapons used against fortified positions should be aimed directly at the aperture. Even if the round or burst misses, enough flaming material enters the position to cause casualties and disrupt the enemy occupants. The M34 WP grenade is difficult to throw far or into a small opening such as a bunker aperture. However, its effects are dramatic when thrown into a room or building.

b. Effects. The M34 is used to ignite and destroy flammable objects, especially wooden structures. It is also used to create an immediate smoke cloud to conceal movement across a narrow open space such as a street. Its smoke is not toxic but can cause choking in heavy concentrations.

(1) The grenade's explosion, bright flash, smoke, and burning WP particles all combine to make the M34 one of the most effective psychological weapons available. The M34 hand grenade throws WP fragments up to 35 meters from the point of detonation. These fragments can attach to clothing or skin and continue burning. Because of its weight, most infantrymen can throw this grenade only 30 to 40 meters.

(2) The Marine must avoid injury from friendly use of the M34. The M34 can ignite if the WP inside is exposed to the air. Bullets and shell fragments have been known to strike and rupture M34 grenades; therefore, grenades must be protected from enemy fire.

(3) The M34 WP grenade is an effective weapon against enemy armored vehicles when used in the close confines of combat in built-up areas. It can be thrown or dropped from upper stories onto enemy vehicles. The M34 can be combined with flammable liquids, detonating cords, blasting caps, and fuze igniters to create the eagle fireball, a field-expedient antiarmor device. (See FM 21-75, *Combat Skills of the Soldier*, Appendix H.)

(4) The M34 is also excellent as a screening device. A grenade can be thrown from behind cover into an open street or plaza. When it explodes, the enemy's observation is temporarily obscured. Thus, friendly forces can quickly cross the open area—if the enemy fires, it is unaimed and presents less of a danger. If screening smoke is used to cover a squad's movement across short open areas, it will reduce expected casualties from small-arms fire by about 90 percent.

8. Hand Grenades. Hand grenades are used extensively during combat in built-up areas. Smoke grenades are used for screening and signaling. Riot control grenades are used to drive the enemy out of deep fortifications. Fragmentation and concussion grenades are used to clear the enemy out of rooms and basements. Hand grenades are the most-used explosive munition during intense combat in built-up areas. In World War II, it was common for a battalion fighting in a city to use more than 500 hand grenades each day.

a. Employment. Smoke and riot-control grenades have similar employment techniques. Fragmentation grenades are used to produce enemy casualties.

(1) The AN-M8 HC grenade produces a dense white or grey smoke. It burns intensely and cannot be easily extinguished once it ignites. The smoke can be dangerous in heavy concentrations because it makes breathing difficult and causes choking. The M8 grenade is normally used for screening. It produces a slowly building screen of longer duration than the M34 WP grenade without the problem of collateral damage caused by scattered burning particles.

(2) The M18-series smoke grenades produce several different colors of smoke that are used for signaling. Yellow smoke is sometimes difficult to see in built-up areas.

(3) The M7A3 CS riot control grenade can be used to drive enemy troops out of fortifications when civilian casualties or collateral damage constraints are considerations. Built-up areas often create variable and shifting wind patterns. When using CS grenades, Marines must prevent the irritating smoke from affecting friendly forces. The CS grenade burns intensely and can ignite flammable structures. Enemy troops wearing even rudimentary chemical protective masks can withstand intense concentrations of CS gas.

(4) The MK3A2 offensive hand grenade, commonly referred to as the concussion grenade, produces casualties during close combat while minimizing the danger to friendly personnel. For this reason, it is the preferred hand grenade for use during offensive operations in a MOUT environment. The grenade produces severe concussion effects in enclosed areas. It can be used for light blasting and demolitions and for creating breach holes in interior walls. The concussion produced by the MK3A2 is much greater than that of the fragmentation grenade. It is very effective against enemy soldiers within bunkers, buildings, and underground passages.

(5) The M67 fragmentation grenade is the most commonly available grenade during combat in built-up areas. It provides suppression during room-to-room or house-to-house fighting, and it is used while clearing rooms of enemy personnel. The M67 grenade has a 3- to 5-second delay fuze. When used at close ranges, it can be cooked-off for two seconds to deny the enemy the time to throw it back. The fragmentation grenade can be rolled, bounced, or ricocheted into areas that cannot be reached by 40-mm grenade launchers.

b. Effects. Each type of hand grenade has its own specific effect during combat in built-up areas.

(1) Both HC smoke and CS gas *displace* oxygen from enclosed areas. Gas masks will not prevent this from occurring because they only filter air, not provide oxygen. Smoke grenades produce dense clouds of colored or white smoke that remain stationary in the surrounding area. They can also cause fires if used indiscriminately. If trapped and concentrated within a small space, smoke grenades and CS gas *can suffocate Marines.*

(2) The fragmentation grenade has more varied effects in combat in built-up areas. It produces a large amount of small, high-velocity fragments that can penetrate sheetrock partitions and are lethal at short ranges (15 to 20 meters). Fragments lose their velocity quickly and are less effective beyond 25 meters.

The fragments from a fragmentation grenade cannot penetrate a single layer of sandbags, a cinder block, or a brick building, but they can perforate wood-framed and tin buildings if exploded close to their walls.

(3) Inside rooms, fragmentation barriers consisting of common office furniture, mattresses, doors, or books can be effective against the fragmentation grenade. For this reason, a room should never be considered safe just because one or two grenades have been detonated inside. Fragmentation grenades detonated on the floor not only throw fragments laterally, but also send fragments and spall downward to lower floors. Predicting how much spall will occur is difficult because flooring material varies, but wooden floors are usually affected the most.

(4) Some foreign grenades throw fragments that are much larger than those of the U.S.-made M67. Light barriers and interior walls would probably be less effective against these grenades than against the M67. A major problem with the U.S.-made fragmentation grenade is its tendency to bounce back off of hard targets. Grenades are often directed at window openings on the ground floor or second floor. At ranges as close as 20 meters, a thrower's chances of missing a standard 1-meter by 1-meter window are high. The fragmentation grenade normally breaks through standard window glass and enters a room. If the grenade strikes at a sharp angle or the glass is thick plate, the grenade could be deflected without penetrating.

(5) Hand grenades are difficult weapons to use. They involve a high risk of fratricide. Commanders should conduct precombat training with hand grenades as part of normal preparations. Marines must be very careful when throwing hand grenades up stairs to avoid having the grenade roll back down on them.

(6) Once pulled, the pins of both fragmentation and concussion grenades can be replaced if the thrower decides not to use the weapon. This pin replacement must be done carefully (see FM 23-30, *Grenades and Pyrotechnic Signals*) and prior to the spoon being released or detaching.

(7) An estimate of the situation (METT-T) and existing ROE will dictate what type of grenade will be used to clear each room. Because of the high rate of expenditure of grenades, units should use assault packs to carry additional grenades of all types. Additional grenades can also be carried in empty ammunition or canteen pouches.

9. 25-mm Automatic Gun. The 25-mm automatic gun mounted on the LAV-25 is an effective weapon during combat in built-up areas. The primary roles of LAV-25s during combat in built-up areas are to provide suppressive fire and to breach exterior walls and fortifications.

a. Obliquity. The 25-mm gun produces its best urban target results when fired perpendicular to the hard surface (zero obliquity). In combat in built-up areas, however, finding a covered firing position that permits low-obliquity firing is unlikely unless the streets and gaps between buildings are wide. Most shots impact the target at an angle, which normally reduces penetration. With the armor-piercing, discarding, sabot with tracer (APDS-T) round, an angle of obliquity of up to 20 degrees can actually improve breaching. The rounds tend to dislodge more wall material for each shot but do not penetrate as deeply into the structure.

b. Target Types. The 25-mm gun has different effects when fired against different urban targets.

(1) Reinforced Concrete. Reinforced concrete walls, which are 12 to 20 inches thick, present problems for the 25-mm gun when trying to create breach holes. It is relatively easy to penetrate, fracture, and clear away the concrete, but the reinforcing rods remain in place. These create a "jail window" effect by preventing entry but allowing grenades or rifle fire to be placed behind the wall. Steel reinforcing rods are normally 3/4 of an inch thick and 6 to 8 inches apart—there is no quick way of cutting these rods. They can be cut with demolition charges, cutting torches, or special power saws. Firing with either APDS-T or high-explosive, incendiary with tracer (HEI-T) rounds from the 25-mm gun will not always cut these rods.

(2) Brick Walls. Brick walls are more easily defeated by the 25-mm gun regardless of their thickness, and they produce the most spall.

(3) Bunker Walls. The 25-mm gun is devastating when fired against sandbag bunker walls. Obliquity has the least effect on the penetration of bunker walls. Bunkers with earth walls up to 36 inches thick are easily penetrated. At short ranges typical of combat in built-up areas, defeating a bunker should be easy, especially if the 25-mm gun can fire at an aperture.

c. Burst Fire. The 25-mm gun's impact on typical urban targets seems to be magnified if the firing is in short bursts. At close ranges, the gunner might need to shift his point of aim in a spiral pattern to ensure that the second and third bursts enlarge the hole. Even without burst fire, sustained 25-mm gunfire can defeat almost all urban targets.

d. Weapon Penetration. The penetration achieved by the two combat rounds (HEI-T and APDS-T) differs slightly, but both are effective.

(1) APDS-T. The APDS-T round penetrates urban targets by retaining its kinetic energy and blasting a small hole deep into the target. The APDS-T round gives the best effects behind the wall, and the armor-piercing core often breaks into two or three fragments that can create multiple enemy casualties. The APDS-T needs as few as four rounds to achieve lethal results behind walls. Table B-10 lists the number of APDS-T rounds needed to create different-sized holes in common urban walls.

Target	Loophole	Breach Hole
3-inch brick wall at 0-degree obliquity	22 rounds	75 rounds
3-inch brick wall at 45-degree obliquity	22 rounds	35[1] rounds
5-inch brick wall at 0-degree obliquity	32 rounds	50[1] rounds
8-inch reinforced concrete at 0-degree obliquity	22 rounds	75 rounds (Note: Reinforcing rods still in place)
8-inch reinforced concrete at 45-degree obliquity	22 rounds	40[1] rounds (Note: Reinforcing rods still in place)

[1] Obliquity and depth tend to increase the amount of wall material removed.

Table B-10. Breaching Effects of APDS-T Rounds

(a) When firing single rounds, the APDS-T round provides the greatest capability for behind-the-wall incapacitation. The APDS-T round can penetrate more than 16 inches of reinforced concrete with enough energy left to cause enemy casualties. It penetrates through both sides of a wood-framed or brick veneer building. Field fortifications are easily penetrated by APDS-T rounds. Table B-11 lists the number of APDS-T rounds needed to create different-sized holes in commonly found bunkers.

(b) The APDS-T round creates a hazardous situation for exposed personnel because of the pieces of sabot that are thrown off of the round.

Type Bunker	Obliquity	Penetration	Loophole	Small Breach Hole
36-inch sand/timber	0 degree	1 round	25 rounds	40 rounds
36-inch sand/6-inch concrete	0 degree	6 rounds	6 rounds	20 rounds

Table B-11. Number of APDS-T Rounds Needed To Create Different-Sized Holes in Bunkers

WARNING

Personnel not under cover forward of the 25-mm gun's muzzle and within the danger zone could be injured or killed by these sabots, even if the penetrator passes overhead to hit the target.

The danger zone extends at an angle of about 10 degrees below the muzzle level, out to at least 100 meters, and about 17 degrees left and right of the muzzle. Figure B-11 shows the hazard area of the APDS-T round.

Figure B-11. APDS-T Danger Zone

(2) HEI-T. The HEI-T round penetrates urban targets by blasting away chunks of material.

(a) The HEI-T round does not penetrate an urban target as well as the APDS-T, but it does strip away a greater amount of material for each round. The HEI-T does more damage to an urban target when fired in multiple short bursts because the cumulative impact of multiple rounds is greater than the sum of individual rounds. Table B-12 lists the number of HEI-T rounds needed to create different-sized holes.

Target	Loophole	Breach Hole
3-inch brick wall at 0-degree obliquity	10 rounds	20 rounds
3-inch brick wall at 45-degree obliquity	20 rounds	25 rounds
5-inch brick wall at 0-degree obliquity	30 rounds	60 rounds

8-inch reinforced concrete at 0-degree obliquity	15 rounds	25 rounds
8-inch reinforced concrete at 45-degree obliquity	15 rounds	30 rounds

Table B-12. Number of HEI-T Rounds Needed To Create Different-Sized Holes in Bunkers

(b) The HEI-T round does not provide single-round perforation or incapacitating fragments when used against any external masonry structural wall. It can create first-round fragments behind wood-framed and brick veneer walls. HEI-T rounds cannot penetrate a bunker as quickly as APDS-T rounds, but they can create more damage inside the bunker once the external earth has been stripped away. Against a heavy bunker, about 40 rounds of HEI-T ammunition are needed to strip away the external earth shielding and breach the inner lining of concrete or timber. The HEI-T round is also used for suppression against known or suspected firing ports such as doors, windows, and loopholes.

10. Tank Cannon. The powerful, high-velocity cannon mounted on the M1A1 tank provides Marines with a key requirement for victory in built-up areas—heavy direct-fire support. Although the infantry assumes the lead role during combat in built-up areas, tanks and infantry work as a close team. Tanks move down streets after the infantry has cleared them of any suspected ATGM positions and, in turn, support the infantry with fire. The tank is one of the most effective weapons for heavy fire against structures. The primary role of the tank cannon during combat in built-up areas is to provide heavy direct fire against buildings and strongpoints that are identified as targets by the infantry. The wall and fortification breaching effects of the 120-mm tank cannon are major assets to Marines fighting in built-up areas.

a. Obliquity. The tank cannon produces its best urban target effects when fired perpendicular to the hard surface (zero obliquity). During combat in built-up areas, however, finding a covered firing position that permits low-obliquity firing is unlikely. Most shots strike the target at an angle that would normally reduce penetration. With tank cannon APDS rounds, obliquity angles of up to 25 degrees have little effect, but angles greater than 45 degrees greatly reduce penetration.

b. Ammunition. Armor-piercing, fin-stabilized, discarding sabot (APFSDS) rounds are the most commonly carried tank ammunition. These rounds work best against armored vehicles. The 120-mm cannon also carries an effective high-explosive, antitank multipurpose (HEAT-MP) round.

c. Characteristics. The 120-mm tank cannon has two specific characteristics that affect its employment in built-up areas: limited elevation and depression and short arming ranges. In addition, the M1A1 tank has another characteristic that is not involved with its cannon but that affects Marines working with or around the tank—*extremely hot turbine exhaust.*

(1) The M1A1 tank can elevate its cannon 20 degrees and depress it 10 degrees. The lower depression limit creates a 35-foot (10.8-meter) dead space around a tank. On a street 16 meters wide, this dead space extends to the buildings on each side (Figure B-12). Similarly, there is a zone overhead in which the tank cannot fire (Figure B-13). This dead space offers ideal locations for short-range antiarmor weapons and allows hidden enemy gunners to fire at the tank when the tank cannot fire back. It also exposes the tank's most vulnerable areas: the flanks, rear, and top. Infantry must move ahead, alongside, and to the rear of tanks to provide close protection. The extreme heat produced immediately to the rear of the M1A1 tank prevents dismounted infantry from following closely, but protection from small-arms fire and fragments is still provided by the tank's bulk and armor. The M1A1 tank also has a blind spot caused by the zero-degree depression available over part of the back deck. To engage any target in this area, the tank must pivot to convert the rear target to a flank target.

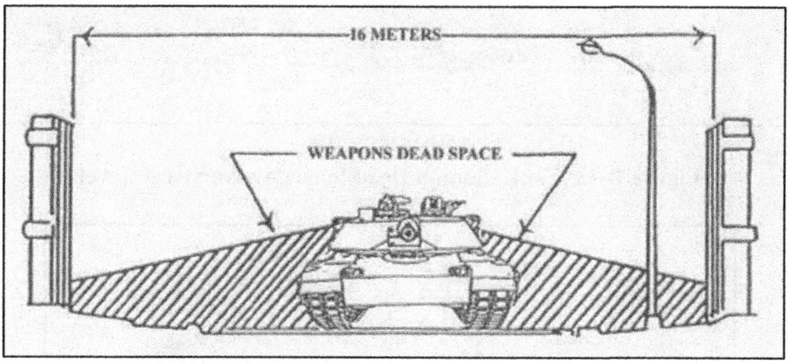

Figure B-12. Tank Cannon Dead Space at Street Level

(2) The 120-mm HEAT-MP round arms at about 36 feet. This arming distance allows the tank to engage targets from short ranges. The armor of the tank protects the crew from both the blowback effects of the round and enemy return fire. The APFSDS round does not need to arm and can, therefore, be fired at almost any range. The discarding portions of the round can be lethal to exposed infantry forward of the tank.

d. Target Effects. HEAT rounds are most effective against masonry walls. The APFSDS round can penetrate deeply into a structure but does not create as large a hole or displace as much spall behind the target. In contrast to lighter HEAT rounds, tank HEAT rounds are large enough to displace enough spall to inflict casualties inside a building. One HEAT round normally creates a breach hole in all but the thickest masonry construction—brick veneer and wood-framed constructions are demolished by a single round. Even the 120-mm HEAT round cannot cut all of the reinforcing rods, which are usually left in place, often hindering entry through the breach hole (Figure B-14). Both HEAT and APFSDS rounds are effective against all field fortifications. Only large earth berms and heavy mass-construction buildings can provide protection against tank fire.

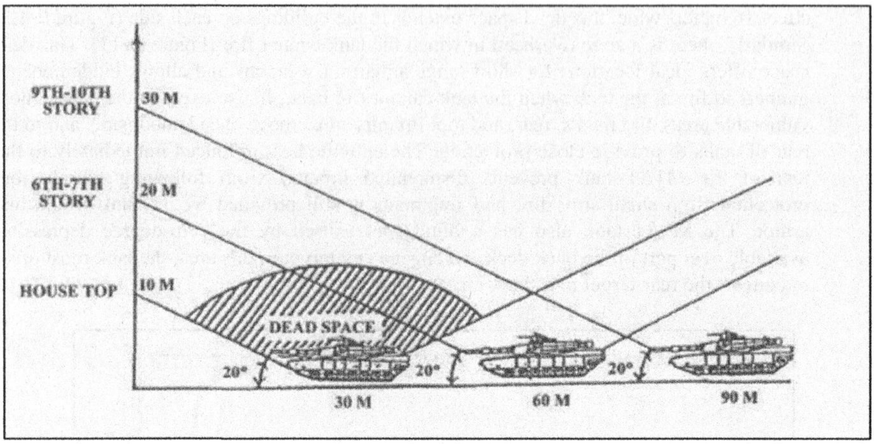

Figure B-13. Tank Cannon Dead Space Above Street Level

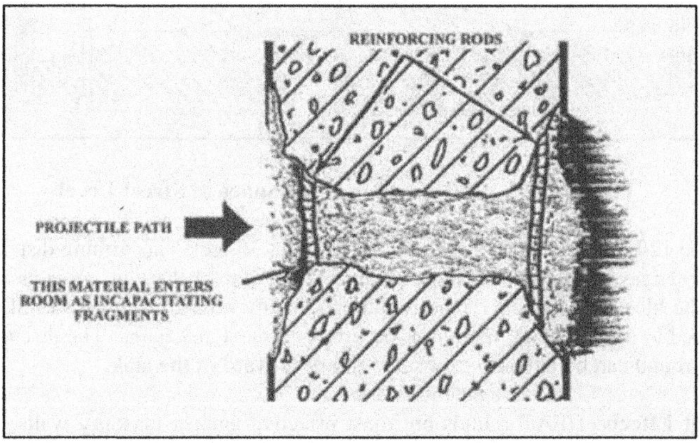

Figure B-14. Tank HEAT Round Effects on Reinforced Concrete Walls

e. Employment. Tank-heavy forces could be at a severe disadvantage during combat in built-up areas, but a few tanks working with the infantry can be most effective, especially if they work well together at the small-unit level. Tank, infantry, and engineer task forces are normally formed to attack a fortified area. The tank platoon or section can work together with rifle squads or platoons.

(1) Tanks need infantry on the ground to provide security in built-up areas and to designate targets. Against targets protected by structures, tanks should be escorted forward to the most covered location that provides a clear shot. On-the-spot instructions by infantry leaders ensure that the tank's fire is accurate and its exposure is limited. The tank commander may have to halt in a covered position, dismount, and reconnoiter his route forward into a firing position.

(2) When the tank main gun fires, it creates a large fireball and smoke cloud. In the confines of a built-up area, dirt and masonry dust are also picked up and added to this cloud. The target is further obscured by the smoke and dust of the explosion. Depending on the local conditions, this obscuration could last as long as two or three minutes. Marines can use this period to reposition or advance unseen by the enemy. Caution must be exercised, however, because the enemy might also move.

(3) A tank cannon creates an overpressure and noise hazard to exposed Marines. All dismounted Marines working near tanks should wear Kevlar helmets and protective vests, as well as ballistic eye protection. If possible, they should also wear earplugs and avoid the tank's frontal 60-degree arc during firing (Figure B-15).

WARNING

- **The overpressure from the tank 120-mm cannon can kill a marine found within a 90-degree arc extending from the muzzle of the gun tube out to 200 meters.**

- **From 200 to 1,000 meters along the line of fire, on a frontage of about 400 meters, dismounted Marines must be aware of the danger from discarding sabot petals, which can kill or seriously injure personnel.**

- **Personnel outside the tank should remain at least 50 meters from the tank in all directions as they may receive damaging effects from firing noise and overpressure.**

- **Personnel must also wear hearing protection when operating within 704 meters of a tank that is firing its main gun.**

TM 9-2350-264-10-2, *Operator's Manual for Operation Under Usual and Unusual Conditions for Tank, Combat, Full-Tracked: 120mm Gun, M1A1 General Abrams*, with Change 4

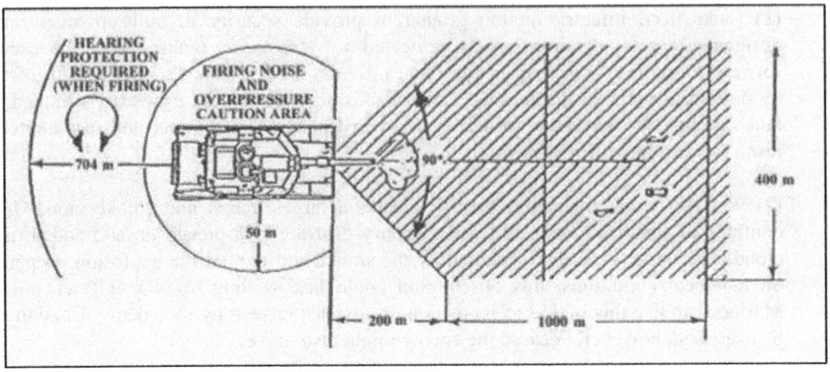

Figure B-15. Danger Areas Around a Tank Firing a 120-mm Gun

(4) Tanks are equipped with powerful thermal sights that can be used to detect enemy personnel and weapons that are hidden in shadows and behind openings. Dust, fires, and thick smoke significantly degrade these sights.

(5) Tanks have turret-mounted grenade launchers that project screening smoke grenades. The grenades use a bursting charge and burning RP particles to create this screen. *Burning particles can easily start uncontrolled fires and are hazardous to dismounted Marines near the tank.* The tank commander and the infantry small-unit leader must coordinate when and under what conditions these launchers can be used. Grenade launchers are a useful feature for protecting the tank but can cause significant problems if unwisely used.

(6) The tank's size and armor can provide dismounted Marines with cover from direct-fire weapons and fragments. With coordination, tanks can provide moving cover for Marines as they advance across small open areas. However, enemy fire that strikes but does not penetrate a tank is a major threat to nearby Marines. Fragmentation generated by antitank rounds and ricochets off of tank armor have historically been a prime cause of casualties while infantry was working with tanks in built-up areas.

11. Artillery. A major source of fire support for infantry forces fighting in built-up areas is field artillery weapons. If the built-up area is near the coast, NGF can be used. Field artillery employment can be in either the indirect- or direct-fire mode.

a. Indirect Fire. Low-angle indirect artillery fire is not effective in attacking targets within walls and masonry structures. It tends to impact on roofs or upper stories rather than on structurally critical wall areas or pillars. High-angle fire can achieve the trajectories necessary to strike targets in urban areas.

Weapons of at least 155 mm are required against thick reinforced concrete, stone, or brick walls. Even with heavy artillery, large expenditures of ammunition are required to knock down buildings of any size. Tall buildings also create areas of indirect-fire dead space, which are areas that cannot be engaged by indirect fire because of a combination of building height and angle of fall of the projectile (Figure B-16). Usually the dead space for low-angle indirect fire is about five times the height of the highest building over which the rounds must pass.

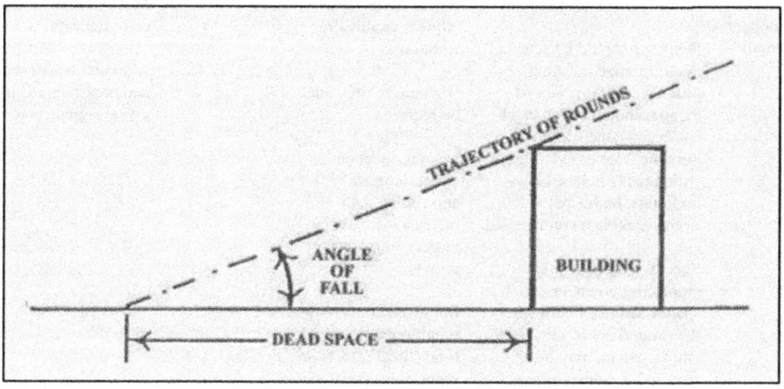

Figure B-16. Indirect-Fire Dead Space (Low Angle)

b. Direct Fire. A likely use of Marine artillery in an urban direct-fire role is to reinforce infantry against tough or important urban targets. Artillery should be used in this role only after an analysis of the need for heavy direct fire and the tradeoff involved in the extreme decentralization of artillery firepower. Artillery has the same need for close security and target designation as tanks.

c. Target Effects. Medium-caliber (155-mm) direct fire has a devastating effect against masonry construction and field fortifications. Smaller artillery (105-mm) pieces in the direct-fire mode are much less destructive than the larger caliber weapons. However, all Marine artillery is towed and, therefore, more difficult to employ in the direct-fire mode.

The 155-mm (M198) towed howitzer is effective because of its rate of fire and penetration. HE rounds can penetrate up to 38 inches of brick and unreinforced concrete. Projectiles can penetrate up to 28 inches of reinforced concrete with considerable damage beyond the wall. HE rounds with concrete-piercing fuzes provide an excellent means of penetrating strong reinforced concrete structures. One round can penetrate up to 46 inches. Five rounds are needed to reliably create a 1.5-meter breach in a wall that is 1 meter thick. About 10 rounds are needed to create the same size breach in a wall that is 1.5 meters thick. Superquick fuzing causes the rubble to be blown into the building, whereas delay fuzing tends to blow the rubble outward into the street. (See Table B-13 for employment considerations for artillery.)

Organization for Combat	Movement/Positioning	Delivery of Fire	Security	Command and Control
Centralized control is required during initial phases; decentralized control is required during later phases to support semi-independent actions of small units.	Movement should occur during night or periods of reduced visibility when possible. There are few displacements, often by platoon or section. Positions should be selected that minimize masking, provide several routes of escape, and afford as much cover and concealment as possible. Use of existing structures (garages, office buildings, highway overpasses) is recommended. Special techniques for emplacing howitzers, such as spades against a curb when the ground is not suitable for emplacement, may be required. Explosives may be required to soften emplacement of howitzers. Reconnaissance, selection, and occupation of position (RSOP) elements should be well armed because they may have to clear areas to be occupied. Extensive route reconnaissance is required. Target acquisition devices are somewhat degraded. Radars should be emplaced to cover likely areas of enemy indirect-fire weapon employment. Radars should not be placed in the midst of an urban area because of masking.	Both direct and indirect fires are delivered for supported units. Destruction of fortifications may require assault fire techniques. High-angle fires may be required. Need for accurate meteorological (MET) and survey data increases because most targets are point targets. Improved conventional munition and variable time (fuze) effects are reduced by structures, although they are effective against personnel on rooftops. HE delay is used for penetration effects. Illumination, chemical incendiary ammunition, and smoke are effective. Ammunition expenditures will be heavy. Lasers and PGMs permit destruction of targets with minimal rubbling of adjacent buildings. Tall buildings may hamper laser use. Batteries must be prepared for hasty survey techniques. Magnetic instruments are impaired.	Positions must be fortified.	Radio communications are impaired by buildings. Wire can usually be run overhead. Make use of civilian communications. A greater use of messengers and prearranged audio and visual signals is required.

Table B-13. Artillery Employment Considerations in Built-Up Areas

12. Mortars. The urban environment greatly restricts low-angle indirect fires because of overhead masking. While all indirect-fire weapons are subject to overhead masking, mortars are less affected than field artillery weapons because of the mortar's higher trajectory. For low-angle artillery fire, dead space is about five times the height of the building behind which the target sits. For mortar fire, dead space is only about one-half the height of the building. Because of these advantages, mortars are even more important than field artillery to the infantry during combat in built-up areas.

a. Employment. Not only can mortars fire into the deep defilade created by tall buildings, but they can also fire out of it. Mortars emplaced behind buildings are difficult for the enemy to accurately locate and even harder to hit with counterfire. Because of their light weight, even heavy mortars can be hand-carried to firing positions that may not be accessible to vehicles.

(1) The 60-mm and 81-mm mortars have limited effects on structural targets. Even with delay fuzes, they seldom penetrate more than the upper stories of light buildings. However, their wide area coverage and multioption fuzes make them useful against an enemy force advancing through streets, through other open areas, or over rubble.

(2) Mortar platoons often operate as separate firing sections during combat in built-up areas. The lack of large open areas can preclude establishing a platoon firing position. Figure B-17 shows how two mortar sections that are separated by only one street can be effective in massing fires and be protected from countermortar fire by employing defilade and dispersion.

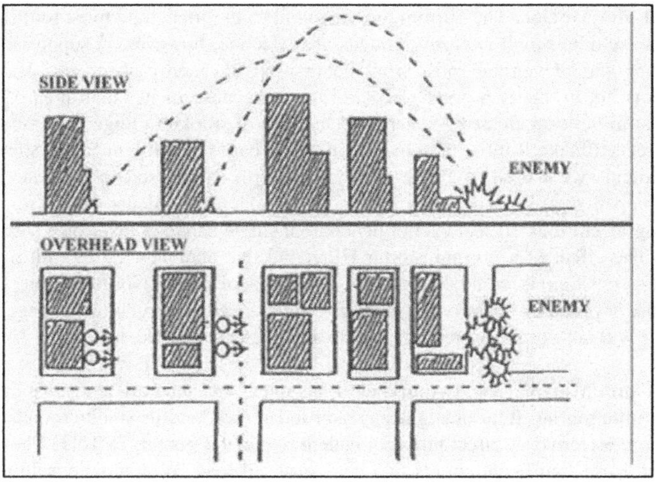

Figure B-17. Split-Section Mortar Operations on Adjacent Streets

(3) All three of the standard mortar projectiles are useful during combat in built-up areas. HE fragmentation rounds are most commonly used. WP is effective for starting fires in buildings and forcing the enemy out of cellars and light-framed buildings, and it is also the most effective mortar round against dug-in enemy tanks. Even near-misses blind and suppress the tank crew, forcing them to button up. Hits are difficult to achieve but are effective.

(4) Because the artificial roughness of urbanized terrain reduces wind speed and increases atmosphere mixing, mortar smoke tends to persist longer and give greater coverage in built-up areas than in open terrain.

(5) Urban masking has an impact on the use of illumination. In built-up areas, it is often necessary to plan illumination behind friendly positions, which places friendly troops in shadows and enemy troops in the light. Illumination rounds are difficult to adjust and are often of limited use because of the deep canyon nature of the urbanized area. Rapidly shifting wind currents in built-up areas also affect mortar illumination, making it less effective.

b. Effects of Mortar Fire. The multioption fuze on newer U.S. mortar rounds makes them effective weapons on urbanized terrain. Delay settings can increase penetration slightly, and proximity bursts can increase the lethal area covered by fragments. Tall buildings can cause proximity-fuzed mortar rounds to detonate prematurely if they pass too closely to the buildings.

(1) 60-mm Mortar. The 60-mm mortar round cannot penetrate most rooftops, even with a delay setting. Small explosive rounds are effective, however, in suppressing snipers on rooftops and preventing roofs from being used by enemy observers. The 60-mm WP round is not normally a good screening round because of its small area of coverage. In combat in built-up areas, however, the tendency of smoke to linger and the small areas to be screened make it more effective. During the battle for Hue in South Vietnam, 60-mm WP rounds were used to create small, short-term smoke screens to conceal movement across open areas such as parks, plazas, and bridges. Fragments from 60-mm HE rounds landing as close as 10 feet cannot penetrate a single sandbag layer or a single-layer brick wall. The effect of a 60-mm mortar HE round that achieves a direct hit on a bunker or fighting position is equivalent to 1 - 2 pounds of TNT. Normally, the blast will not collapse a properly constructed bunker, but it can cause structural damage. The 60-mm mortar will not normally create craters in a hard-surfaced road.

(2) 81-mm Mortar. The 81-mm mortar has much the same effect against urban targets as the 60-mm mortar. It has a slightly greater lethal area, and its smoke rounds (WP and RP) are more effective. A direct hit is equivalent to about 2 pounds of TNT. The 81-mm round cannot cause significant craters in a hard-surfaced road. With a delay setting, the 81-mm round can penetrate the roofs of light buildings.

13. Naval Gunfire. The primary NGF weapon, the 5-inch/54, has a high rate of fire and is roughly equivalent to the 155-mm howitzer in target effect. Target engagement considerations are the same as for artillery; however, the 5-inch/54's flat trajectory is even more affected by terrain masking and a constantly changing gun-target line. NGF can achieve greater target engagement by firing reduced charges at high angles; however, range will be reduced.

14. Aerial Weapons. Both rotary- and fixed-wing aircraft can quickly deliver large volumes of firepower over large built-up areas. Specific targets are hard to distinguish from the air. Good ground-to-air communications are vital in successfully employing aerial firepower. Planners have historically tended to overestimate the effects of HE on defenders. Modern, large buildings are remarkably resistant to damage from bombs and rocket fire.

a. Rotary-Wing Aircraft. Attack helicopters can be used to engage targets in built-up areas. Enemy armored vehicles in small parks, boulevards, or other open areas are good targets for attack helicopters.

(1) The HELLFIRE missile has a larger warhead and greater range than the TOW, but it too is a shaped-charge warhead and is not specifically designed for use against masonry targets. Laser target designation for the HELLFIRE may be difficult as a result of laser reflections off of glass and shiny metal surfaces. The use of attack helicopters to deliver ATGMs against targets in the upper stories of high buildings is sometimes desirable.

(2) The 2.75-inch folding-fin aerial rocket and the 20-mm cannon on Marine attack helicopters are good area weapons to use against enemy forces in the open or under light cover. They are usually ineffective against a large masonry target. The 20-mm cannon produces many ricochets, especially if armor-piercing ammunition is fired into built-up areas.

b. Fixed-Wing Aircraft. Close air support to ground forces fighting in built-up areas is a difficult mission for fixed-wing aircraft. Targets are hard to locate and identify, enemy and friendly forces could be intermingled, and enemy short-range air defense weapons are hard to suppress.

(1) Because enemy and friendly forces may be separated by only one building, accurate delivery of ordnance is critical. Marking panels, lights, electronic beacons, smoke, or some other positive identification of friendly forces is needed.

(2) General-purpose bombs that weigh from 500 to 2,000 pounds are moderately effective in creating casualties among enemy troops located in large buildings. High-dive-angle bomb runs increase accuracy and penetration but also increase the aircraft's exposure to antiaircraft weapons. Low-dive-angle bomb runs using high-drag (retarded) bombs can be used to get bombs into upper stories. Penetration is not good with high-drag bombs. Sometimes aerial bombs pass completely through light-clad buildings and explode on the outside.

(3) Aerial rockets and 20-mm cannons are only moderately effective against enemy soldiers in built-up areas because rockets lack the accuracy to concentrate their effects. The 20-mm cannon rounds penetrate only slightly better than the .50-caliber round, 20-mm armor-piercing rounds can ricochet badly, and tracers can start fires.

(4) The AC-130 aircraft has weapons that can be most effective during combat in built-up areas. This aircraft can deliver accurate fire from a 20-mm Vulcan cannon, 40-mm rapid-fire cannon, and 105-mm howitzer. The 105-mm howitzer round is effective against the roofs and upper floors of buildings. The AC-130 is accurate enough to concentrate its 40-mm cannon and 105-mm howitzer fire onto a single spot to create a rooftop breach that allows fires to be directed deep into the building.

(5) Laser and optically guided munitions can be effective against high-value targets. Specially developed, heavy, laser-guided bombs can penetrate hardened targets. However, problems associated with dense smoke and dust clouds hanging over the built-up area and laser scatter can restrict their use. If the launching aircraft can achieve a successful laser designation and lock-on, these weapons have devastating effects, penetrating deep into reinforced concrete before exploding with great force. If launched without a lock-on, or if the laser spot is lost, these weapons are unpredictable and can travel long distances before they impact.

15. Demolitions. Combat in built-up areas requires the extensive use of demolitions. All Marines, not just engineers, should be trained to employ demolitions. See FM 5-250, *Explosives and Demolitions*, for specific information on the safe use of demolitions.

a. Demolitions. Demolitions come in two types: TNT and C4. Exposed Marines must take cover or move at least 300 meters away from bulk explosives that are being used to breach walls.

(1) TNT comes in 1/4-, 1/2-, and 1-pound blocks. About 5 pounds of TNT are needed to breach a nonreinforced concrete wall that is 12 inches thick if the explosives are laid next to the wall and are not tamped. If the explosives are tamped, about 2 pounds are sufficient.

(2) C4 comes in many different-sized blocks. Normally, it is found in 1 1/4- and 2 1/2-pound blocks within prepackaged satchel charges. About 10 pounds of C4 placed between waist and chest high will blow a hole in the average masonry wall large enough for a man to walk through.

b. Shaped Charges. There are two sizes of shaped charges: a 15-pound M2A3 and a 40-pound M3A3. The M3A3 is the shaped charge that is most likely to be used in built-up areas. It can penetrate 5 feet of reinforced concrete. The hole tapers from 5 inches down to 2 inches. The amount of spall thrown behind the target wall is considerable. There is also a large safety hazard area for friendly forces.

c. Satchel Charges. There are two standard satchel charges: the M183 and the M37. Both come in their own carrying satchel with detonators and blasting cords. Each weighs 20 pounds. The M183 has 16 individual 1 1/4-pound blocks that can be used separately. When used untamped, a satchel breaches a 3-foot-thick concrete wall. Satchel charges are very powerful. Debris is thrown great distances. Friendly forces must move away and take cover before detonation.

d. Cratering Charges. The standard cratering charge is a 43-pound cylinder of ammonium nitrate. This explosive does not have the shattering effect of bulk TNT or C4. It is more useful in deliberate demolitions than in hasty ones.

Appendix C

Attacking and Clearing Buildings

Offensive operations on urbanized terrain will require Marines to attack and clear buildings.

1. **Introduction.** At the rifle platoon and squad level, the major offensive tasks for combat in built-up areas are attacking and clearing buildings. This involves:

- Suppressing defensive fires
- Advancing Marine assault forces
- Assaulting buildings
- Clearing buildings
- Reorganizing assault forces.

To accomplish these tasks, the assaulting unit task organizes into assault, security, and support forces or elements. (See Appendix A)

2. **Requirements.** There are five interrelated requirements for attacking a defended building: *fire support, movement, assault, clearing,* and *reorganization.* Proper application and integration of these requirements reduce casualties and hasten accomplishment of the mission. The proper integration of these requirements are determined by METT-T. For example, larger cities may have high-rise apartment buildings and industrial complexes that are separated by parking areas or parks. Increased fire support may be required to suppress and obscure enemy gunners covering the open terrain and spaces between buildings. Conversely, a built-up area with few open areas can provide numerous covered routes that may decrease fire support requirements.

3. **Fire Support.** Fire support necessary to advance the assault force may include:

- Suppressing and obscuring enemy gunners within the objective building(s) and adjacent structures
- Isolating the objective building(s) with fires to prevent enemy withdrawal, reinforcement, or counterattack
- Breaching walls en route to and in the objective structure
- Destroying enemy positions with direct-fire weapons

- Securing cleared portions of the objective
- Providing replacements for the assault force
- Providing resupply of ammunition and explosives
- Evacuating casualties and prisoners.

a. The size of the support force is determined by the type and size of the objective building(s), whether the adjacent terrain provides open or covered approaches, and the organization and strength of enemy defenses.

b. The support force for a Marine rifle squad may consist of one fire team that could be reinforced with heavy machine guns, antitank weapons, and/or armored vehicles. In situations involving a larger assault force, a rifle platoon or company reinforced with armored vehicles, engineers, and/or artillery may be required to support the movement and assault of an adjacent platoon or company.

c. After seizing objective buildings, the assault force reorganizes and may be required to provide supporting fires for a subsequent assault. Each weapon is assigned a target or area to cover. Individual small-arms weapons place fires on likely enemy weapon positions—loopholes, windows, and roof areas. Snipers are best employed in placing accurate fire through loopholes or engaging long-range targets. M203 fires are directed through windows or loopholes.

d. AT4s and demolitions are employed to breach walls, doors, barricades, and window barriers on the ground level of structures. The tank cannon or LAV-mounted 25-mm gun engages first-floor targets and breaches walls for attacking Marines. Armored vehicle- or wheeled vehicle-mounted machine guns can engage suspected enemy positions on upper floors and in adjacent structures. In addition to destroying or weakening structures, the tank's main gun rounds cause casualties by their explosive effects and by hurling debris throughout the interior of structures.

e. NGF, artillery, and mortars should initially use time fuzes to initially clear exposed enemy personnel, weapons, OPs, and radio sites from rooftops. They should then use delayed fuze action to cause casualties among the defenders inside the structure as a result of shrapnel and falling debris. They may also provide smoke to cover the movement and assault phases of the attack. Artillery can also be used in the direct-fire mode.

f. In support of the movement and assault phases of the attack, buildings can be attacked by fixed-wing aircraft or by attack helicopters.

4. Movement. The assault force minimizes the effects of enemy defensive fires during movement by:

- Using covered routes
- Moving after defensive fires have been suppressed or obscured
- Moving at night or during other periods of reduced visibility
- Selecting routes that will not mask friendly suppressive fires
- Crossing open areas (e.g., streets and spaces between buildings) quickly under the concealment of smoke and suppressive fires
- Moving on rooftops that are not covered by enemy direct fires.

a. The requirement for speed may dictate movement through streets and alleys without clearing all buildings. When moving in this manner, the maneuver force should employ infantry to lead the column, and supported by AAVs, LAVs, wheeled vehicle-mounted machine guns, or tanks following.

b. Marines should move along each side of a street with the leading squad staying abreast of armored vehicles. When not accompanied by armored vehicles, Marines move in a single file along one side of the street under cover of fires from supporting weapons. Individual Marines should be detailed to observe and cover a specific area, such as second-floor windows on the opposite side of the street.

5. Assault. Assault forces quickly and violently execute their assault and subsequent clearing operations. Momentum is maintained to prevent the enemy from organizing a more determined resistance on other floors or in other rooms. Small-unit leaders should keep the assault force moving without allowing the operation to become disorganized.

a. An assault in a built-up area involves the elementary skills of close combat. Leaders must:

- Be trained in the required techniques to defeat the enemy in a face-to-face encounter
- Keep themselves in excellent physical condition
- Have confidence in their abilities and those of their unit.

b. The composition of the assault force will vary depending on METT-T; however, its basic organization at the squad level consists of two assault elements carrying a fighting load of equipment and as much ammunition as possible, especially grenades.

6. Clearing. Entering at the top and fighting downward is the preferred method of clearing a building. Clearing a building is easier from an upper story because gravity and building construction become assets to the assault elements when throwing hand grenades and moving from floor to floor. However, this method is only feasible when access to an upper floor or rooftop can be gained from the windows or roofs of adjoining buildings and assault elements can be transported to the rooftops by helicopter (Figure C-1). Helicopters should land only on those buildings that have special heliports or parking garages on the roofs. When these are unavailable, Marines can rappel or fastrope onto the roof or dismount as the helicopter hovers a few feet above the roof. Assault elements then breach the roof or common walls with explosives and use ropes to enter the lower floors. Stairs are guarded by a covering team or security force when not in use to protect the assault force from a flank attack.

Figure C-1. Helicopters Used To Clear Buildings

a. Although the top-to-bottom method is preferred for clearing a building, assaulting the bottom floor and clearing upward is a common method. When using this technique, the assault force closes on the flank(s) or rear of the building. It then clears each room on the ground floor and then, moving upward, begin a systematic clearance of the remaining floors.

b. Entry can be gained through walls breached by explosives or gunfire. Assault elements should avoid windows and doors because they are usually covered by fire or boobytrapped. If tanks are attached to the company, they can breach the wall by firing their main gun (Figure C-2).

Figure C-2. Main Gun Used To Breach Exterior

c. Shortly before the assault, suppressive fires on the objective should be increased until masked by the advancing assault force. Once masked, fires are shifted to upper windows and continued until the assault force has entered the building. At that time, fires are shifted to adjacent buildings to prevent enemy withdrawal or reinforcement.

d. Hand grenades should be used to clear the space before assault element members enter through a breached wall. Immediately after the explosion, the assault element clearing team enters and uses three round bursts to fire into the interior of the room, concentrating on areas that are possible enemy positions.

e. Once inside the building, the assault force must cover staircases and seize rooms that overlook approaches to the building. These actions are required to isolate enemy forces within the building and to prevent reinforcement. Assault elements should clear each ground floor room and then the basement. Sequence for the assault is that:

(1) The assault force leader determines which room(s) to clear first.

(2) The support force provides suppressive fire while the assault force systematically clears the building. It also provides suppressive fire on adjacent buildings to prevent enemy reinforcements or withdrawal. The support force destroys any enemy trying to exit the building.

(3) The security force isolates the immediate area.

(4) After assault elements establish a foothold in the building, the assault force leader proceeds with the systematic clearing of the building.

f. When clearing a building from the top floor down, assault forces should gain entrance through a common wall or the roof of an adjoining building (Figure C-3). Demolition charges are used to breach the wall and gain entrance to the top floor. Access to lower floors and rooms may be gained by breaching holes in the floor and having the Marines jump or slide down ropes to the lower floors. Stairs can be used if they are first cleared.

g. When using the top-to-bottom method of clearing, requirements remain the same as for other methods. After the floor is breached to gain access to a lower floor, a grenade is allowed to cook-off and is dropped to the lower room. A Marine then sprays the lower room with gunfire using three-round bursts and drops through the mousehole.

Figure C-3. Upper Floors Secured

h. Marines should avoid predictability when entering and clearing rooms. They should vary entry through doors and mouseholes to keep the enemy off guard (Figure C-4). As rooms are cleared, doors should be left open, and a predetermined mark (cloth, tape, spray paint) should be placed on the doorjamb or over the door, as established by unit SOPs.

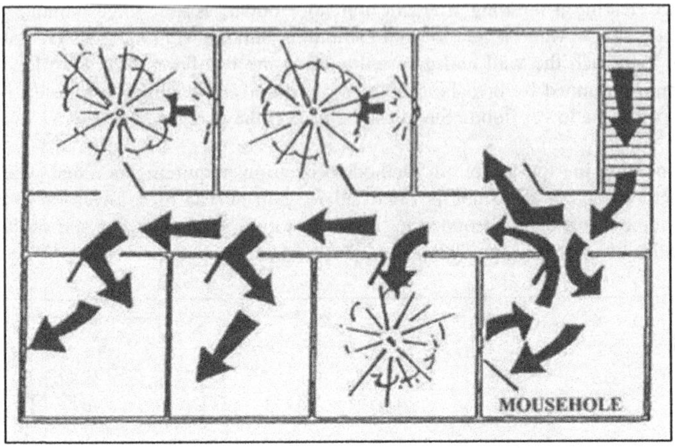

Figure C-4. Varying Techniques for Clearing Rooms

i. If there is a basement, it should be cleared as soon as possible, preferably at the same time as the ground floor. The procedures for clearing a basement are the same as for any room or floor. However, basements often contain entrances to tunnels such as sewers and communications cable tunnels. These should be cleared and secured to prevent the enemy from infiltrating back into cleared areas.

j. Types of buildings that may be cleared include brick buildings and houses, box-wall buildings, heavy-clad framed buildings, and light-clad framed buildings.

(1) Brick Buildings and Houses. The best way to enter a brick building is to blow a breaching hole in the side by using a tank firing HEAT ammunition (Figure C-5). If tanks are not available, a door or window in the rear of the building usually provides better cover and concealment for entry than one in the front. If there is enough cover and concealment, the assault force should enter the rear of the building at an upper level by using a fire escape or grappling hook.

Brick houses usually have similar floor plans on each floor (Figure C-6); therefore, ground floors are cleared the same way as upper floors.

Converely, the floor plans in larger brick buildings typically are different on ground-floor levels than on upper levels (Figure C-7).

Figure C-5. Entering a Brick Building

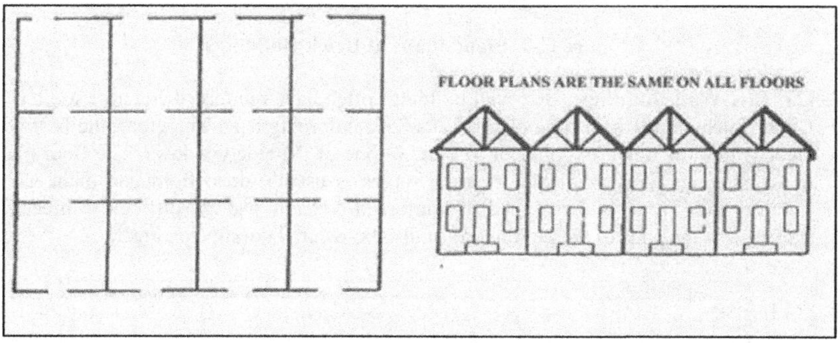

Figure C-6. Brick Houses Have Similar Floor Plans

To clear from building to building, the best method is to move from rooftop to rooftop because the roofs of brick buildings are usually easy to breach. The walls between buildings are at least three bricks thick (total of six bricks between buildings) and require large quantities of demolitions to breach. Walls are normally easier to breach on an upper floor than on a lower floor because the walls are thinner on upper floors. If rooftops are covered by fire and there are not enough demolitions to breach walls between buildings, clearing from rear to rear of buildings is safer than clearing from front to front.

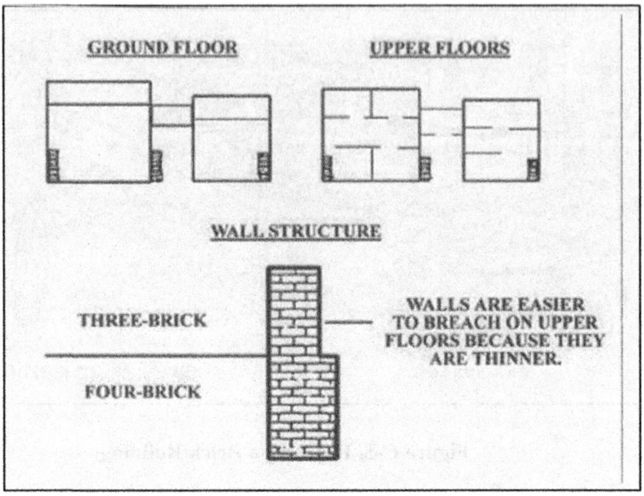

Figure C-7. Floor Plans of Brick Buildings

(2) Box-Wall Buildings. Box-wall buildings often have reinforced concrete walls (Figure C-8), which are difficult to breach because of reinforcing bars. Therefore, the best way to enter is to blow down the door or to blow in one of the side windows. The floor plans of these buildings are predictable; clearing rooms is usually done from one main hallway. Interior walls are also constructed of reinforced concrete and are difficult to breach. The stairways at the ends of the building must also be secured during clearing.

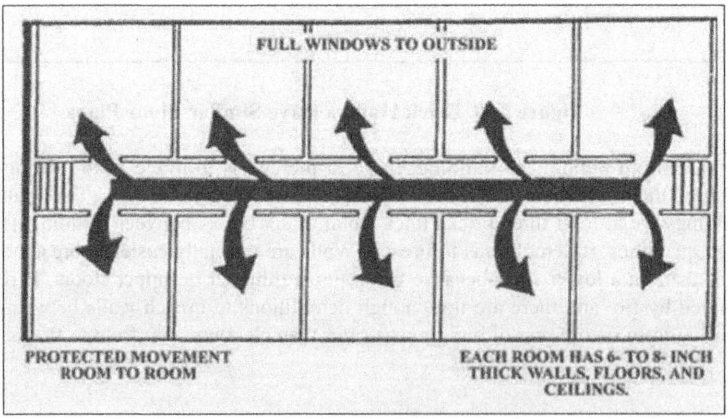

Figure C-8. Box-Wall Buildings

(3) **Heavy-Clad Framed Buildings.** Tanks can be used to breach a hole in heavy-clad framed buildings. Their floor plans are normally oriented around a stairway or elevator. this must be secured during clearing (Figure C-9). The interior walls of these buildings can be breached, although they may require use of demolitions.

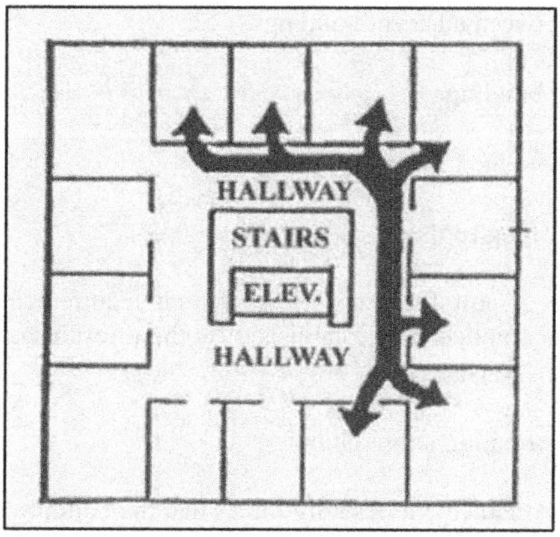

Figure C-9. Heavy-Clad Framed Buildings

(4) **Light-Clad Framed Buildings.** Light-clad framed buildings (Figure C-10), can be cleared in the same manner as heavy-clad buildings. Secure the central stairway and clear in a circular pattern. Walls are usually thin enough to be breached with an ax.

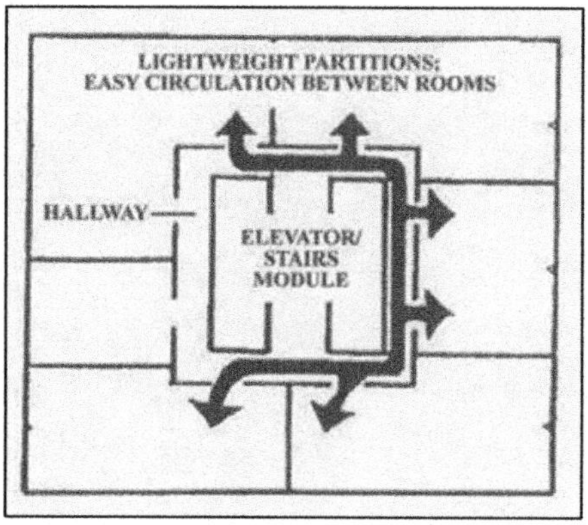

Figure C-10. Light-Clad Framed Buildings

7. **Consolidation and Reorganization.** Consolidation and reorganization of the assault force in a cleared building must be accomplished quickly to repel enemy counterattacks and to prevent the

enemy from infiltrating back into the cleared building. After securing a floor (bottom, middle, or top), selected members of the assault force are assigned to cover potential enemy counterattack routes to the building. Those sentinels alert the assault force and place a heavy volume of fire on enemy forces approaching the building. They should cover:

- Enemy mouseholes between adjacent buildings
- Covered routes to the building
- Underground routes into the basement
- Approaches over adjoining roofs.

As the remainder of the assault force completes search requirements, defensive positions are assigned and tasks are accomplished as established by the priorities of work in the defense. The following actions may then be taken:

- Resupplying and redistributing ammunition
- Marking the building to indicate to friendly forces that the building has been cleared
- Assuming an overwatch mission and supporting an assault on another building
- Treating and evacuating wounded personnel
- Further developing a defensive position if the building is to be occupied for any period.

Appendix D

Fighting Positions

A critical defensive task in combat in built-up areas is the selection and preparation of fighting positions.

1. Considerations. Leaders should consider the following factors when establishing fighting positions:

a. Protection. Leaders should select buildings that provide protection from direct and indirect fires. Reinforced concrete buildings with three or more floors provide suitable protection, while buildings constructed of wood, paneling, or other light materials require reinforcement to gain sufficient protection. One- to two-story buildings without strongly constructed cellars are vulnerable to indirect fires and require construction of overhead protection for each firing position.

b. Dispersion. Normally, a position should not be established in a single building when it is possible to occupy two or more buildings that permit mutually supporting fires. A position in one building without mutual support is vulnerable to bypass, isolation, and subsequent destruction from any direction.

c. Concealment. The fighting position should be concealed. Buildings that are obvious defensive positions should not be selected. The occupation of exposed buildings may be necessary because of requirements for security and fields of fire. Therefore, reinforcement of the structure must be accomplished to provide suitable protection within the building.

d. Fields of Fire. Positions should be mutually supporting and have interlocking fields of fire in all directions. Clearing fields of fire may require the destruction of adjacent buildings by using explosives, engineer equipment, and field expedients. Care should be taken to avoid highlighting the building as a defensive position.

e. Covered Routes. Defensive positions should have at least one covered route that permits resupply, MEDEVAC, reinforcement, or withdrawal from the building. The route can be established by one of the following means:

- Through walls to adjacent buildings
- Through underground systems
- Through communications trenches
- Behind protective buildings.

f. Observation. The building should permit observation of enemy avenues of approach and adjacent defensive sectors.

g. Fire Hazard. Avoid selecting positions in buildings that are a fire hazard. If flammable structures must be occupied, the danger of fire can be reduced by wetting the structure, laying an inch of sand on the floors, and providing fire extinguishers and firefighting equipment. Routes of escape should also be prepared in case of fire.

h. Time. The time available to prepare the defense could be the most critical factor. If enough time is insufficient, buildings that require extensive preparation should not be used. Conversely, buildings located in less desirable areas that require little improvement may be a better choice for a defensive position.

2. **Preparation.** Preparation of fighting positions depends on proper selection and construction.

a. Selecting Positions. Each weapon should be assigned a primary sector of fire to cover enemy approaches. Alternate positions that overwatch the primary sector should also be selected. These positions are usually located in an adjacent room on the same floor (Figures D-1A, D-1B, and D-1C). Each weapon must be assigned a supplementary position (to engage attacks from other directions) and an FPL.

b. Building Positions. There are many ways to establish a fighting position in a building.

(1) Window Position. Marines should kneel or stand on either side of a window. To fire downward from upper floors, tables or similar objects can be placed against the wall to provide additional elevation, but they must be positioned to prevent the weapon from protruding through the window (Figure D-2). Leaders should inspect positions to determine the width of sector that each position can engage.

(2) Loopholes. To avoid establishing a pattern of always firing from windows, loopholes should be prepared in walls. Marines should avoid firing directly through loopholes to enhance individual protection.

(a) Several loopholes are usually required for each weapon (primary, alternate, and supplementary positions). The number of loopholes should be carefully considered because they can weaken walls and reduce protection. Engineers should be consulted before numerous loopholes are made. Loopholes should be made by punching or drilling holes in walls and should be placed where they are concealed. Blasting loopholes can result in a large hole that can easily be seen by the enemy.

(b) Loopholes should be cone-shaped to obtain a wide arc of fire, to facilitate engagement of high and low targets, and to reduce the size of the exterior aperture (Figure D-3). The edges of a loophole splinter when hit by bullets; therefore, protective linings such as an empty sandbag held in place by wire mesh will reduce spalling effects. When not in use, loopholes should be covered with sandbags to prevent the enemy from firing into or observing through them.

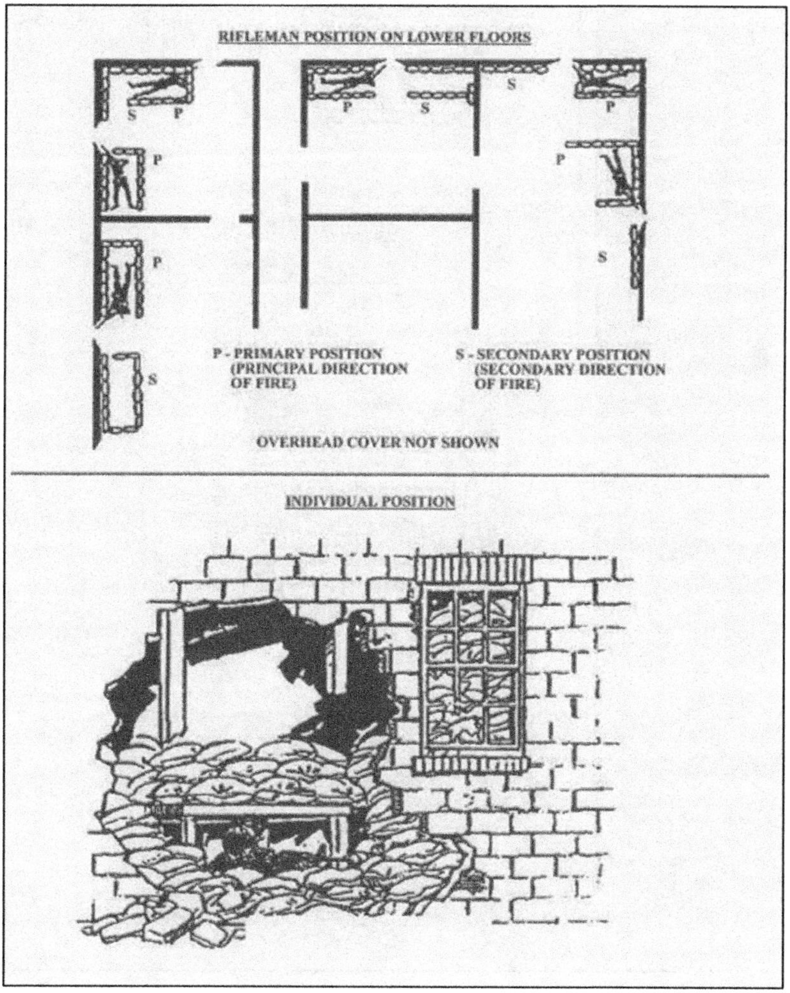

Figure D-1A. Weapon Positions

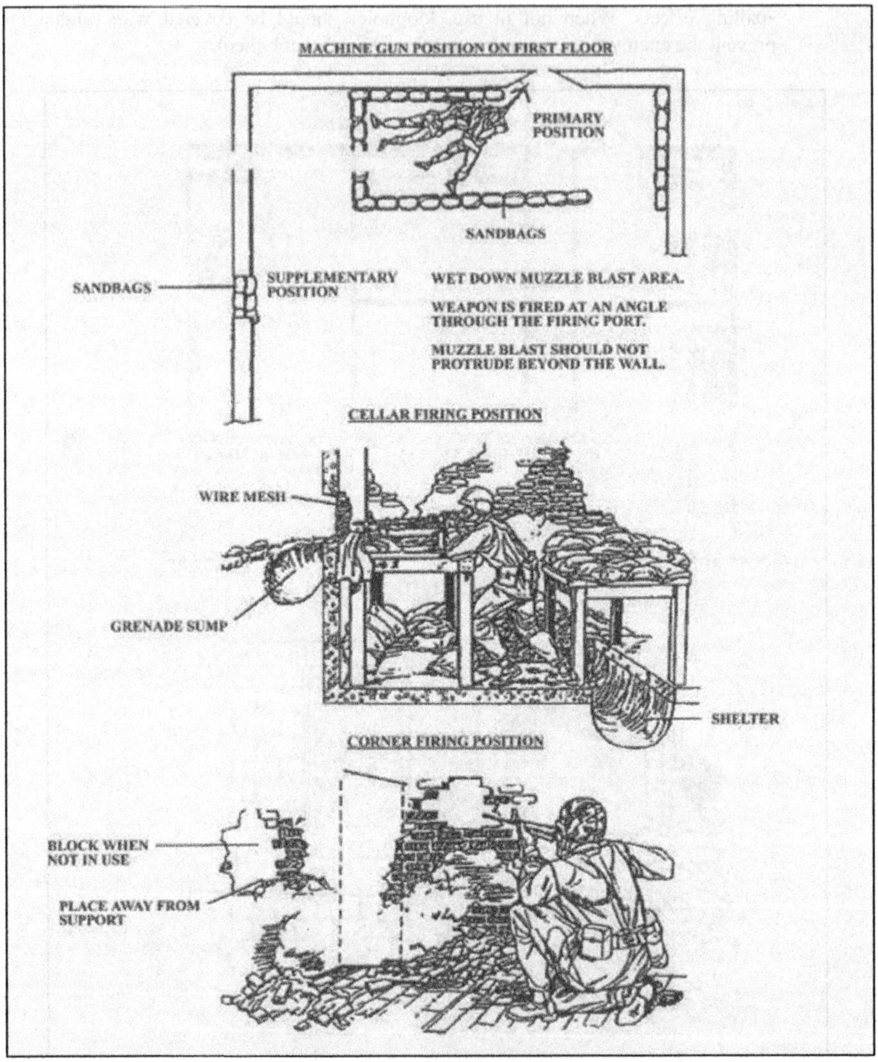

Figure D-1B. Weapon Positions (Continued)

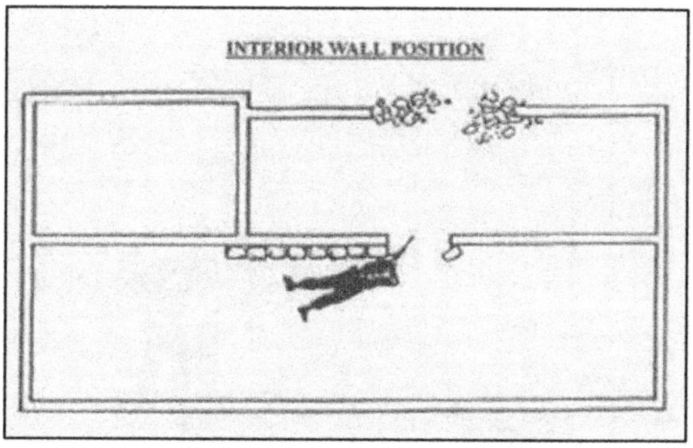

Figure D-1C. Interior Wall Position

Figure D-2. Window Position

(c) Loopholes should also be prepared in interior walls and ceilings of buildings to permit fighting within the position. Interior loopholes should enable overwatch of stairs, halls, and unoccupied rooms and be concealed by pictures, drapes, or furniture. Loopholes in floors permit the defender to engage enemy personnel on lower floors with small-arms fire and grenades.

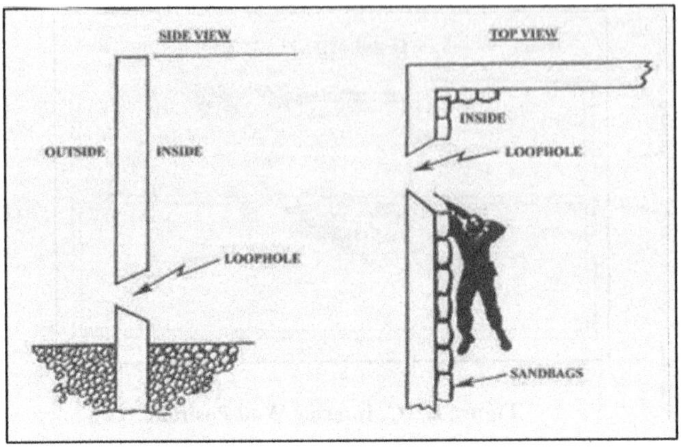

Figure D-3. Cone-Shaped Loopholes

(d) Although walls provide some frontal protection, they should be reinforced with sandbags, furniture filled with dirt, or other expedients. Each position should have overhead and all-around protection (Figure D-4).

Figure D-4. Position With Overhead and All-Around Protection

c. Other Construction Tasks. Other construction tasks may need to be performed. These include work on:

(1) Basements and Ground Floors. Basements require preparation similar to that of the ground floor. Any underground system that is not used by the defender but that could provide enemy access to the position must be blocked.

(a) Doors. Unused doors should be locked, nailed shut, blocked, and reinforced with furniture, sandbags, or other field expedients. Outside doors can be boobytrapped by engineers or other personnel.

(b) Hallways. If not required for the defender's movement, hallways should be blocked with furniture and tactical wire (Figure D-5). If authorized, boobytraps should be employed to impede enemy movement.

(c) Stairs. Defenders should block stairs not used by the defense with furniture and tactical wire (Figure D-5) or remove them. If possible, all stairs should be blocked, and ladders should be used to move from floor to floor and then removed when not being used. Boobytraps should also be employed on stairs.

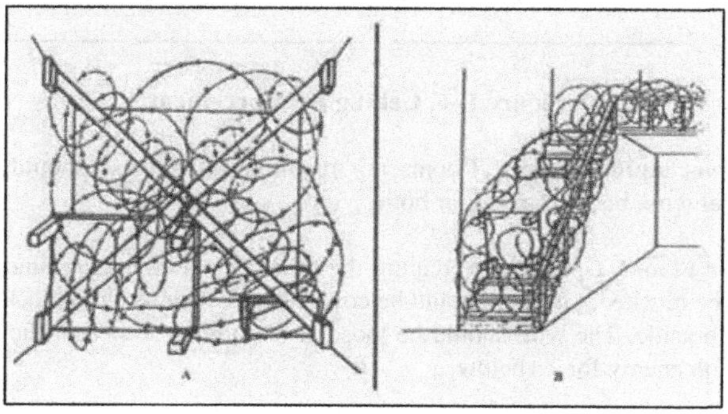

Figure D-5. Blocking Hallways and Stairs

(d) Windows. All glass should be removed. Windows not being used should be blocked with boards or sandbags.

(e) Floors. Fighting positions should be constructed on the ground or basement floors. If there is no basement, fighting positions on the first floor must be constructed to provide additional protection from heavy direct-fire weapons.

(f) Ceilings. Supports that can withstand the weight of rubble from upper floors should be placed under ceilings to provide overhead cover (Figure D-6).

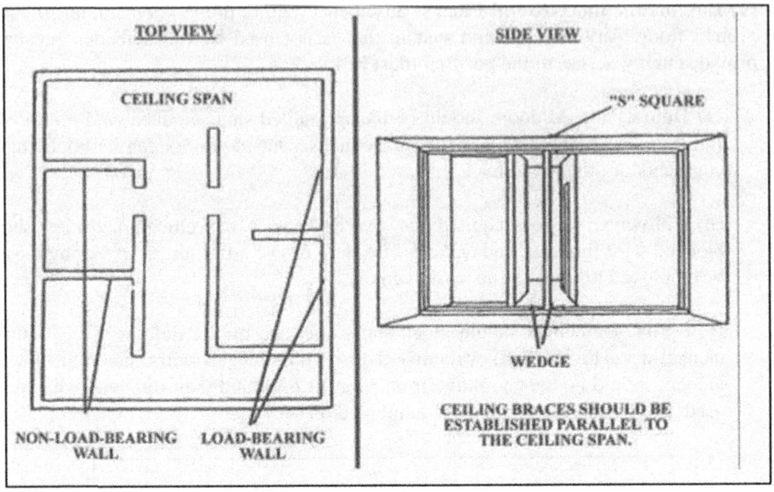

Figure D-6. Ceiling Reinforcement

(g) Unoccupied Rooms. Rooms not required for defense should be blocked with tactical wire, boobytrapped, or both.

(2) Upper Floors. Upper floors require the same preparation as ground floors. Windows need not be blocked, but they should be covered with wire mesh to block grenades thrown from the outside. The wire should be loose at the bottom to permit the defender to drop grenades on enemy forces below.

(3) Interior Routes. Routes are required that permit defending forces to move within the building to engage enemy forces from any direction or reinforce other forces (Figure D-7). Escape routes should also be planned and constructed to permit rapid evacuation of a room or the building. Small holes (called mouseholes) should be made through interior walls to permit movement between rooms. Once the defender has withdrawn to another level, such holes should be clearly marked for both day and night identification. All personnel must be briefed as to where the various routes are located. Rehearsals should be conducted so that everyone becomes familiar with the routes.

(4) Fire Prevention. Buildings that have wooden floors and raftered ceilings require extensive fire prevention measures. The attic and other wooden floors should be covered with about 1 inch of sand or dirt, and buckets of water should be positioned for immediate use. Firefighting materials (dirt, sand, fire extinguishers, and blankets) should be placed on each floor for immediate use. Water basins and bathtubs should be filled as a reserve for firefighting. All electricity and gas should be turned off. Fire breaks can be created by destroying buildings adjacent to the defensive position.

Figure D-7. Movement Between Floors

(5) Communications. Telephone lines should be laid through adjacent buildings or underground systems or buried in shallow trenches. Radio antennas can be concealed by placing them among civilian television antennas, along the sides of chimneys and steeples, or out windows that direct FM communications away from enemy electronic warfare sources and ground observation. Telephone lines within the building should be laid through walls and floors.

(6) Rubbling. Rubbling parts of the building may provide useful cover and concealment for weapons emplacements. Rubbling should be performed only by trained engineers.

(7) Rooftops. Positions in flat-roofed buildings require obstacles that restrict helicopter landings (See Figure F-2). Rooftops that are accessible from adjacent structures should be covered with tactical wire or other expedients and must be guarded. Entrances to buildings from rooftops can be blocked if compatible with the overall defensive plan. Any structure on the outside of a building that could assist in scaling the buildings to gain access to upper floors or to the rooftop should be removed or blocked.

(8) Obstacles. Obstacles should be positioned adjacent to buildings to stop or delay tanks and infantry.

(9) Fields of Fire. Fields of fire should be improved around the defensive position. Selected buildings may be destroyed to enlarge fields of fire. Obstacles to ATGMs, such as telephone or electrical wires, should be cleared. Dead space should be covered with mines and obstacles.

3. Tank and Other Armored Vehicle Positions. Fighting positions for tanks and other armored vehicles are essential to a complete and effective defensive plan in built-up areas.

a. Armored Vehicle Positions. Armored vehicle positions are selected and developed to obtain the best cover, concealment, observation, and fields of fire while retaining the vehicle's ability to move.

(1) If fields of fire are restricted to streets, hull-down positions should be used to gain cover and to fire directly down streets (Figure D-8). From those positions, armored vehicles are protected and can rapidly move to alternate positions. Buildings collapsing from enemy fires are a minimal hazard to armored crews.

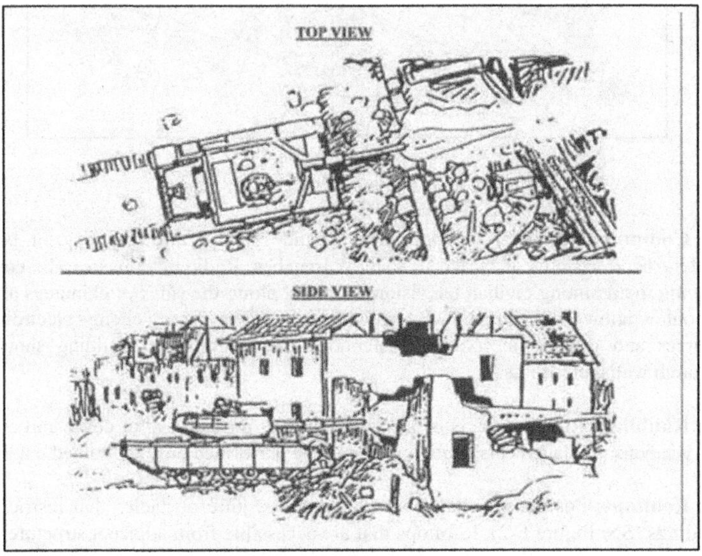

Figure D-8. Hull-Down Position

(2) The hide position (Figure D-9) covers and conceals the vehicle until time to move into position for engagement of targets. Because the crew will not be able to see advancing enemy forces, an observer from the vehicle or a nearby infantry unit must be concealed in an adjacent building to alert the crew. The observer acquires the target and signals the armored vehicle to move to the firing position and fire. After firing, the vehicle moves to an alternate position to avoid compromising one location.

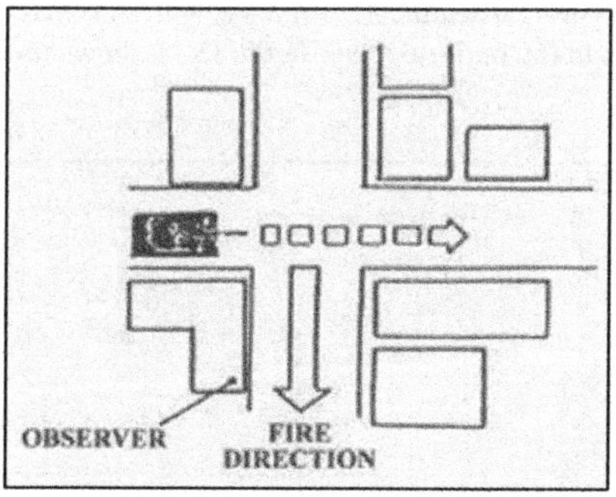

Figure D-9. Hide Position

(3) The building hide position (Figure D-10) conceals the vehicle inside a building. If basement hide positions are inaccessible, engineers can evaluate the building's floor strength and prepare for the vehicle. Once the position is detected, it should be evacuated to avoid enemy fires.

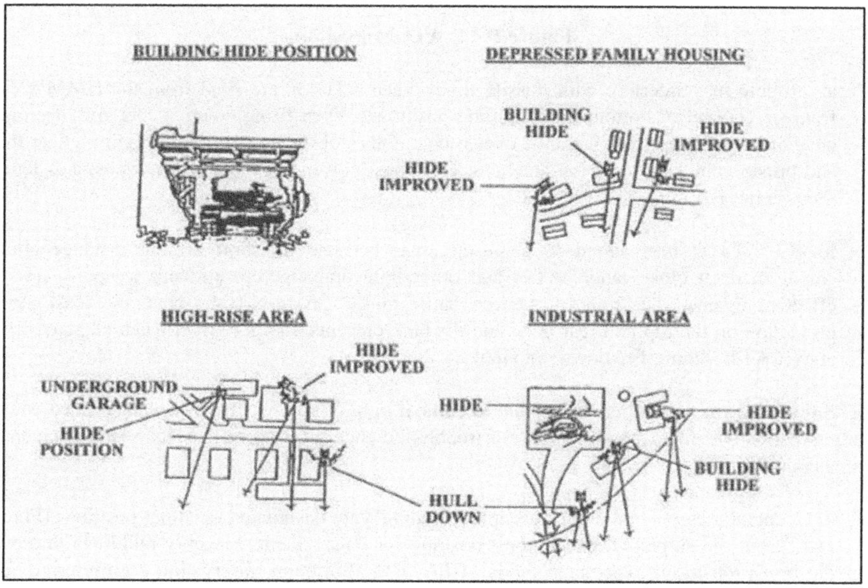

Figure D-10. Building Hide Position

4. Antitank Guided Missile Positions. HMMWVs with ATGMs are employed in areas that maximize their capabilities in the built-up area. Figure D-11 shows mounted and dismounted firing positions.

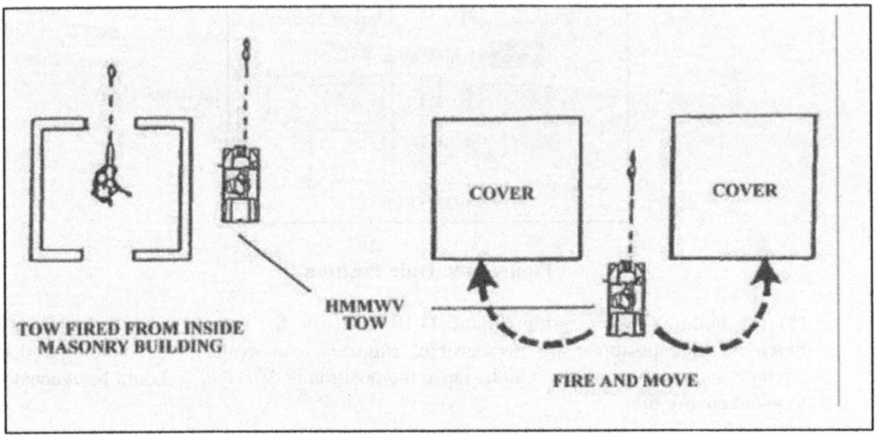

Figure D-11. ATGM Positions

a. Rubble may interfere with missile flight when ATGMs are fired from the HMMWV or from street-level or bottom-floor fighting positions. When firing down streets, missiles must have at least 1 meter of clearance over rubble. Other obstacles to missile flight include trees and brush, vehicles, television antennas, buildings, power lines and wires, walls, and fences (See Figure B-7).

b. An AT4 is best suited for built-up areas because its short arming distance allows employment at close range. AT4s and other light and medium antitank weapons are not effective against the front of modern battle tanks. Because tanks have the least armor protection on the top and rear deck and the tank presents a larger target when engaged from above, AT4s should fire down onto tanks.

5. Sniper Positions. Snipers contribute to combat in built-up areas by firing on selected enemy targets. Effective sniper employment can trouble the enemy far more than its relative personnel numbers reflect.

a. General areas (a building or group of buildings) are designated as sniper positions (Figure D-12), but the sniper selects the best position for engagement. Masonry buildings that offer the best protection, long-range fields of fire, and all-around observation are preferred. The sniper also selects several secondary and supplementary positions to cover his areas of responsibility.

b. Engagement priorities for snipers are determined by the relative importance of the targets to the effective operations of the enemy. Sniper targets may include tank commanders, crewmen of direct fire support weapons, crewmen of crew-served weapons, officers, FOs, and radiotelephone operators.

c. The urban environment often limits snipers to firing down or across streets. Snipers can be employed to cover rooftops, obstacles, dead space, and gaps in final protective fires.

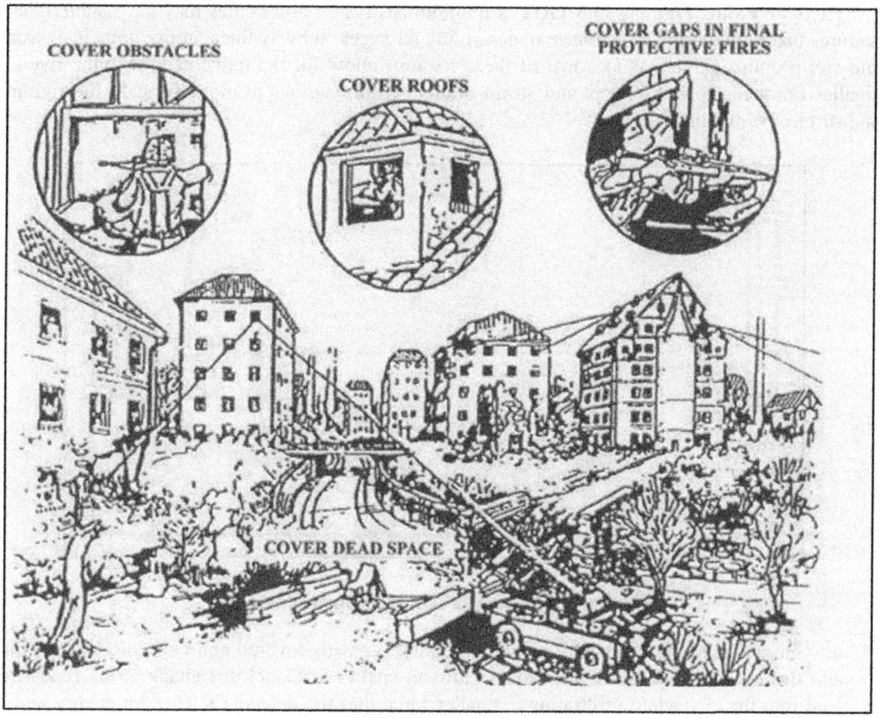

Figure D-12. Sniper Positions

Appendix E

Subterranean Operations

Knowledge of the nature and location of underground facilities is valuable to both the urban attacker and defender.

1. Tactical Value. Fighting in MOUT is multidimensional. Larger cities may have subterranean features that include sunken garages, underground passages, subway lines, utility tunnels, sewers, and storm drains (Figure E-1). Most of these features allow for the movement of units. Even in smaller European towns, sewers and storm drains permit Marines to move beneath the fighting and surface behind the enemy.

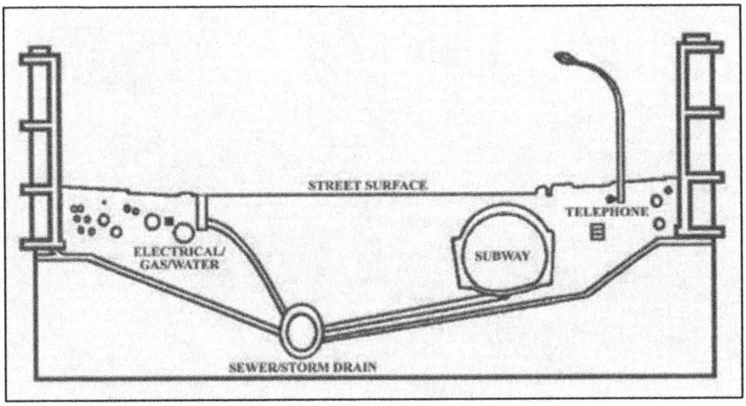

Figure E-1. Potential Underground Passages

a. Subterranean passages may provide the attacker with covered and concealed routes into and through built-up areas. This may allow an enemy to launch his attack along roads that lead into the city while infiltrating a smaller force into the defender's rear. An enemy unit in the defender's rear area may disrupt his defense and obstruct the avenues of withdrawal for his forward defense.

b. An attack along a subterranean avenue of approach could become the main effort. Even if the subterranean effort is not immediately successful, it forces the defender to fight on atleast two levels and to extend resources to more than just street-level fighting.

c. The existence of subterranean passages forces the defender to defend the built-up area above and below ground. While these passages are usually more of a disadvantage to the defender than to the attacker they may offer some advantages in the defense. When thoroughly reconnoitered and controlled by the defender, subterranean passages provide excellent covered and concealed routes to move reinforcements or to launch counterattacks. They also provide ready-made covered lines of communication for the movement of supplies and evacuation of casualties and provide places to cache supplies for forward companies. Subterranean passages also offer the defender a ready-made conduit for communications wire, with protection from armored vehicles and indirect fires.

2. Denial to the Enemy. Subterranean passages are useful to the defender only to the extent that the attacker can be denied their use. The defender has an advantage in that, given the confining, dark environment of these passages, a small group of determined Marines in a prepared position can defeat a numerically superior force.

a. Tunnels afford the attacker little cover and concealment except for the darkness and any manmade barriers. The passageways provide tight fields of fire and amplify the effects of grenades. Obstacles at intersections in tunnels support excellent ambush sites and may turn the subterranean passages into a maze. These obstacles can be created quickly by using chunks of rubble, furniture, and parts of abandoned vehicles interspersed with M18 claymore mines.

b. A thorough reconnaissance of the subterranean or sewer system is required. To be effective, obstacles must be located at critical intersections in the passage network so that they trap attackers in a kill zone but allow defenders freedom of movement (Figure E-2).

3. Subterranean Reconnaissance Techniques. The mission of local reconnaissance for a platoon- or company-sized area of responsibility should be given to a squad-sized element.

a. The squad leader tasks one fire team with providing forward security. One fire team follows in support, and one fire team provides rear security (Figure E-3). The squad leader moves directly behind the pointman and records data collected by the patrol. If available, an engineer should be with the lead fire team. The rear security fire team protects the ground- and subterranean-level point of entry. This fire team is responsible for detecting enemy movement on the patrol's rear and serves as the communications link between the squad leader and higher headquarters. Organization of the patrol may vary based on METT-T.

b. The squad leader should carry a map, compass, street plan, and notebook in which he has detailed the information that he must gather. One Marine is assigned to carry the tools needed to open manhole covers. If the patrol is to move more than 200 to 300 meters, or if the platoon leader directs, the squad may carry a sound-powered telephone (TA-1) and wire dispenser (XM-306A) for communications. Radios are often unreliable in this environment. Marines (especially the pointman, squad leader, engineer, and fire team leaders) should be equipped with night vision devices to maintain surveillance within the sewer. Thermal night devices are also useful as there may not be enough ambient light for accurate visibility with night vision devices.

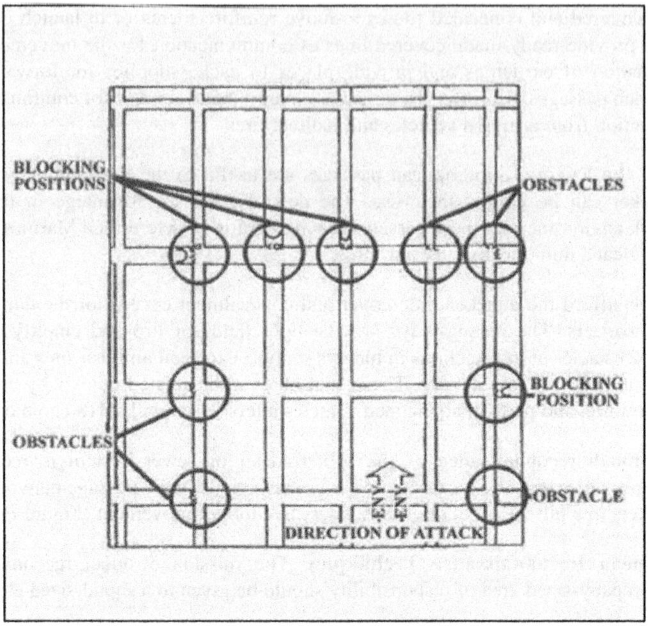

Figure E-2. Defense of a Sewer System

Figure E-3. Organization of the Patrol Unit

c. Marines entering the sewer should carry a sketch of the sewer system that includes magnetic north, azimuths, distances, and manholes. They should also carry protective masks, flashlights, gloves, and chalk (or chemiluminescent lights) for marking features along the route. The patrol should also be equipped with a safety rope to which each man is tied. To improve their footing in slippery sewers and storm drains, the squad members may wrap chicken wire or screen wire around their boots.

d. A constant concern to units conducting a subterranean patrol is chemical defense. Enemy chemical agents used in tunnels may be suddenly encountered in dense concentrations with little dispersion. The pointman should carry an automatic chemical agent alarm system to provide instant warning of the presence of chemical agents. M8 detector paper can also be used to detect chemical agents. At the first indication that harmful gases are present, the squad should mask.

e. In addition to enemy chemical agents, noxious gases from decomposing sewage can also pose a threat. These gases are not detected by chemical agent alarm systems, nor are they completely filtered by the protective mask. Physical signs that indicate their presence in harmful quantities are nausea and dizziness. The squad leader should be constantly alert to these signs and know the shortest route to the surface for fresh air. Some sewer gases deplete oxygen, a situation against which the field protective mask provides no protection.

f. Once the squad is organized and equipped, it moves to the entrance of the tunnel. This may be a manhole. With the manhole cover removed, the squad waits 15 minutes before entry to allow any gases that may be present to dissipate. The pointman then descends into the tunnel to determine whether the air is safe to breathe and whether movement is restricted. The pointman should remain in the tunnel for 10 minutes before the rest of the squad follows. If he becomes ill or is exposed to danger, he can be pulled out by the safety rope.

g. When the squad is moving through the tunnel, the pointman moves about 10 meters in front of the squad leader. Other squad members maintain 5-meter intervals. If water in the tunnel is flowing faster than 2.5 meters per second or the sewer contains slippery obstacles, those intervals should be increased to prevent all patrol members from falling if one man slips. All squad members should stay tied into the safety rope so that they can easily be retrieved from danger. The last Marine in the patrol marks the route with chalk or chemiluminescent lights so other units can find the squad.

h. The squad leader should record the azimuth and pace count of each turn he takes in the tunnel. When he encounters a manhole to the surface, the pointman should open it and determine the location, which the patrol leader then records. The use of recognition signals (Figure E-4) prevents friendly forces from accidentally shooting the pointman as he appears at a manhole.

Figure E-4. Recognition Signals

i. Once the patrol has returned and submitted its report, the platoon leader must decide how, or whether, to use the tunnel. In the offense, a tunnel might provide a covered route to move behind the enemy's defenses. In the defense, a tunnel might provide a covered passage between positions. In any case, the squad members that provided the reconnaissance should act as guides along the route.

j. If the tunnel is to be blocked, the platoon may emplace concertina wire, early warning devices, and antipersonnel mines. A security force should be established at the entrance of the sewer (Figure E-5) to provide warning against any enemy trying to approach the platoon's defense. It should be equipped with command-detonated illumination. The overpressure from grenades, mines, and boobytraps exploding in a sewer or tunnel can have adverse effects on friendly forces, including ruptured eardrums and wounds from flying debris. Also, gases found in sewers can be ignited by the blast effects of these munitions. For these reasons, small-arms weapons should be employed in tunnels and sewers. Friendly personnel should be outside of tunnels or out of range of the effects when mines or demolitions are detonated.

4. Psychological Considerations. Combat operations in subterranean passages are similar to night combat operations. The psychological factors of night operations reduce confidence, cause fear, and increase a feeling of isolation. This feeling of isolation is further magnified by the tight confines of the tunnels. The layout of tunnels could require greater dispersion between positions.

Figure E-5. Security Position Established at the Entrance to a Sewer

a. Leaders must enforce measures to help dispel the feelings of fear and isolation experienced when operating in tunnels. These measures include leadership training, physical and mental fitness, sleep discipline, and stress management.

b. Leaders maintain communication with Marines manning positions in tunnels either through personal visits, messengers, or field telephone. Communication increases situational awareness and provides insight into the mental state of Marines. Marines manning positions below ground should be given as much information as possible on the organization of the tunnels and the importance of the mission. They should be briefed on contingency plans and alternate positions should their primary positions become untenable.

c. Physical and mental fitness can be maintained by periodically rotating Marines out of tunnels so they can stand and walk in fresh air and sunlight. Stress management is also a factor of operations in tunnels. Historically, combat in built-up areas has been one of the most stressful forms of combat. Fighting in continuous darkness and restricted maneuver space causes even more stress to Marines than street fighting.

Appendix F

Obstacles, Mines, and Demolitions

Obstacles are used extensively in combat in built-up areas to allow the defender to canalize the enemy, impede his movement, and disrupt his attack.

Section I

Obstacles

1. Introduction. Obstacles are designed to prevent movement by personnel, to separate infantry from tanks, and to slow or stop vehicles. (See MCRP 3-17A, *MAGTF Breaching Operations.*)

2. Types. Antipersonnel mines, barbed wire, boobytraps, and exploding flame devices are used to construct antipersonnel obstacles (Figure F-1). (See FMFRP 13-5/FM 5-34, *Engineer Field Data*; FM 5-250, *Explosives and Demolitions*; and FM 3-11, *Flame, Riot Control Agent, and Herbicide Operations,* for more detailed information.) These obstacles are used to block the following infantry approaches:

- Streets
- Buildings
- Roofs
- Open spaces
- Dead space
- Underground systems.

a. The approval authority to boobytrap buildings is normally the GCE commander. This authority may be delegated to subordinate commanders.

b. The three types of wire obstacles used in defensive operations are protective, tactical, and supplementary. Wire obstacles should be integrated with other obstacles and covered by fire and observation.

(1) Protective wire is usually located beyond hand-grenade range (40 to 100 meters) from the defensive position.

(2) Tactical wire is positioned to increase the effectiveness of friendly weapons fire. Tactical wire is usually positioned on the friendly side of machine gun FPLs.

(3) Supplementary wire is used to break up the pattern of tactical and protective wire to prevent the enemy from easily locating friendly weapons.

c. Obstacles may be placed in dead spaces to restrict infantry movement in areas that cannot be observed and that are protected from direct fires.

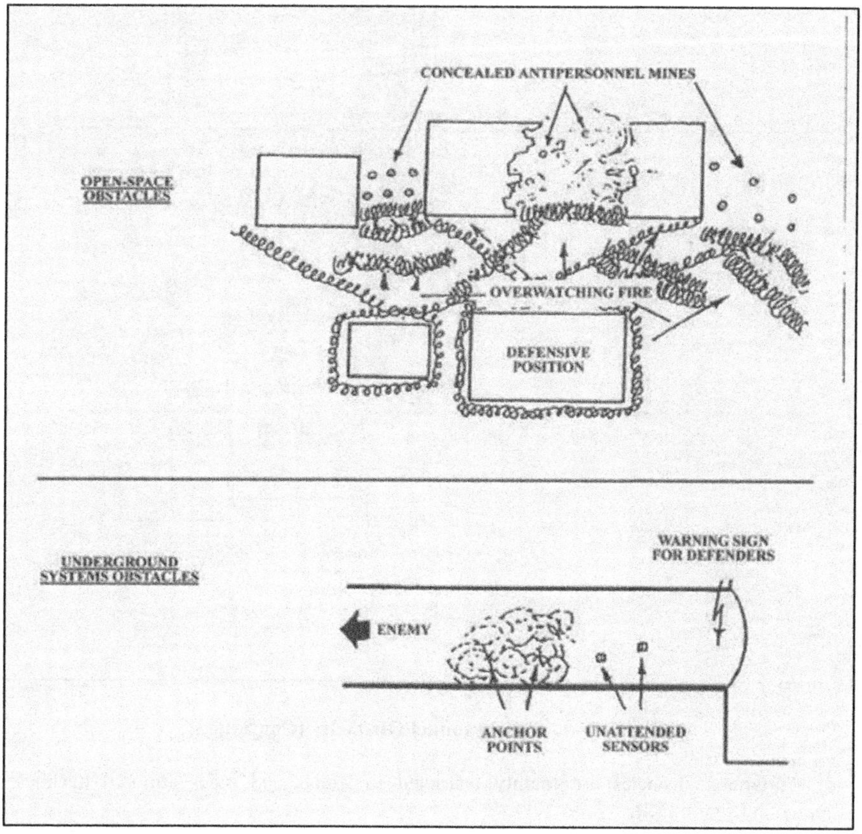

Figure F-1. Antipersonnel Obstacles

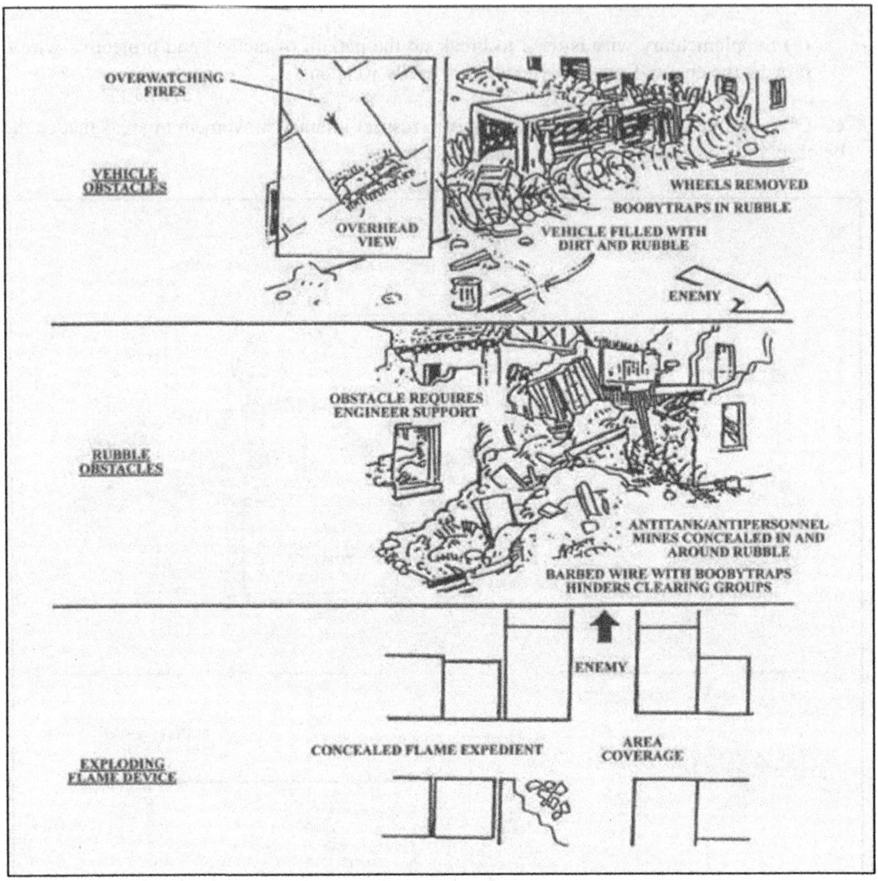

Figure F-1. Antipersonnel Obstacles (Continued)

d. Antiarmor obstacles are mainly restricted to streets and other armored avenues of approach (Figure F-2).

e. Some antitank mines may be sited and employed as antihelicopter mines to cover probable HLZs (Figure F-2).

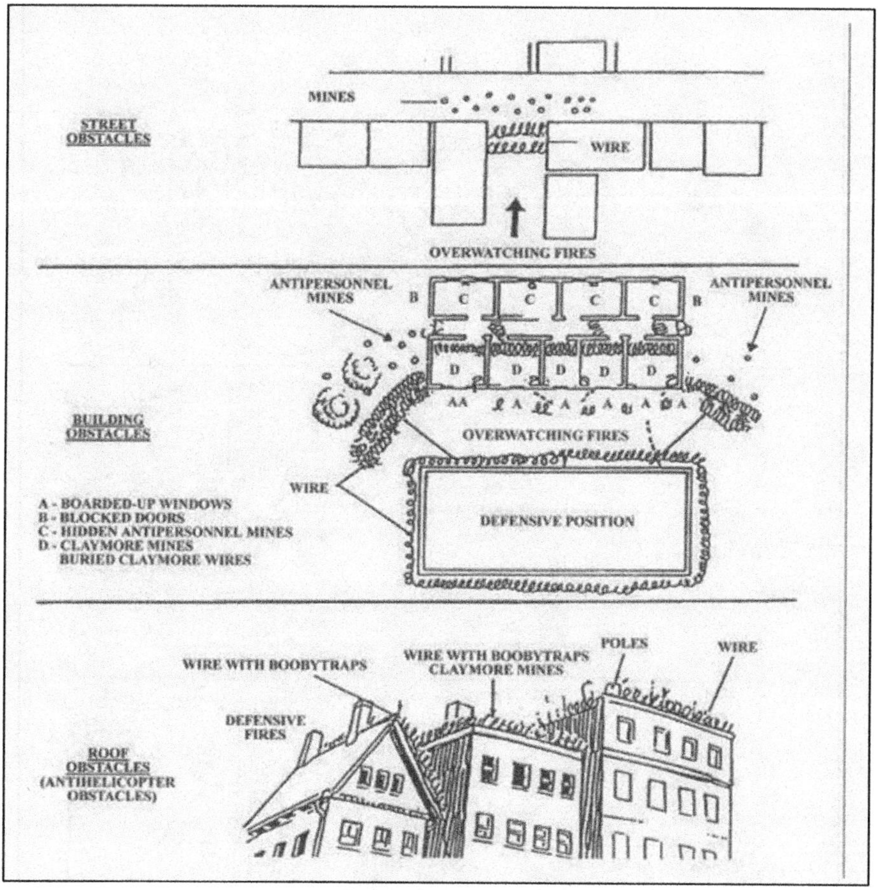

Figure F-2. Antiarmor Obstacles

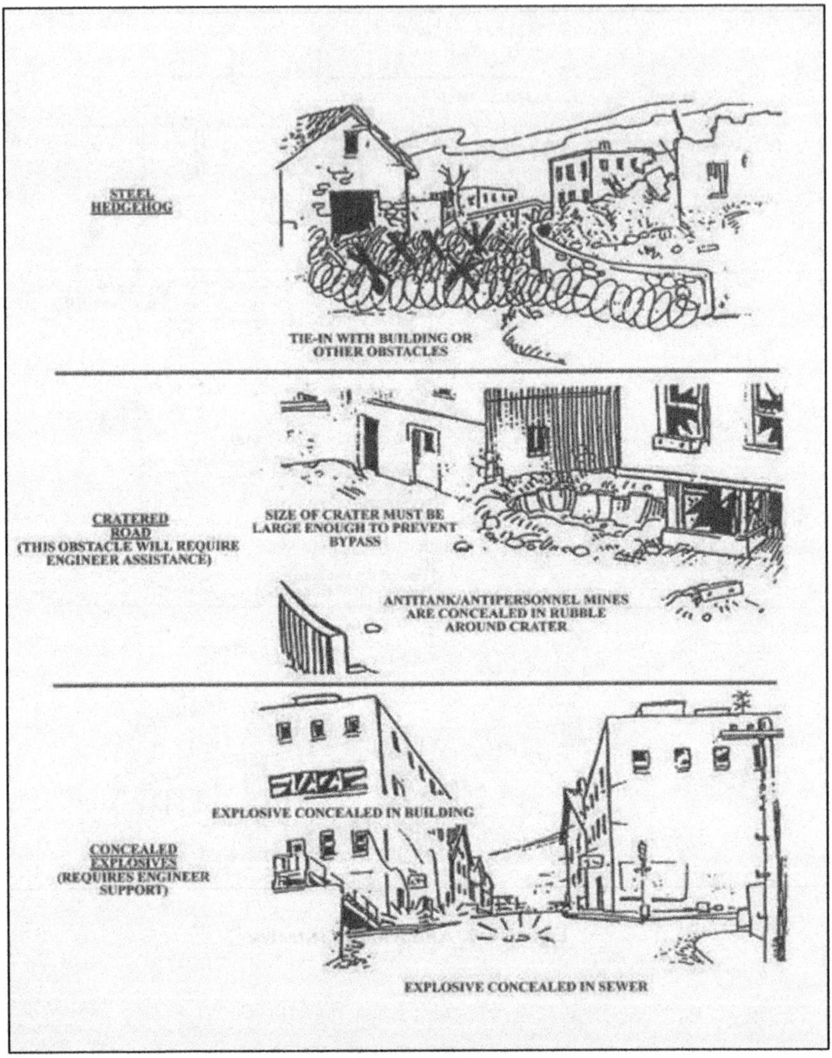

Figure F-2. Antiarmor Obstacles (Continued)

3. Construction of Obstacles. Obstacles are constructed in buildings to deny enemy infantry covered routes and weapons positions that are close to friendly defensive positions. They can be constructed of explosives, by using wire, or by using boobytraps within buildings. The building may be prepared as an explosive or flame trap for command detonation after enemy occupation.

Section II

Mines

4. Introduction. The employment or discovery of mines in built-up areas should be recorded on a building sketch (Figure F-3). (See FM 20-32.) The sketch should include the number of the building (taken from a city map, if available) and all floor plans. It should also include the type of mine and firing device. When possible, mined buildings should be marked on the friendly side (Figure F-4). Clearing areas or buildings that have been mined is extremely difficult and dangerous. Therefore, such areas should be considered "no go" areas for infantry. This factor must be carefully considered when planning and authorizing the placement of mines. (See Table F-1 for notional approving authority for minefields, but with the understanding that information in that table is subject to change with international agreements and ROE.)

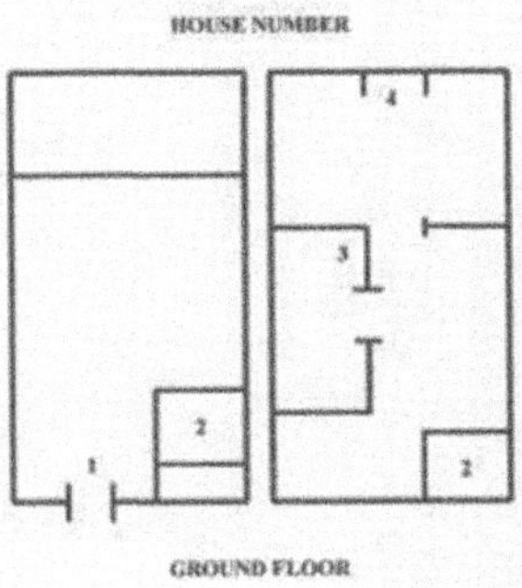

MINE	LOCATION	TYPE	FIRING DEVICE
1	FRONT DOOR	CLAYMORE	COMMAND
2	STAIRS	CLAYMORE	COMMAND
3	CLOSET	CLAYMORE	COMMAND
4	DRESSER	CLAYMORE	COMMAND

Figure F-3. Building Sketch Showing Mines

Figure F-4. Marking Mined Buildings

	Minefields (Type)		FASCAM (Duration)		Other Obstacles	
Level	Tactical	Protective	Long	Short	Tactical	Protective
MEF	X	X	X	X	X	X
Division	X	X	X	X	X	X
Regiment	X	X	X	X	X	X
Battalion	X	X		X	X	X
Company		X				X

Table F-1. Notional Minefield Employment Authority Matrix

5. **Types.** Several types of mines may be encountered or employed in built-up areas.

a. The M14 mine (a plastic type of antipersonnel landmine) cannot currently be used by American forces due to the Leahy Amendment of 1995 and the Certain Conventional Weapons Convention, of which the United States is a signatory nation. It may, however, be encountered in MOUT or be employed by coalition or enemy forces. For this reason, the following information is given as a guide for what U.S. forces may encounter. This type of antipersonnel landmine is normally used in conjunction with metallic antipersonnel, antitank, or chemical mines to confuse and hinder enemy breaching attempts. When encountered, it must be carefully handled because its light weight makes it easy to displace (Figure F-5). However, its size makes it ideal for covering obscure places such as stairs and cellars.

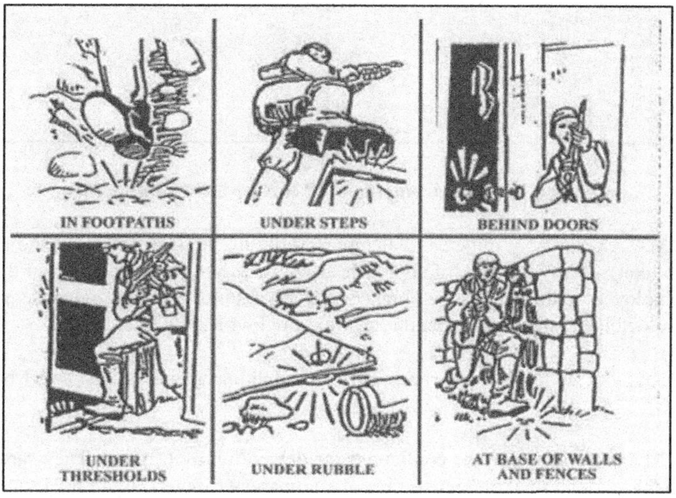

Figure F-5. Typical Employment of the M14 Antipersonnel Mine

b. The M16A2 antipersonnel landmine is subject to the same legal limitations as the M14. It is a boundary type of antipersonnel landmine, which makes it ideal for covering large areas such as rooftops, backyards, parks, and cellars. It can be expediently rigged for command detonation by attaching a rope or piece of communications wire to the pull ring (Figure F-6).

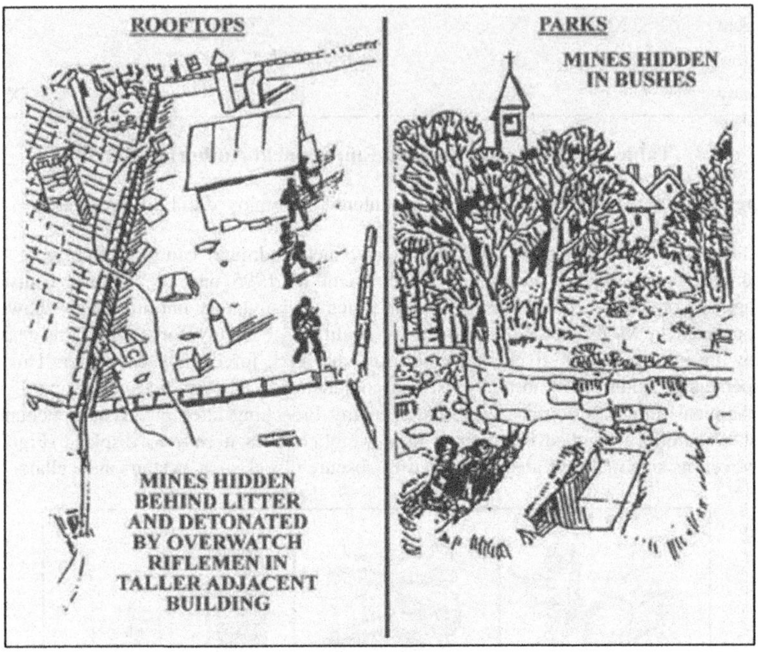

Figure F-6. Emplacements of M16A2 Antipersonnel Mine

c. The M18A1 claymore mine can be employed during the reorganization and consolidation phase on likely enemy avenues of approach. It does not have to be installed in the street, but can be employed on the sides of buildings or any other sturdy structure. The claymore mine, in command-detonated mode, is currently legal for employment by U.S. forces.

(1) Claymore mines can also be used for demolition against thin-skinned buildings and walls.

(2) Claymore mines may be configured for detonation by tripwire. They can help fill the dead space in the final protective fires of automatic weapons (Figure F-7).

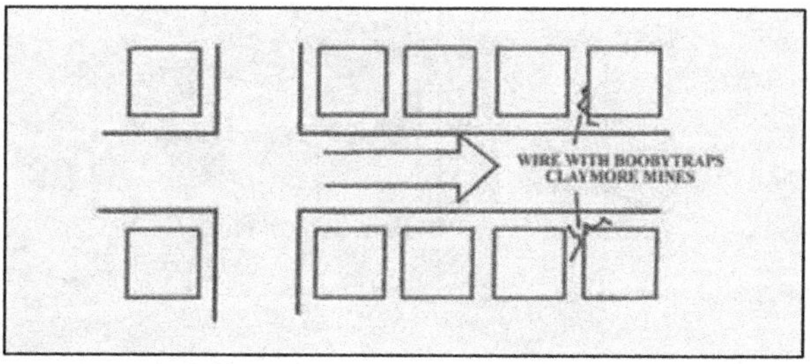

Figure F-7. Claymore Mine Used To Cover Dead Space of Automatic Weapons

(3) Claymore mines can be used in several ways in the offense. For example, if friendly forces are advancing on a city, claymore mines can be used in conjunction with blocking positions to cut off enemy avenues of escape (Figure F-8).

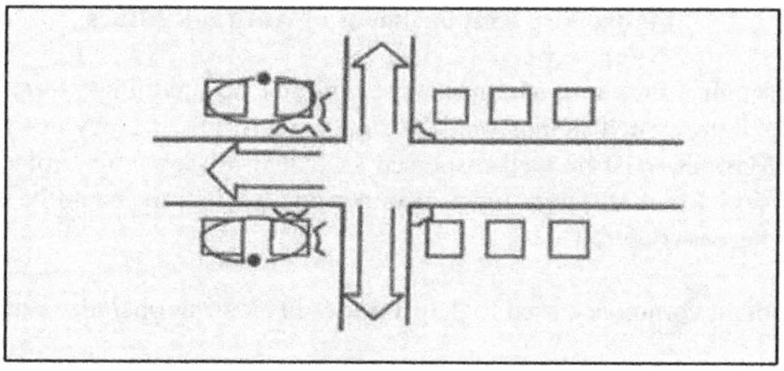

Figure F-8. Claymore Mines Used To Block Enemy Avenues of Escape

d. The M15, M19, and M21 antitank mines are used to stop or turn enemy armor and should be employed (Figure F-9):

- In conjunction with other manmade obstacles and covered by observation and fire
- In streets or alleys to block routes of advance in narrow defiles
- As command-detonated mines with other demolitions.

6. Enemy Mines and Boobytraps. Buildings contain many areas and items that are potential hiding places for boobytraps, for example, doors, windows, telephones, stairs, books, canteens, and so on.

Figure F-9. Emplacement of Antitank Mines

When moving through a building, Marines must not pick up anything—food, valuables, enemy weapons, or other items. Such items could be rigged with pressure devices that explode when they are moved. Marines must be well-dispersed so that if a boobytrap explodes, the number of casualties will be few. Many different types of mines and boobytraps could be encountered during combat in built-up areas (Figure F-10).

a. The equipment commonly used to detect mines in clearing operations includes:

- Mine detectors
- Probes
- Grappling hooks
- Ropes
- Explosives
- Flak vests
- Eye protection
- Engineer tape.

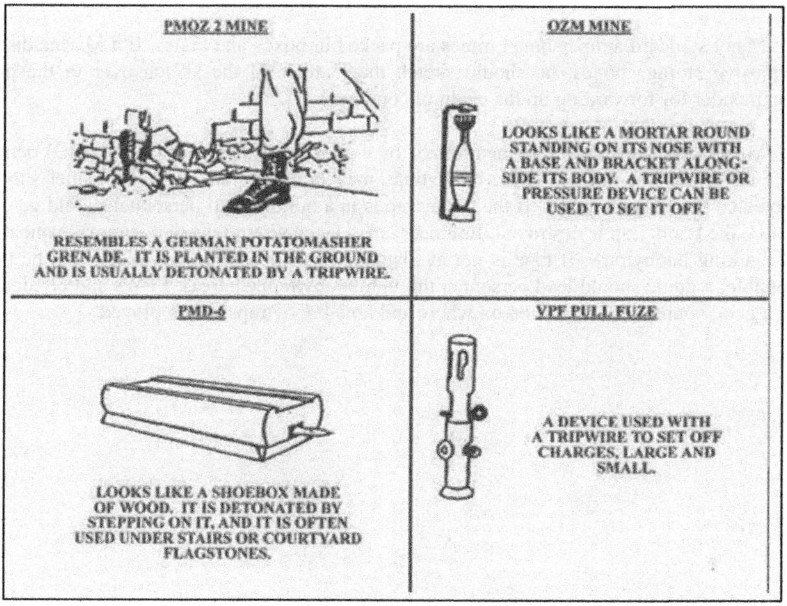

Figure F-10. Common Threat Antipersonnel Mines and Boobytraps

b. To detect tripwires, a Marine can use a 10-foot pole with 5 feet of string tied on one end. A small weight attached to the loose end of the string will snag on the tripwire. This allows the lead Marine to easily detect a tripwire (Figure F-11). Marines can also use a flashlight to cause reflections and shadows to find tripwires. A wand with a dangling thread can be swept over the ground ahead with the thread being watched for any disturbance by a tripwire.

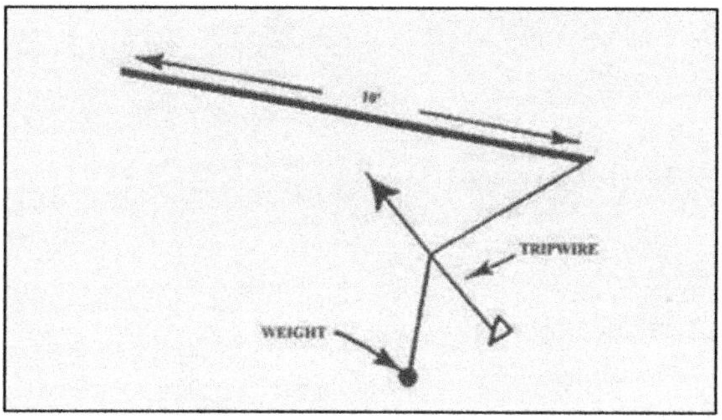

Figure F-11. Tripwire Detection

c. Many standard antipersonnel mines are packed in boxes and crates. If a Marine discovers explosive storage boxes, he should sketch them and turn the sketch over to the platoon commander for forwarding up the chain of command.

d. Most boobytraps should be neutralized by explosive ordnance disposal (EOD) personnel. If EOD teams are not available, boobytraps may be blown in place. Personnel should be protected by adequate cover. If the boobytrap is in a building, all personnel should go outside before the boobytrap is destroyed. Engineer tape placed around the danger area is one method of marking boobytraps. If tape is not available, strips ripped from bedsheets can be used. If possible, a guide should lead personnel through boobytrapped areas. EPWs and civilians may be a good source of information on where and how boobytraps are employed.

Section III

Demolitions

7. Introduction. Demolitions are used more often during combat in built-up areas than during operations in open natural terrain. Demolition operations should be directed by the engineers that support the GCE. However, if engineers are involved in the preparation and execution of the barrier plan, other trained Marines can prepare mouseholes, breach walls, and rubble buildings themselves if assisted and advised by attached engineers.

8. Offensive Use. When assaulting or clearing a built-up area, demolitions may enable the GCE commander to create an avenue of approach through buildings.

a. Every other man in an assault force should carry demolitions, and other selected personnel should carry blasting caps. The same man should *not* carry both the explosives and the blasting caps because of the danger of sympathetic detonation. As the demolitions are expended by the assault force, they should be replaced by explosives carried by the support force.

b. One of the most difficult breaching operations faced by the assault force is the breaching of masonry and reinforced concrete walls. When demolitions must be used, composition C4 is the ideal explosive to use. Normally, building walls are 15 inches thick or less. Assuming that all outer walls are constructed of reinforced concrete, a rule of thumb for breaching is to place 10 pounds of C4 against the target between waist and chest height. When detonated, this normally blows a hole large enough for a man to go through. The amounts of TNT and C4 required to breach concrete are shown in Table F-2.

c. Mouseholes provide the safest method of moving between rooms and floors. They can be created with C4. Because C4 comes packaged with an adhesive backing or can be emplaced using pressure-sensitive tape, it is ideal for this purpose. When using C4 to blow a mousehole in a lath and plaster wall, one block or a strip of blocks should be placed on the wall from neck to knee height. Charges should be primed with detonating cord or electrical blasting caps to obtain simultaneous detonation, which will blow a hole large enough for a man to fit through.

9. Defensive Use. The use of demolitions in defensive operations is similar to their use in offensive operations. When defending a built-up area, demolitions are used to create covered and concealed routes through walls and buildings that can be used for withdrawals, reinforcements, or counterattacks. Demolitions are also used to create obstacles and clear fields of fire.

a. Marines can use demolitions for creating mouseholes and constructing command-detonated mines. Expedient C4 satchel charges can be concealed in firing positions and along movement routes for detonation in the vicinity of the enemy.

Thickness of Material	TNT	C4	Size of Opening
Reinforced Concrete			
Up to 10 cm (4 inches)	.792 lbs	.591 lbs	10 to 15 cm (4 to 6 inches)
10 to 20 cm (4 to 8 inches)	3.4 lbs	2.54 lbs	15 to 30 cm (6 to 12 inches)
Nonreinforced Concrete Masonry			
Thickness of Material	**TNT**	**C4**	**Size of Opening**
Up to 35 cm (14 inches)	3.66 lbs	2.73 lbs	35 cm (14 inches)
35 to 45 cm (14 to 18 inches)	7.5 lbs	5.6 lbs	45 cm (18 inches)
45 to 50 cm (18 to 20 inches)	10.2 lbs	7.61 lbs	50 cm (20 inches)

Table F-2. Amounts of TNT and C4 Required To Breach Concrete

b. The engineers should furnish technical assistance for selective rubbling. Normally, buildings can be rubbled by using shaped charges or C4 on the supports and major beams.

c. Charges should be placed directly against the surface to be breached unless a shaped charge is used. Whenever possible, demolitions should be tamped to increase their effectiveness. Tamping materials could be sandbags, rubble, or desks and chairs (Figure F-12).

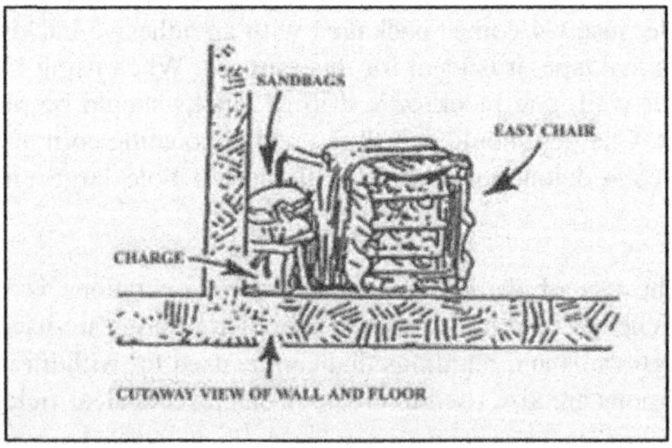

Figure F-12. Chair Used To Tamp Breaching Charge

d. For exterior walls, tamping of breaching charges may be impossible because of enemy fire. Thus, the untamped charge normally requires twice the explosive charge to produce the same effect as a tamped charge (Figure F-13).

METHODS OF PLACEMENT				
THICKNESS OF CONCRETE				
FEET	POUNDS OF TNT	POUNDS OF C4	POUNDS OF TNT	POUNDS OF C4
2	14	11	28	21
2 1/2	27	21	54	41
3	39	30	78	59
3 1/2	62	47	124	93
4	93	70	185	138
4 1/2	132	99	263	196
5	147	106	284	211
5 1/2	189	141	376	282
6	245	186	490	366

Figure F-13. Breaching Reinforced Concrete

e. When enemy fire prevents an approach to the wall, the breaching charge can be attached to a pole and slid into position for untamped detonation at the base of the wall (Figure F-14).

Figure F-14. Charge Placement When Small-Arms Fire Cannot Be Suppressed

f. The internal walls of most buildings function as partitions rather than as load-bearing walls. Therefore, smaller explosive charges can be used to breach them. In the absence of C4 or other military explosives, internal walls can be breached by using a claymore mine (Figure F-15). The claymore should be tamped to increase its effectiveness and to reduce the amount of explosive force directed to the rear.

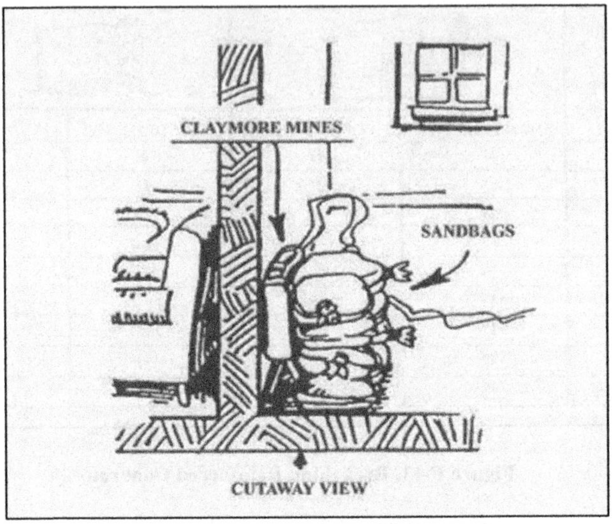

Figure F-15. Tamping a Claymore Mine To Breach Internal Walls

g. The Molotov cocktail (Figure F-16) is an expedient device for disabling both wheeled and tracked vehicles. It is easy to make because the materials are readily available. The results are most effective because of the close engagement in built-up areas. The objective is to ignite a flammable portion of the vehicle such as the fuel or ammunition that it is transporting. The following materials are needed to make a Molotov cocktail:

- Bottle or other glass container
- Gas (60 percent)
- Oil (40 percent)
- Rag for use as a wick.

The gas and oil are mixed thoroughly (60 percent gas to 40 percent oil). The rag is soaked with the mixture, then the mixture is placed into the bottle. The rag is then inserted in the opening of the bottle as a wick. When a target is sighted, the wick is lit and the bottle is thrown hard enough to break.

Figure F-16. Molotov Cocktail

WARNING

Ensure that a safe distance is maintained when throwing a Molotov cocktail.

h. The bunker bomb (Figure F-17) is an expedient, explosive flame weapon that is best used against fortified positions or rooms. This expedient munition should be used with a nonelectrical firing system. The following materials are required to make a bunker bomb:

- 1 metal small-arms ammunition container
- 1 gallon of gasoline
- 50 feet of detonating cord
- 1 nonelectrical blasting cap
- 1 M60 fuze igniter
- 7 1/2 feet of M700 time fuze
- 1 M49 trip flare or M34 WP grenade.

(1) Step 1. Fill the ammunition can 3/4 full with fuel and secure the lid.

(2) Step 2. "Hasty whip" the device with 15 turns around the center of the container using 44 feet of detonating cord. Leave 2-foot "pigtails" for attaching the igniter and fuze igniter.

(3) Step 3. Tape the igniter (M49 trip flare or M34 WP grenade) to the container handle.

(4) Step 4. Place one detonating cord pigtail end under the igniter spoon handle. Tape it in place.

(5) Step 5. Attach the M60 fuze igniter and the nonelectrical blasting cap to the M700 time fuze.

(6) Step 6. Attach the nonelectrical firing system to the other pigtail by making a loop in the detonating cord and attaching the blasting cap to it.

(7) Step 7. Remove the safety pin from the igniter (M49 trip flare or M34 WP grenade). The device is ready to be fired.

WARNING

Never carry the device by the handle or igniter. Remove the igniter safety pin only when it is time to use the device. Use extreme care when handling or carrying nonelectrical firing systems. Protect blasting caps from shock and extreme heat. Do not allow the time fuze to kink or become knotted. Doing so may disrupt the powder train and may cause a misfire. Prime detonating cord and remove the time fuze igniter safety pin only when it is time to use the device.

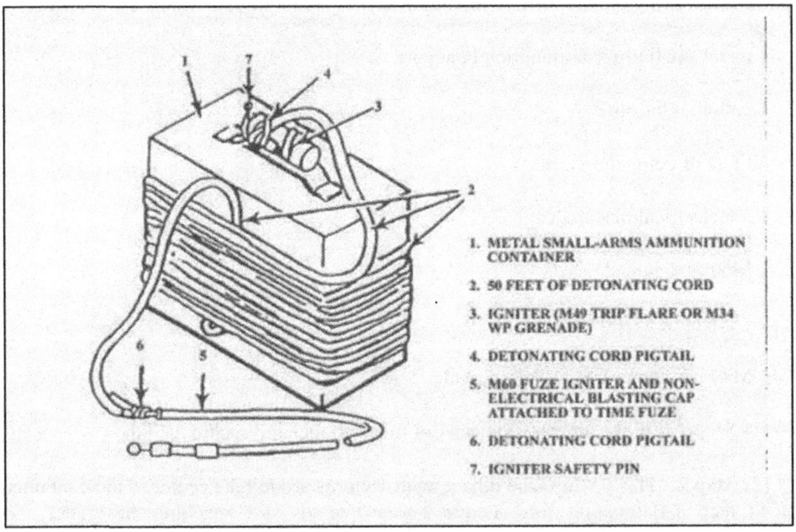

Figure F-17. Bunker Bomb Made From an Ammunition Can

10. Safety. The greatest danger to friendly personnel from demolitions is the debris thrown by the explosion. Leaders must ensure that protective measures are enforced. The safe distance listed in Table F-3 indicates the danger of demolition effects.

Pounds of Explosive	Safe Distance in Meters[1]	Pounds of Explosive	Safe Distance in Meters[1]
1 to 29	300	150	534
30	311	175	560
35	327	200	585
40	342	225	609
45	356	250	630
50	369	275	651
60	392	300	670
70	413	325	688
80	431	350	705
90	449	375	722
100	465	400	737
125	500	425	750

[1]These distances will be modified in combat when troops are in other buildings, around corners, or behind intervening walls.

Table F-3. Minimum Safe Distances for Personnel in the Open

a. The following are the guidelines for using demolitions:

- Keep the blasting machine under the control of a noncommissioned officer.
- Wear helmets and flack jackets at all times while firing explosives.
- Handle misfires with extreme care.
- Clear the room and protect personnel when blowing interior walls.

b. Some charges should be prepared beforehand, without detonators, to save time. Examples include 10- or 20-pound breaching charges of C4 and expedient-shaped charges in No. 10 cans.

- Use C4 to breach hard targets (masonry construction).

- Do not take chances.
- Do not divide responsibility for demolition work.
- Do not mix demolitions and detonators until ready to use.
- Do not carry demolitions and caps together.

Appendix G

Nuclear, Biological, and Chemical Considerations

Current U.S. policy regarding lethal or incapacitating agents is that their use against an armed enemy requires approval at the NCA level. Potential enemies may not operate under the same restrictions. Field commanders must be prepared to assume an adequate NBC defensive posture when engaged in urban fighting.

1. Protection From Nuclear, Biological, and Chemical Weapons. Generally, the lowest floor or basement of a reinforced concrete or steel-formed building offers good protection from nuclear hazards and initial liquid chemical contamination although some chemical agents tend to collect in lower areas. Tanks and LAVs/AAVs also provide protection. (See FMFM 11/MCWP 3-37 series manuals for greater detail.)

a. Biological attacks are difficult to detect and recognize. Biological agents can be disseminated by using aerosols, vectors, and covert methods. Because biological agents can be sprayed or dropped in bomblets, personnel who observe such indicators should promptly report them. Prompt reporting and treatment of the sick speed the employment of medical countermeasures. Although buildings and shelters provide some protection against spraying, they provide little protection against biological agents.

b. Chemical agents cause casualties by being inhaled or absorbed through the skin. They may afford Marines a few seconds to mask. Buildings have a channeling effect and tend to contain the effects of an agent, causing great variations in chemical concentration from room to room or from building to building. Chemical agents usually settle in low places, making sewers and subways hazardous hiding places. A prepared defender should include some collective protective measures in the defensive network. Personnel using fans may be able to put enough overpressure into tunnels to keep some chemical agents from entering. The protective mask and battle dress overgarment provide the best protection against chemical agents.

c. Personal hygiene is a critical defensive measure against infection and disease. Unfortunately, built-up areas are characterized by sophisticated sanitation systems. When those systems are destroyed, the resulting sanitary conditions become much worse than those in areas where sanitation facilities do not exist. Commanders must ensure that personnel employ sanitation measures and that their immunizations are current.

d. Commanders should plan their mission-oriented protective posture (MOPP) with the realization that increased logistics demands in built-up areas also apply to NBC equipment. Protective clothing, detection and decontamination equipment, and sealed containers of food and water must be stockpiled the same as other critical supplies. When operating in protective clothing, commanders must make allowances for the strenuous activities normally associated with combat in built-up areas.

(1) Detection. After an NBC attack, battalions should dispatch their detection and survey teams. Detection in built-up areas is complicated by the numerous rooms within buildings. Teams should conduct tests and surveys of major streets, intersections, and buildings in their area for inclusion in initial NBC reports. A systematic survey of all buildings, rooms, and underground facilities must be accomplished before occupation by unmasked personnel. All data should be forwarded using the appropriate NBC report.

(2) Decontamination. Personnel begin decontamination operations as soon after an NBC attack as the mission allows. Personnel should conduct individual decontamination of themselves and their personal equipment. Unit commanders determine the need for MOPP gear exchange and the requirements for a hasty or deliberate decontamination operation.

(a) Nuclear. Personnel should wear wet-weather gear for certain decontamination operations (hosing down buildings) to prevent radioactive material from touching the skin.

(b) Chemical and Biological. Roads, sidewalks, and other hard surfaces are best decontaminated by weathering, if time permits. Agents can also be covered with several inches of dirt or sand to provide protection. Fragment testing should be conducted periodically to ensure that the agent has not seeped through the covering. For critical sections of roads, a truck-mounted M12A1 (power-driven decontamination apparatus (PDDA)) can be used to spray supertropical bleach (STB) slurry; this aids rapid decontamination. Buildings are difficult to decontaminate, especially wooden ones. Some techniques for their decontamination are scrubbing with STB slurry; washing with hot, soapy water; washing or spraying with a soda solution; and airing.

2. Smoke Operations. The use of smoke is an integral part of either offensive or defensive operations in built-up areas. In the offense, smoke can support the maneuver of combat units and deception operations. Smoke employed in the defense obscures enemy air and ground observation, limiting the accuracy of enemy fires and intelligence collection.

a. Smoke should not be used when it degrades the effectiveness of friendly forces. Likewise, an extremely dense concentration of smoke in a closed area displaces the oxygen, suffocating Marines even when they are wearing protective masks.

b. Smoke pots, generators, or artillery smoke munitions should be used to cover the withdrawal of defending forces or the movement of attacking forces. Artillery-delivered WP can be effective against enemy forces by screening, in addition to causing casualties and fires. Consideration should be given to the incendiary effects of WP, smoke (HC and M825), and illumination munitions on the litter and debris of built-up areas.

c. Smoke grenades can be massed to provide a hasty screen for concealing personnel moving across streets and alleys. Smoke grenades can also be used for signaling; those launched by an M203 can be used to mark targets for fixed and rotary-wing air support.

d. The use of smoke in built-up areas is affected by the complex wind patterns caused by buildings. When covering a built-up area with a smoke haze or blanket, personnel must include all buildings. Failure to obscure tall buildings, towers, and steeples provides enemy observers with reference points for fire placement within the built-up area.

3. Riot Control Agents. When authorized, riot control agents such as CS and CN can be used to drive enemy forces from positions or to deny them areas for occupation. Riot control agents are incapacitating but generally have no lasting effects. They may be appropriate when preventing civilian casualties is a planning consideration. However, riot control agents may not be effective against an enemy that is well trained in chemical defense.

Appendix H

MOUT Under Limited-Visibility Conditions

To be successful, leaders must use limited-visibility conditions to their advantage.

1. Advantages. When fighting in built-up areas during night or periods of limited visibility, attacking or defending forces have several advantages.

a. In many cases, Marines may have a technological advantage in thermal imagery and light intensification over their opponents. This enables Marines to identify, engage, and destroy enemy targets before being detected by the enemy.

b. Possessing the capability to conduct operations as well at night or during periods of reduced visibility as during daylight gives Marines a tremendous advantage over an enemy with reduced capability. It also aids the generation of tempo and allows Marines to gain or retain the initiative.

c. Generally, ranges of direct-fire engagements are greatly reduced in the MOUT environment. During periods of limited visibility, effective unaided target acquisition ranges are further reduced. This enables attacking forces to close to shorter ranges, thus increasing the lethality and accuracy of weapons. Attacking forces can also take advantage of the enemy's reduced ability to see and can more effectively engage the enemy before being detected with thermal imagery or light intensification devices.

d. Helicopterborne assaults are best conducted during periods of limited visibility because the enemy's air defenses are degraded.

e. Attacking during periods of limited visibility gives the attacker a greater chance of achieving surprise.

2. Disadvantages. When fighting in built-up areas during periods of limited visibility, attacking and defending forces also face some disadvantages.

a. Command and control is difficult in any operation in a built-up area; periods of limited visibility increase this difficulty.

b. Marines have an instinctive tendency to move close together during periods of limited visibility. Constant attention must be given to prevent Marines from "bunching up."

c. Marines may become disoriented easily because of the combined effects of low visibility and the characteristics of built-up areas.

d. Target identification becomes more difficult in limited-visibility conditions. Depending on the training of the individual, the Marine may fire at anything seen or may hesitate too long before firing. Improper target identification is one of the leading causes of fratricide, so leaders must pay close attention to individual target engagement.

3. Fratricide Avoidance. The risk of fratricide is much greater during periods of limited visibility. The key to avoiding fratricide is increased situational awareness by leaders and individuals, coupled with realistic training in target identification. (See *Fratricide: Reducing Self-Inflicted Losses*, Newsletter, No. 92-4, April 92, Center for Army Lessons Learned, U.S. Army Combined Arms Command, Fort Leavenworth, Kansas.) Other considerations include:

a. Graphic control measures should be clearly defined and obvious. Examples include distinctive buildings, large boulevards, rivers, and so forth.

b. Leaders must exercise firm control when engaging targets. Movements should also be coordinated and controlled.

c. Cleared rooms and buildings should be distinctly marked to identify cleared areas and friendly forces to base-of-fire elements supporting the maneuver.

d. Visible markers (for example, glint tape or thermal strips) should be attached to individual Marines for rapid identification.

e. Far and near recognition symbols should be coordinated and used properly.

f. Units employing close air support must exercise firm control. Failure to do so may lead to the pilot becoming disoriented and engaging friend and foe alike.

4. Urban Environmental Effects on Night Vision Devices. Built-up areas affect standard night vision devices and sights differently than do open areas. This may cause some confusion because the images Marines receive through their night vision devices are unusual compared to those they may be used to.

a. Most built-up areas have electric power; therefore, street lights and/or building lights may "white out" some light intensification devices unless the power is disrupted or lights are turned off.

b. Fires may be burning in the urban area. Open flames cause problems not only for light intensification devices, but also for thermal devices.

c. Subterranean areas and the interiors of buildings will not have ambient light if the power is off. Passive night vision devices must then use an artificial light source, such as infrared radiation, to provide enough ambient light for the devices to work effectively.

d. The many reflective surfaces found in built-up areas may cause false images, particularly for laser range finders and laser target designators.

e. Thermal imaging devices may not be able to see through large amounts of dust particles suspended in the air.

f. Smoke and dust degrade the effectiveness of night vision devices.

g. Fog degrades long-range target acquisition from thermal sights.

h. Weapons flashes within enclosed areas appear to be much brighter. This causes Marines to lose their night vision and washes out light intensification devices.

5. Considerations. The environment of built-up areas presents special challenges and considerations during periods of limited visibility.

a. The use of glint tape, thermal tape, or chemiluminescent lights is an important consideration. These can be used to mark the FLOT, casualties, cleared buildings and rooms, weapons positions, and individual Marines. Their use must be clearly addressed in the unit's SOP. When markers are used for extended periods, their meanings should change because the enemy may be able to capture or manufacture and use these marking devices to their advantage.

b. The use of tracer and incendiary ammunition may be restricted to prevent fires. The light of fires whites out some night vision devices and may interfere with or confuse thermal devices.

c. The control of power stations may be essential to operations during periods of limited visibility. Control of power stations enables friendly forces to control, to a degree, background illumination. Shutting off the power to the street lights is much easier than shooting out the lights. Commanders must balance the tradeoff between force protection and maintaining law and order after the battle is over. During cold weather, the control of power stations may be critical to the welfare of the civilian population.

d. Identifying friendly units, noncombatant civilians, and enemy troops becomes more difficult during limited-visibility operations.

e. Locating the source of sounds becomes more difficult because of the natural echoing in built-up areas and the tendency of sounds to carry farther at night.

f. Locating boobytraps and obstacles also becomes more difficult at night. Movement rates are normally slower than during periods of normal visibility.

6. Special Equipment. Fighting during periods of limited visibility requires some specialized equipment to maximize advantages.

a. As a rule, thermal imaging devices such as the AN/PAS-7 infrared viewer and the AN/TAS-5 Dragon infrared sight are better for limited-visibility operations than are light intensification devices such as the AN/PVS-7B. Light intensification devices are more easily washed out by background light, weapons flashes in enclosed areas, and fires. Thermal devices, while also affected by fires, are not as easily washed out.

b. The AN/PAQ-4A infrared aiming device is similar to its off-the-shelf laser aiming sight counterparts, except it is not visible to the naked eye. Infrared pen lights can also be attached to weapons and, when used in conjunction with night vision devices, can provide a quick sight picture, illuminate rooms and hallways, identify obstacles and boobytraps, and identify friendly forces.

c. Other night sights for weapons include the AN/TVS-5 crew-served weapon night vision sight, the AN/PVS-4 individual weapons night vision sight, and the AN/UAS-12 night vision sight, which mounts to the modular universal laser equipment (MULE) and the TOW missile system.

d. Trip flares, flares, illumination from mortars, and artillery and spotlights (visible light or infrared radiation) can be used to blind enemy night vision devices or to artificially illuminate the battlefield (Figure H-1). (See FMFM 6-8, *Supporting Arms Observer, Spotter, and Controller*, for more information on illumination from mortars and artillery.)

e. Spare batteries for the night vision devices should be carried to keep the devices operational. Soft, clean rags should be used to clean the lenses.

7. Combat Support. Coordinating employment of forces and fire support in limited-visibility conditions is a major concern to leaders during MOUT. The concentration of forces and fires at the point of decision is facilitated by the technological edge that night vision devices provide.

a. Any degradation of accuracy in artillery fire will likely be a result of the limitations of target acquisition assets. While FOs and FACs may have thermal sights and laser range finders, most Marines on the battlefield do not yet have devices that will enable them to acquire targets accurately. The following are some devices and techniques to improve target acquisition for indirect fires:

(1) If the target is within LOS, tanks and LAVs can rapidly identify the target and provide an accurate range.

(2) Preregistered targets are effective if the target reference point can be observed and the observer has clear communications with the firing unit.

(3) Fixed-wing and rotary-wing aircraft can be used to identify targets and adjust supporting arms.

Figure H-1. Use of Indirect-Fire Illumination During MOUT

8. Combat Service Support. Maneuver unit commanders and their Marines are not the only individuals who must adjust to combat under limited-visibility conditions in MOUT. Logisticians at every level must anticipate requirements for these conditions.

a. Units conducting resupply operations during periods of limited visibility should consider:

- Issuing drivers and vehicle commanders night vision devices so that the vehicles going to and from logistic release points do not require illumination (This also prevents the enemy from acquiring resupply locations by following vehicles with blackout lights on.)

- Maintaining strict noise and light discipline

- Following a clearly marked route to avoid obstacles and prevent the resupply vehicle(s) from becoming disoriented

- Providing radios to resupply vehicles

- Issuing each vehicle a map of the AO (preferably a city map with the street names).

b. Combat units operating for an extended time under conditions of limited visibility should have enough batteries to keep the night vision devices functioning at optimum power and sensitivity.

c. A large operational readiness float of night vision devices should be maintained.

d. Casualty collection during periods of limited visibility is more difficult. Clear methods for marking any casualties must be established before the operation begins.

e. CSS operations located in existing structures should not be visible from a distance. This includes limiting vehicle traffic to an absolute minimum, sealing doors and windows to prevent light leakage, and dispersing assets as much as possible.

9. Operational Considerations. Marine units conduct attacks during periods of limited visibility to gain or sustain the momentum of the attack. Before conducting a limited-visibility attack, the commander must balance the risks and ensure that every Marine understands the mission, intent, and control measures. Rehearsals and strict command and control reduce casualties and greatly enhance the chances for mission accomplishment.

a. To reduce confusion, Marines should clear buildings and rooms using the same techniques they use during the day.

b. Movement rates are slower in the dark. Each Marine must remain alert for mines, boobytraps, and enemy positions.

c. Rifle squads and fire teams should be equipped with a mixture of both thermal imaging and light intensification devices whenever possible. This enables the squads and fire teams to obtain a better picture of the night environment and enables Marines to balance the strengths and weaknesses of each type of night vision device for maximum results.

d. When moving through buildings, assault forces mark cleared rooms and buildings and communicate with the support forces. Marking cleared rooms is especially critical if more than one assault force or element is in the same building. Communications with supporting forces are imperative to avoid fratricide.

e. If flashlights or chemiluminescent lights are used, they should be held away from the head or chest area. This will make it harder for enemy soldiers firing at the light to kill the Marine holding the flashlight or chemiluminescent light.

f. Units must know where everyone is during offensive operations. This not only reduces the risk of fratricide, but also decreases the time spent identifying, locating, and treating casualties. Also, it greatly reduces the chance of Marines becoming disoriented and separated from the unit.

g. Assault units should be aware of adjacent fires that may diminish the effectiveness of night vision devices. Weapons flashes within small rooms cause Marines to lose their night vision and can wash out light intensification devices. Also, enemy soldiers may use flares inside and outside of buildings to deliberately render night vision devices ineffective.

h. Leaders must ensure that all Marines follow the ROE and the laws of land warfare. This is critical if the enemy is intermixed with the local civilian population. Also, leaders must follow all control measures.

i. Enemy forces can be expected to use periods of limited visibility to the same advantages as Marine forces do.

j. Enemy forces may have access to sophisticated night vision devices manufactured in Europe, the United States, Japan, Korea, and the former Soviet Union.

Appendix I

Urban Building Analysis

As in other types of operations, success in urban combat depends largely on the ability to analyze the military aspects of terrain. Marines must be able to recognize certain terrain features when evaluating urbanized terrain.

1. Types of Mass-Construction Buildings. Mass-construction buildings are those in which the outside walls support the weight of the building and its contents. Additional support, especially in wide buildings, comes from using load-bearing interior walls, strongpoints (called pilasters) on the exterior walls, cast-iron interior columns, and arches or braces over the windows and doors (Figure I-1). Modern types of mass-construction buildings are wall and slab structures, such as many modern apartments and hotels, and "tilt-up" structures commonly used for industry or storage. Mass-construction buildings are built in many ways:

- The walls can be built in place by using brick, block, or poured-in-place concrete.
- The walls can be prefabricated and then tilted up, or they can be reinforced-concrete panels.
- The walls can be prefabricated and assembled like boxes.

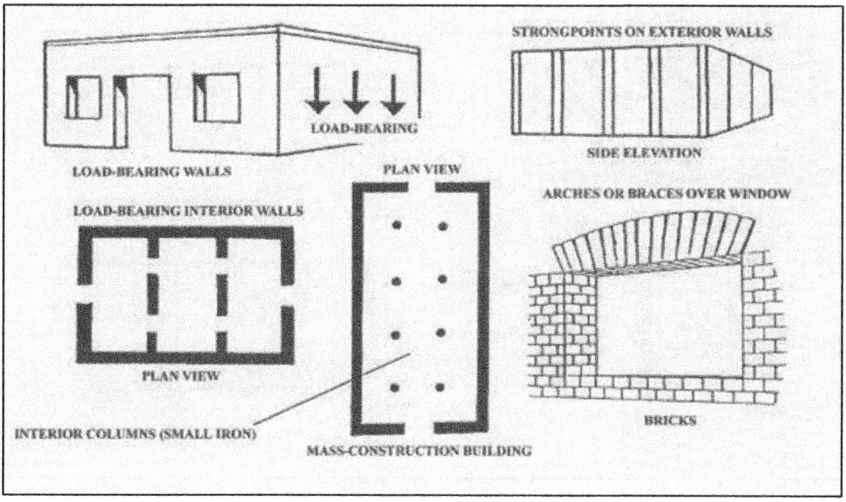

Figure I-1. Mass-Construction Buildings

a. Brick buildings are, in some regions, the most common and most important of the mass-construction buildings. In Europe, brick buildings are commonly covered with a stucco veneer so that bricks do not show (Figure I-2). One of the most common uses of brick buildings is the small store. These buildings are found in all built-up areas but are most common in the core periphery (Figure I-3).

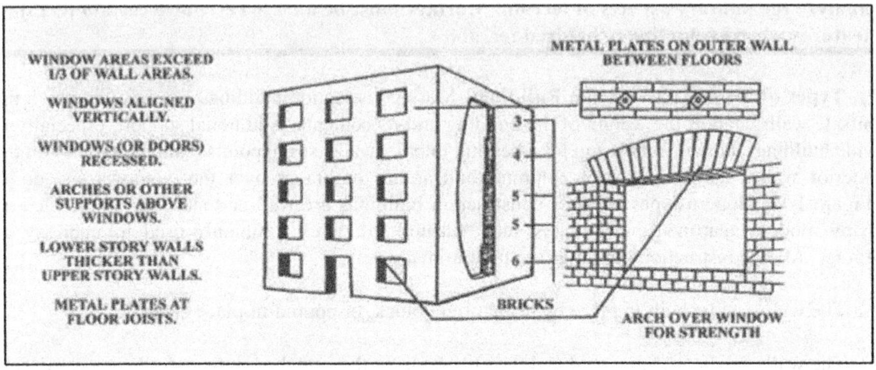

Figure I-2. Brick Buildings

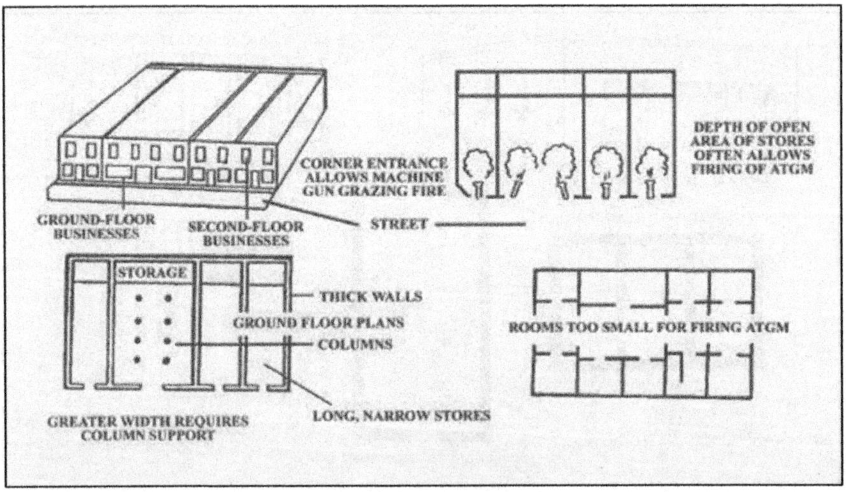

Figure I-3. Brick Store

b. Another common mass-construction building found mainly in industrial areas and along commercial ribbons is the warehouse. It is built of poured-in-place concrete reinforced with steel bars or of prefabricated tilt-up walls. The walls of warehouses provide good cover, although the roofs are generally light and vulnerable. The warehouses' large open bays permit firing of ATGMs or surface-to-air missiles and, because they are normally found in outlying areas, often afford adequate fields of fire for ATGMs. These buildings are built on slabs that can normally support the weight of vehicles and can provide excellent cover and concealment for tanks (Figure I-4).

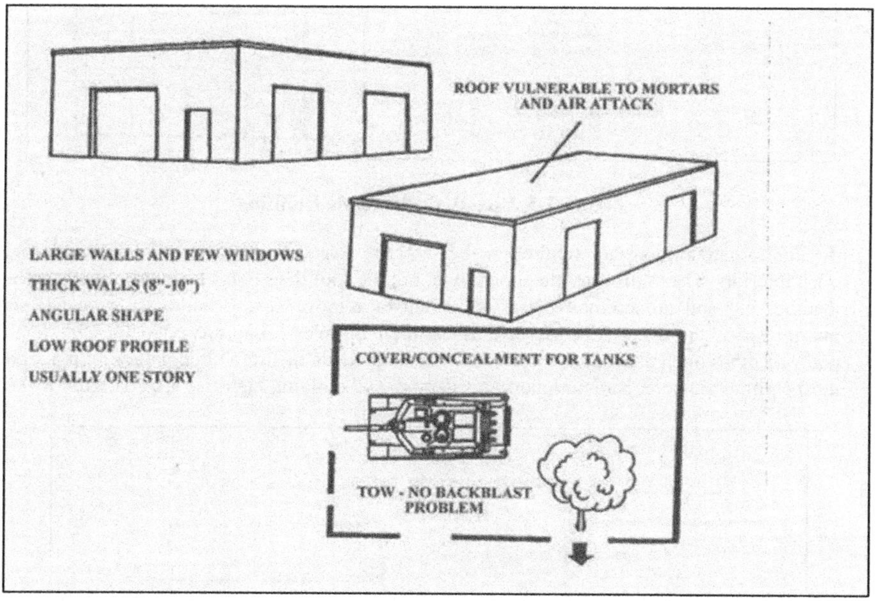

Figure I-4. Warehouse

c. Another type of mass-construction building is the box-wall-principle type. It is made from prefabricated panels of 6- to 8-inch-thick reinforced concrete. The outside wall is often glass. The box-wall-principle building provides good cover, except at the glass wall. The rooms are normally too small for ATGMs to be fired. A good circulation pattern exists from room to room and from floor to floor. These buildings are commonly used as hotels or apartments and are located mainly in residential and outlying areas (Figure I-5).

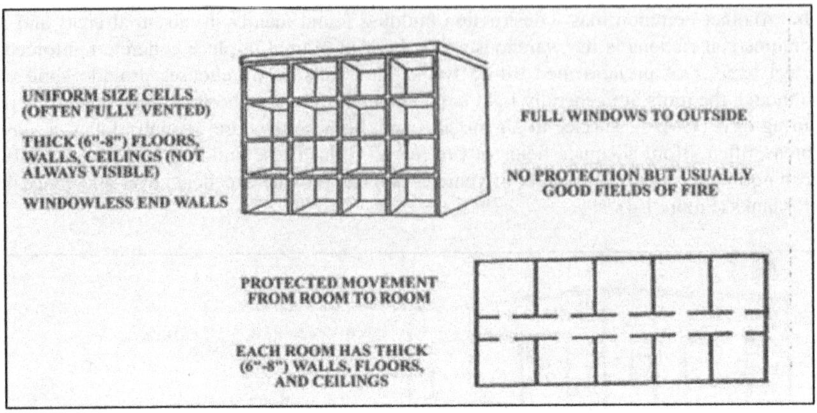

Figure I-5. Box-Wall-Principle Building

d. Public gathering places (churches, theaters) are mass-construction buildings with large, open interiors. The walls provide good cover, but the roof does not. The interior walls are not load-bearing and are normally easy to breach or remove. These buildings have adequate interior space for firing ATGMs or surface-to-air missiles. They are often located next to parks or other open areas and, therefore, have long fields of fire. Public gathering places are most common in core, core periphery, residential, and outlying high-rise areas (Figure I-6).

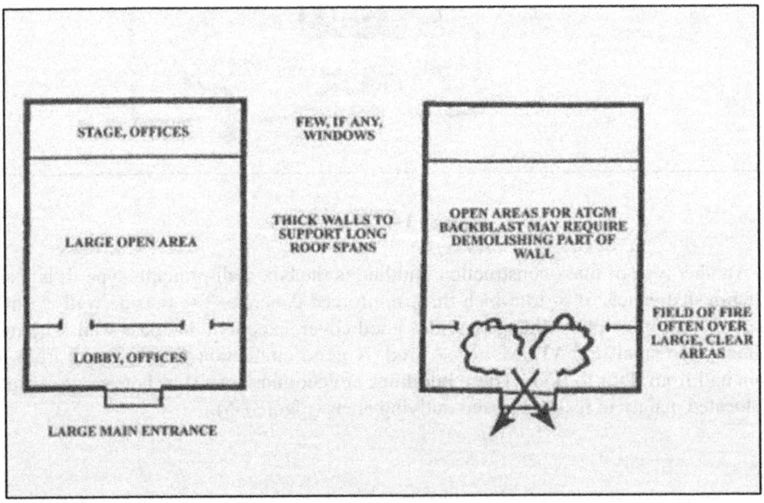

Figure I-6. Public Gathering Places

2. Types of Framed Buildings. Framed buildings are supported by a skeleton of columns and beams and are usually taller than frameless buildings (Figure I-7). The exterior walls are not load-bearing and are referred to as either heavy clad or light clad. Another type of framed building often found in cities is the garage, which generally has no cladding.

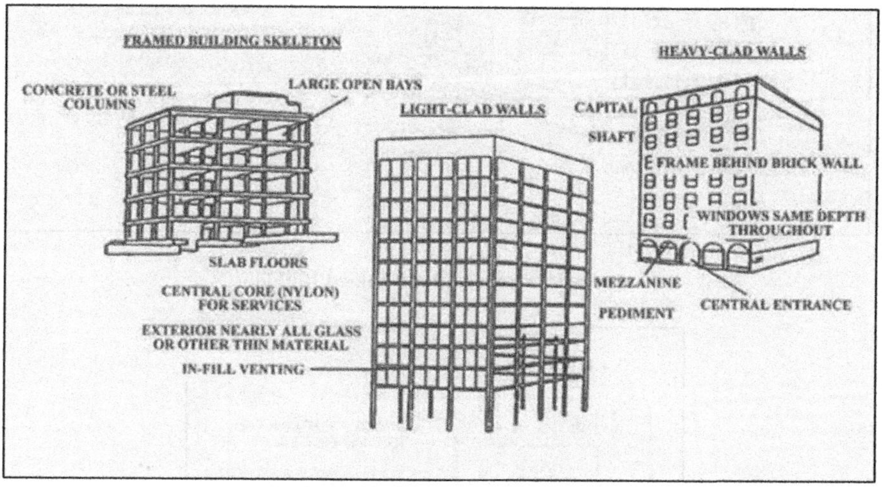

Figure I-7. Framed Buildings

a. Heavy-clad buildings were common when framed buildings were first introduced. Their walls are made of brick and block that are sometimes almost as thick as frameless brick walls, although not as protective. Heavy-clad framed buildings are found in core and core periphery areas. They can be recognized by a classic style or architecture in which each building is designed with three sections: the pediment, shaft, and capital. Unlike the brick building, the walls are the same thickness on all floors, and the windows are set at the same depth throughout. Often the frame members (the columns) can be seen, especially at the ground floor. The cladding, consisting of layers of terra cotta blocks, brick, and stone veneer, does not provide as good a cover as the walls of brick buildings. It protects against small-arms fire and light shrapnel but does not provide much cover against heavy weapons (Figure I-8).

(1) The floor plans of these buildings depend on their functions. Office buildings normally have small offices surrounding an interior hall. These offices have the same dimensions as the distance between columns (some large offices are as large as two times the distance between columns). These rooms are too small to permit firing of ATGMs but do provide some cover for snipers or machine gunners (Figures I-9 and I-10).

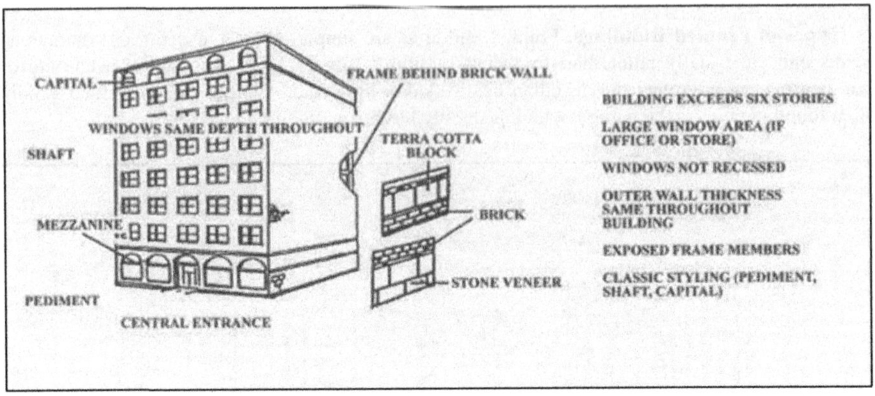

Figure I-8. Heavy-Clad Framed Building

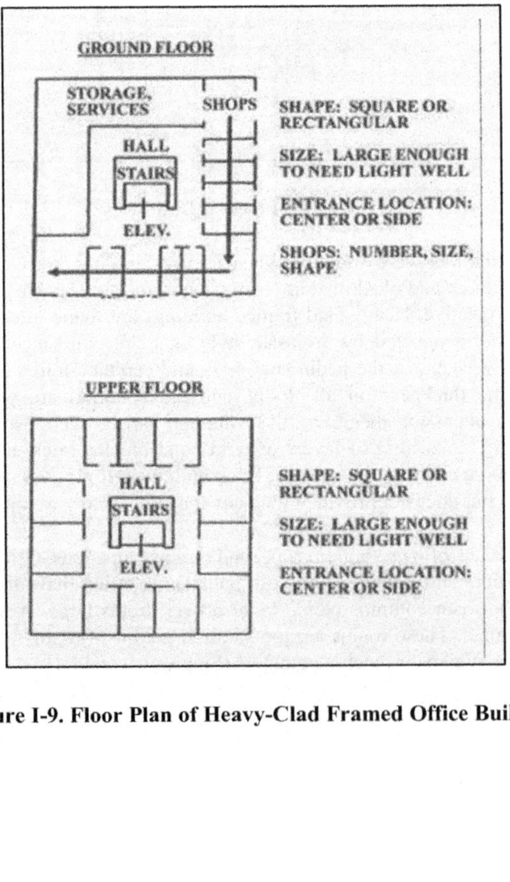

Figure I-9. Floor Plan of Heavy-Clad Framed Office Building

(2) Department stores normally have large, open interiors (Figure I-11). Such areas permit firing of ATGMs or man-portable surface-to-air missiles (if there are adequate fields of fire). Often a mezzanine level with a large backblast area permits firing down onto tanks. Steel fire doors often exist between sections of the store. The steel fire doors are activated by heat. Once closed, they are difficult to breach or force open, but they effectively divide the store into sections (Figure I-12).

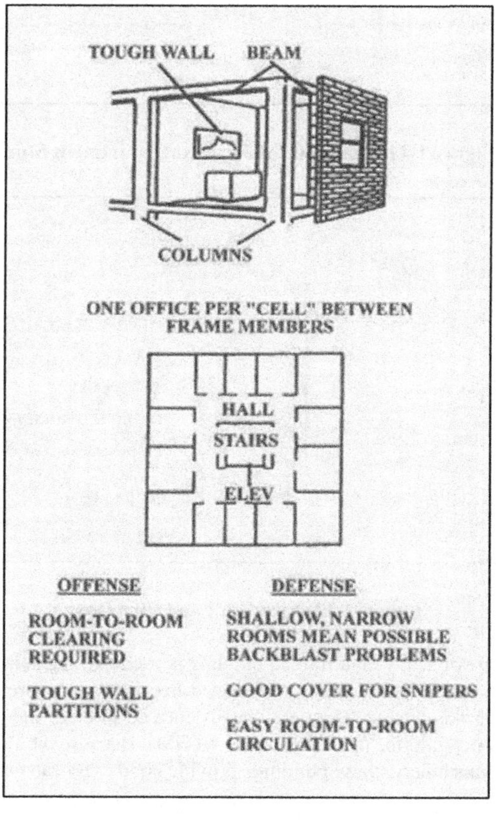

Figure I-10. Heavy-Clad Framed Office

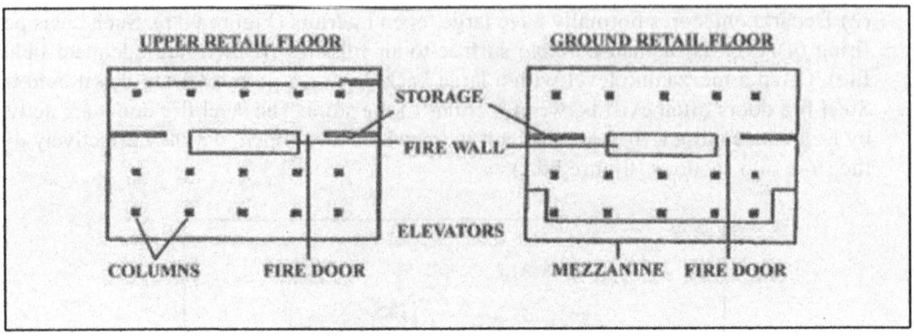

Figure I-11. Heavy-Clad Framed Department Store

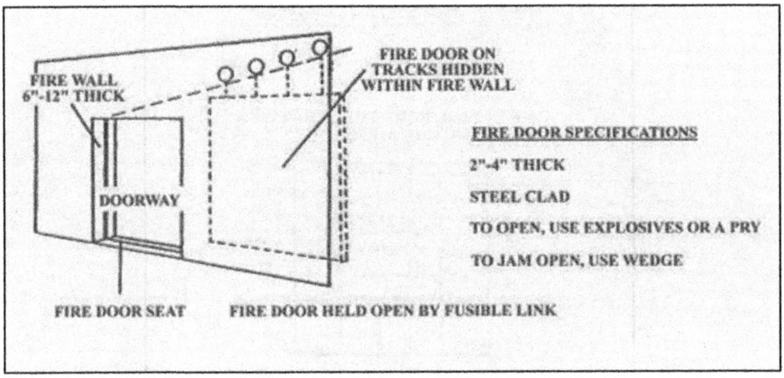

Figure I-12. Fire Wall and Fire Door

(3) Another type of heavy-clad framed building is used as a high-rise factory (Figure I-13). Such buildings are normally easily recognized because the concrete beams and columns are visible from the outside. They are usually located in older industrial areas. The large windows and open interior favor the use of ATGMs. Because the floors are often made to support heavy machinery, these buildings provide good overhead cover.

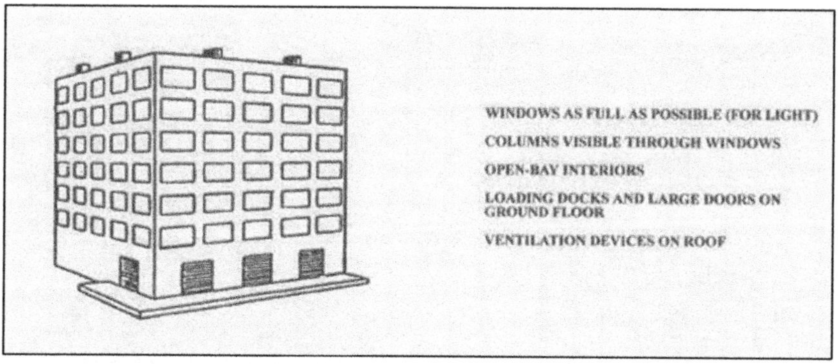

Figure I-13. High-Rise Factory

b. Light-clad buildings are more modern and may be constructed mostly of glass (Figure I-14). Most framed buildings built since World War II are light-clad buildings. They are found in both core and outlying high-rise regions. Their walls consist of a thin layer of brick, lightweight concrete, or glass. Such materials provide minimal protection against any weapon. However, the floors of the buildings are much heavier and provide moderate overhead cover. The rooms in light-clad framed buildings are much bigger than those in heavy-clad buildings. This feature, along with the fact that the buildings usually stand detached from other buildings, favors the employment of ATGMs. The interior partitions are thin, light, and easy to breach (Figure I-15).

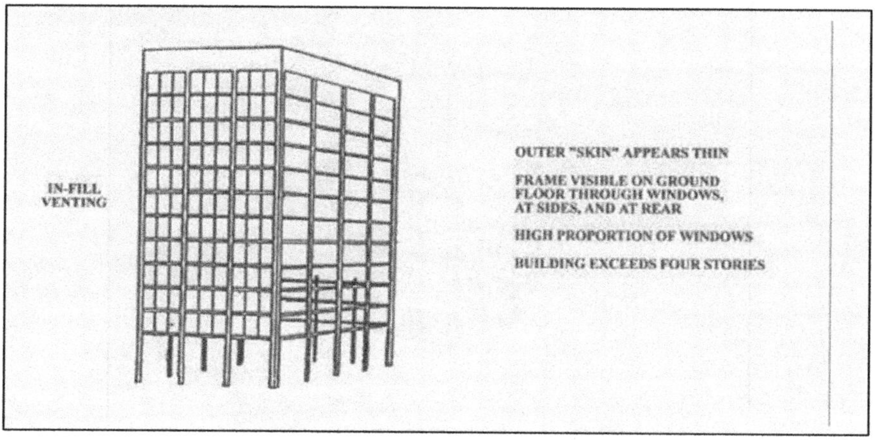

Figure I-14. Light-Clad Framed Building

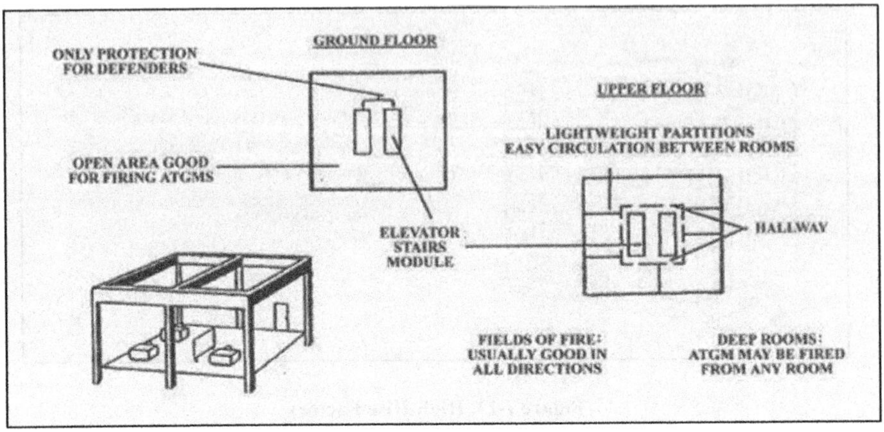

Figure I-15. Light-Clad Framed Room

c. The garage is one of the few buildings in an urban area in which all floors support vehicles. It provides a means to elevate vehicle-mounted TOWs, and the open interiors permit firing of ATGMs. Garages are normally high enough to provide a 360-degree field of fire for antiaircraft weapons. For example, a Stinger could hide under the top floor of the garage, come out to engage an aircraft, and then take cover again (Figure I-16).

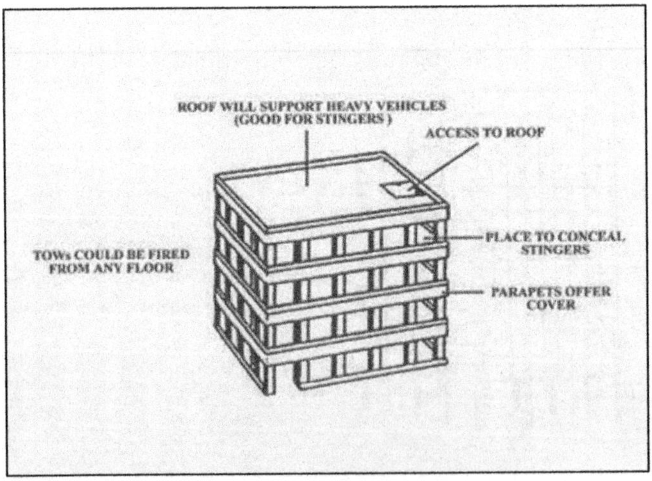

Figure I-16. Garage

3. Floor Plans. Floor plans in buildings follow predictable patterns. One of the factors that determines floor plans is building shape (Figure I-17). The basic principle governing building shape is that rooms normally have access to outside light. This principle helps Marines to analyze and determine the floor plans of large buildings.

4. Residential Areas. The two basic types of houses in the western world are located in and around cities and in rural areas. City-type houses are normally mass-construction brick buildings. Rural-type buildings in the continental United States, South America, and Southeast Asia are commonly made of wood, while in continental Europe, Southwest Asia, and sub-Saharan Africa, where wood is extremely scarce, rural buildings are normally constructed of concrete blocks. Rural-type buildings can also be found within cities (Figure I-18).

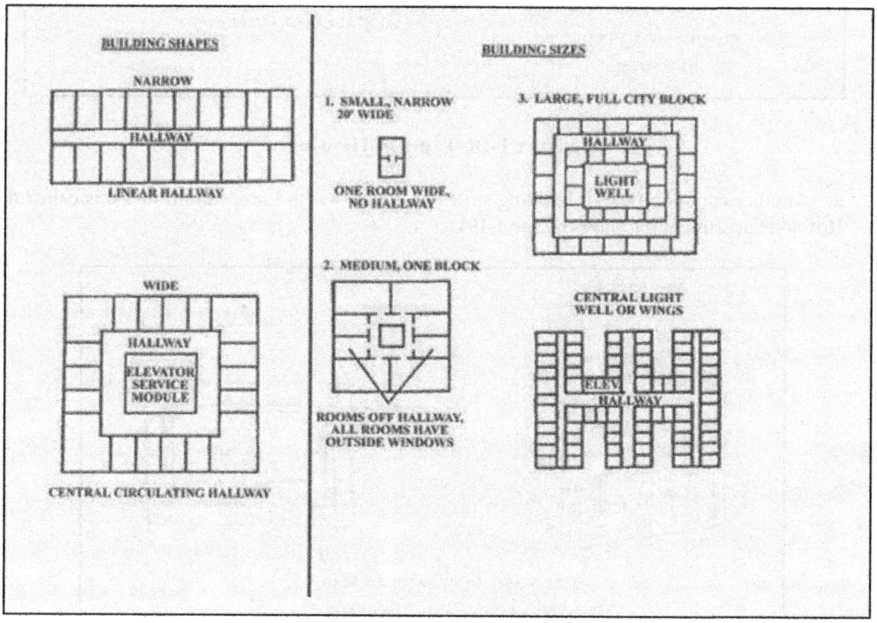

Figure I-17. Building Shapes and Sizes

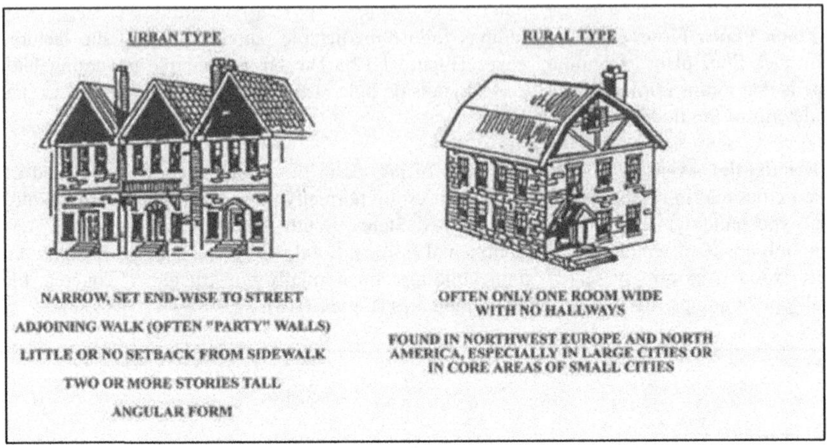

Figure I-18. Types of Housing

a. Another common type of building structure in cities with European influence is called the Hof-style apartment building (Figure I-19).

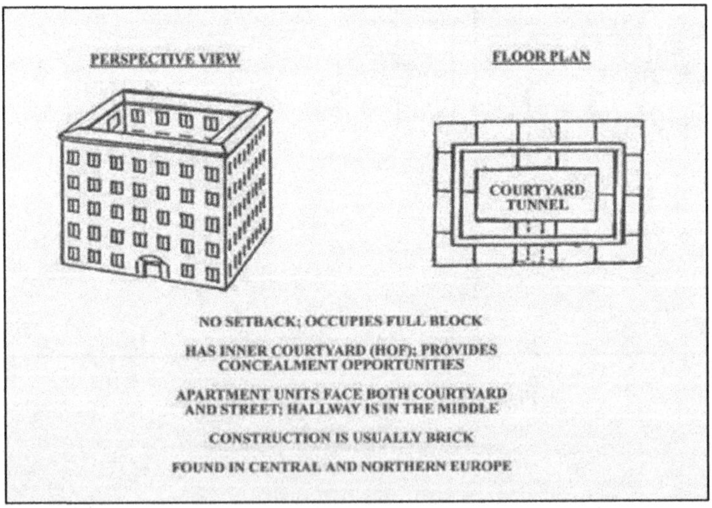

Figure I-19. Hof-Style Apartment Building

b. In the Middle East and tropical regions, the most common type of housing is the enclosed courtyard. Houses are added to one another with little regard to the street pattern. The result is a crooked, narrow maze that is harder to move through or fire in than dense European areas. (Figure I-20).

Figure I-20. Enclosed Courtyard

5. Characteristics of Buildings. Certain characteristics of both mass-construction and framed buildings can be helpful in analyzing a built-up area. Leaders can use Table I-1 to determine how to defend or attack a certain building given the unit's available weapons systems.

Type of Construction	Building Material	Height (Stories)	Average Wall Thickness (cm)
Mass	Stone	1 to 10	75
Mass	Brick	1 to 3	22
Mass	Brick	3 to 6	38
Mass	Concrete block	1 to 5	20
Mass	Concrete wall and slab	1 to 10	22 to 38
Mass	Concrete tilt-ups	1 to 3	18
Framed	Wood	1 to 5	3
Framed	Steel (heavy cladding)	3 to 50	30
Framed	Concrete/steel (light cladding)	3 to 100	2 to 8

Table I-1. Characteristics of Buildings

6. Distribution of Building Types. Certain types of buildings dominate certain parts of a city; this establishes patterns within a city. Analysis of the distribution and nature of these patterns has a direct bearing on military planning and weapon selection (Figure I-21).

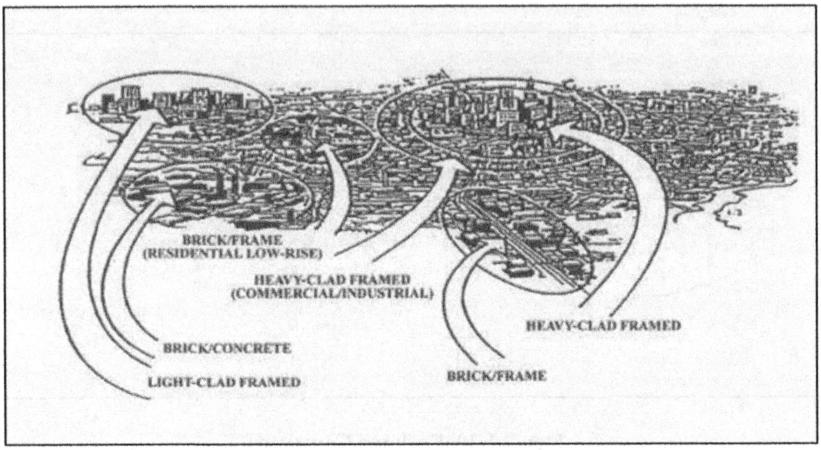

Figure I-21. Distribution of Building Types

a. Mass-construction buildings are the most common structures in built-up areas, forming about two-thirds of all building types. Brick structures account for nearly 60 percent of all buildings, especially in Europe.

b. Steel- and concrete-framed multistory buildings have an importance far beyond their one-third contribution to total ground floor area. They occupy core areas—a city's most valuable land—and as centers of economic and political power, they have a high potential military significance.

c. Open space accounts for about 15 percent of an average city's area. Many open spaces are grass-covered and are used for parks, athletic fields, and golf courses; some are broad, paved areas. The largest open spaces are associated with suburban housing developments where large tracts of land are recreation areas.

d. Streets serving areas consisting of mostly one type of building normally have a common pattern. In downtown areas, for example, high land values result in narrow streets. Street widths are grouped into three major classes: 7 to 15 meters, located in medieval sections of European cities; 15 to 25 meters, located in newer, planned sections of most cities; and 25 to 50 meters, located along broad boulevards or set far apart on large parcels of land. When a street is narrow, observing or firing into windows of a building across the street can be difficult because an observer is forced to look along the building rather than into windows.

When the street is wider, the observer has a better chance to look and fire into the window openings (Figure I-22).

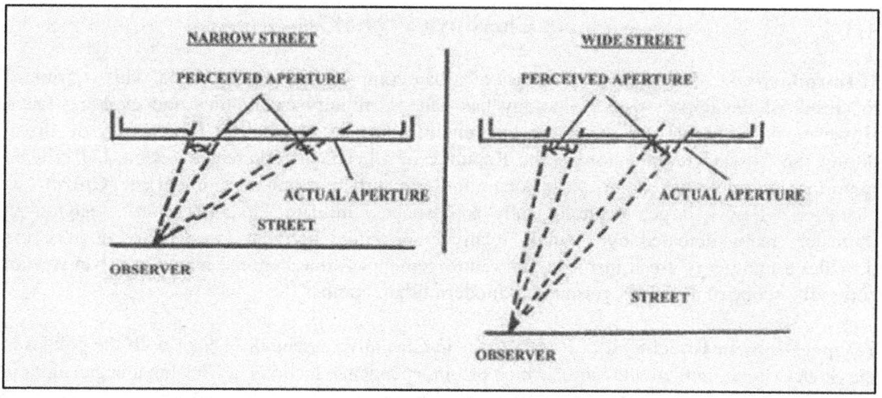

Figure I-22. LOS Distances and Angles of Obliquity

Appendix J

Lessons Learned from Russian Military Operations in Chechnya 1994-1996

1. Introduction. Many historical studies of urban combat have become dated. This is primarily the result of the impact that technology has had on military capabilities and evolving tactics. However, certain tenets of urban combat remain constant. The battles for the city of Grozny during the Russian intervention in the Republic of Chechnya represent a recent and critically important example of large scale operations in urban combat. Combat in Grozny was characterized by a large, technologically sophisticated military force (Russian) engaging and ultimately being defeated by a small, relatively primitive irregular force (Chechen). Grozny provides a number of fresh insights, and reinforcement of time honored tenets of urban warfare, across the scope of activities germane to modern urban combat.

2. Operations in Chechnya. Russia's war in Chechnya essentially began with the collapse of the Soviet Union, immediately after which pro-independence factions in Chechnya began agitating to free the Republic from Russian domination. This eventually led to civil war with pro-Russian factions within the Republic. Russia committed troops to Chechnya on 11 December 1994, for what many in the Russian government and military thought would be a quick campaign to restore Russian primacy. What followed was a bitter, costly, and protracted struggle that ended with the complete withdrawal of Russian military forces from Chechnya.

A primary focus of Russian operations in Chechnya was the capture of the capital city, Grozny. The initial attack on Grozny was conducted by a mechanized force, consisting of 6,000 troops mounted in tanks, BMP infantry fighting vehicles, and BTR armored personnel carriers. The Russians anticipated light opposition, but instead met a determined and heavy resistance from Chechens armed with a large number of antitank weapons. The attack was repulsed with significant Russian losses, both in personnel and armored vehicles (105 of 120 vehicles were lost). A second mechanized attack conducted on New Years Eve, 1994, was also repulsed with the loss of 140 of 200 tanks employed. The Russians captured Grozny some two months later, primarily through the use of excessive, overwhelming firepower, and at great cost to themselves and the local Chechen population.

The capture of Grozny marked the end of the first phase of Russian combat operations in Chechnya. Beginning in March 1995, the Russians became heavily engaged in anti-partisan operations as they tried to gain control of the country. Though the Russians controlled several major Chechen cities and a portion of the countryside by May 1995, they were never able to fully isolate the Republic. Chechen separatist fighters continued to receive a steady flow of arms and supplies from neighboring countries. Consequently, the Chechen fighters were able to maintain the initiative throughout this period, engaging Russian forces whenever and wherever they chose.

Even late in the campaign, Russian intelligence failed to detect Chechen fighters infiltrating Grozny, which enabled the Chechens to launch a major attack against the Russian-controlled city in early August 1996. The Chechens succeeded in capturing the city less than two weeks later.

.3

The recapture of Grozny was a significant loss for the Russians, precipitating a general cease-fire and the withdrawal of Russian forces from the Republic.

Besides the military aspects of combat in Chechnya, other issues such as ethnic, cultural, and religious divides impacted upon the nature of the conflict. The Russians failed to culturally orient their troops, which resulted in many serious cultural mistakes when dealing with the Chechen civilians. Once insulted or mistreated, the Chechen civilians became active participants in the struggle against Russia. The war eventually degenerated into one of abject ferocity and brutality on both sides. The Chechens are reported to have executed Russian prisoners, decapitated bodies, and boobytrapped the dead. Russian wounded and bodies were hung upside down outside of Chechen defensive positions in buildings, causing the Russians to have to shoot through their own people to engage the Chechens. The Russians, initially hesitant to cause civilian casualties, increasingly engaged Chechen units with a growing disregard for the safety of the local populace. As Russian discipline degraded, the incidences of Russian troops engaging in looting, arson, torture, even the summary execution of civilians, rose significantly. The result was the deaths of tens of thousands of civilians.

3. Lessons Learned. The intervention in Chechnya and the battles for Grozny generated many important lessons across the warfighting functions of combat in urbanized terrain. The following information is primarily from the Russian point of view, though some Chechen capabilities are included. The lessons learned are arrayed as warfighting functional considerations.

a. Command and Control.

(1) Russian military commanders lacked clear policy guidance. The Russian military's ability to attain and maintain steady, logical operational progress was hampered, if not negated, by a lack of clear governmental policy. The Russian government was split at the highest levels over the wisdom of committing forces to Chechnya, and the conduct of the intervention.

(2) Poor Russian command and control structure throughout all levels. There was no direct, unified chain of command for the operations in Chechnya. Operational directives often flowed from a number of governmental agencies directly to local unit commanders, often bypassing intermediate military commanders. This resulted in a lack of overall situational awareness, and exceptionally poor coordination among tactical forces.

(3) Russians were unable to contain/prohibit Chechen operations against Russia. When the Russian forces finally began to achieve progress in late May 1995, the Chechens initiated attacks against targets inside Russia. The result was a major propaganda victory for the Chechens.

(4) Excessive senior officer (flag level) interference contributed to operational confusion. There were eight major changes at the senior command level during the course of the intervention. Additionally, there were approximately one hundred general officers on the operational scene, about one general officer per 3,000 troops, which resulted in confusion as they got in each others way.

(5) Poor small unit leadership, particularly at the NCO level, was a primary cause of Russian tactical failures in Grozny. While the Russians recognize that urban combat is small unit intensive, they do not possess a professional NCO corps. Additionally, Russian forces lacked some 12,000 platoon leaders when the intervention began. This had a crippling effect on small unit operations.

(6) Russian inability to isolate Chechnya/Grozny allowed a constant flow of reinforcements and supplies into the country/city. Russian control of any given area was at best, porous. The Chechens continued to receive outside support throughout the campaign, as well as, a large supply of arms and equipment from poorly disciplined Russian troops inside Chechnya.

(7) Both sides used commercial off-the-shelf communication assets. The information war was as vitally important as any other aspect of the campaign. The Chechens had to make do with commercially available communication equipment. They fully exploited the use of cellular telephones, Motorola radios, improvised television stations, video cameras, and the internet. The Russians lacked sufficient military communication equipment, and had to supplement with commercial assets.

(8) Russian tactical communication was very difficult in Grozny. Urban structures in Grozny interfered with Russian military radios, severely hindering tactical communications. The Russians had to establish ground-based and aircraft-based relay stations to overcome the interference.

(9) Boundaries between units were tactical weak points. Units had to think of boundaries in both a horizontal and vertical context. There were many instances when the Chechens held the third floor and above, while the Russians held the first two floors and maybe the roof. This caused a lack of coordination, and mutual support. Additionally, due to poor command and control and coordination between units, boundaries became seams through which the Chechens could pass unopposed.

b. Intelligence.

(1) Lack of high quality intelligence made operations much more difficult and dangerous. Russian intelligence collection and dissemination capabilities were poor throughout the intervention. In March 1996, the Russian Minister of the Interior complained that poor reconnaissance and intelligence allowed Chechen fighters to move in and out of Grozny without opposition, which contributed to the heavy Russian losses.

(2) Directives to minimize civilian casualties ultimately put the Russians at a disadvantage. An advantage accrues to the side with less concern for the safety of the civilian population. Initially, the Russians attempted to limit civilian casualties. The Chechen

.3

fighters took advantage of this reticence, often using civilians, schools, and apartment buildings as a shield. While this may seem to be cold-blooded of the Chechen fighters, what must be understood is that the majority of the populace in Grozny were actually ethnic Russians, and that the ethnic Chechens had no remorse in using them in this manner.

(3) Russian concerns for the safety of the Chechen populace and property declined. As the Russian troops became increasingly frustrated in their attempts to separate Chechen fighters from the local populace, and their casualties began to mount, concerns for civilian safety waned, resulting in greater civilian casualties and further alienating the Chechen people.

(4) Rigorous communications security is essential, even against less sophisticated enemies.Much of the Russian tactical radio traffic was broadcast in the clear, which allowed the Chechens to intercept it. The Chechens were able to redirect Russian airstrikes onto Russian targets, and engage Russian artillery forward observers who broadcast their own position coordinates.

(5) The Russians made extensive use of RPVs for reconnaissance. The Chechen campaign saw the first Russian use of unmanned aerial vehicles for reconnaissance and intelligence gathering. There were primarily used by Russian airborne forces.

c. Maneuver.

(1) Inadequate training in the most basic maneuver and combat skills inhibited Russian operations. Overall, poor Russian combat performance could be traced to a lack of training in fundamental military skills, a situation that was then exacerbated by an additional lack of training in specific urban combat skills. Individual skills training was so inadequate that some units were entirely incapable of undertaking combat operations. The Russian's ability to conduct large scale operations had been seriously eroded by a complete cessation of exercises at the division level in the 2 years prior to the Chechen campaign. Even regimental and battalion level exercises had been curtailed by 75 percent or more. Additionally, The air forces were not any better trained or prepared for their operational role in Chechnya.

(2) Lack of cohesiveness among the Russian forces led to confusion and antagonism. The Russian ground forces consisted primarily of regular army units and Ministry of Interior Internal Troops, which were essentially a police force. With no real unified command and control, the two groups were usually conducting uncoordinated missions, which led to confusion and fratricide.

(3) The nature of cities channels combat operations along narrow lanes of activity. Combat conditions in Grozny were characterized by narrow fields of view, limited fields of fire, and constricted avenues of approach. For the Russians, this meant movement along approaches well known and heavily defended by the Chechens.

(4) Night fighting was the most difficult activity for Russian infantry. The Russian infantry, at all levels, was inadequately trained for night operations, and lacked night vision equipment.

(5) Armored vehicles cannot operate in cities without extensive dismounted infantry support. The Chechens formed antitank hunter-killer teams equipped with RPG-7 and RPG-18 antitank missiles. They engaged Russian tanks with volley fire from above, behind, and the sides. Armor columns not supported by dismounted infantry suffered great losses.

(6) Armored engineer vehicles are indispensable for removing obstacles. The Russians found that armored engineer vehicles were critical to restoring or improving mobility in Grozny. Eventually, the Russians formed special assault groups that included armored engineer vehicles in each group.

(7) The Chechens preferred small unit, hit-and-run ambush tactics. The Chechens relied on small units (15 to 20 troops) to conduct quick, sharp ambushes, from which they would disengage before the Russians could bring significant firepower to bear. Ambushes were often conducted in three tiers, with Chechens in the underground, on the ground floor, and on the roof, with each element having a different task in the ambush. They also used ambush tactics against Russian helicopters.

(8) Urban combat is extremely manpower intensive, and results in significant attrition of personnel and materiel for the attacker. The Russians discovered that a 5 to 1 manpower ratio was often not enough, due to the high casualty rates consistent with urban combat and the requirement to guard virtually every building taken.

(10) Tracked vehicles are preferable to wheeled vehicles in the urban environment. Tracked vehicles provided better mobility than wheeled vehicles due to their significantly greater ability to negotiate obstacles and rubble.

(11) Armored vehicles required additional protection (i.e., reactive armor, sandbags, wire mesh, etc.) to negate attacks from the sides, rear, and top. Russian tank design emphasized greater armor protection on the frontal arc, and less on other areas of the vehicle. This proved inadequate for urban combat, where attacks can come from any direction. The majority of lethal hits came from above, easily penetrating turrets and engine decks, and from the rear.

(12) The spatial qualities and perspectives of urban combat are more "vertical" than non-urban combat. The Chechens regularly fought from basements, rooftops, and upper-floor rooms, making them difficult to locate and engage. Russians soldiers trained in open areas, had to reorient to a confined, highly vertical environment.

(13) Standard Russian military unit configurations were inappropriate for urban combat. Typical Russian army unit organizations, oriented to open spaces and mechanized operations, proved to be inadequate for urban combat. They eventually had to adopt specially configured assault teams that included, among other assets, significantly greater firepower and engineer assets. Essentially, they moved towards a better combined arms team.

(14) The Russians found their helicopters to be far too vulnerable to rooftop snipers and ambushes in the urban setting. While Russian doctrine had long specified the capture of a building from the top-down, it became apparent in Grozny that this tactic was too expensive in terms of helicopter losses. The Chechens made effective use of snipers and RPG-equipped ambush teams to damage or destroy helicopters supporting Russian ground forces.

d. Fires.

(1) Artillery was invaluable in the direct fire mode. The Russians found artillery, to include rocket launchers and mortars, to be very effective for direct fire against point targets , usually at ranges of 150 to 200 meters.

(2) Obscurants (smoke, white phosphorous, tear gas) are especially useful in urban combat. The Russians used significant amounts of smoke and white phosphorus to mask the movement of forces. Every fourth or fifth artillery or mortar round was either smoke or WP. WP also had a toxic effect, easily penetrating Chechen protective masks.

(3) The Chechens effectively used RPGs against helicopters. The Chechens made effect use of ambush teams equipped with RPGs to engage Russian helicopters, either by damaging or destroying the helicopter or simply prohibiting helicopterborne operations.

(4) Air defense guns (i.e., ZSU-23-4) were valuable in suppressing ground targets. The Russians found air defense guns effective against multi-story buildings because they had sufficient elevation to hit targets in the upper stories. Air defense guns became a primary asset in the Russian special assault groups.

(5) The Russians determined that overwhelming firepower can make-up for organizational and tactical deficiencies in the short run if collateral damage is disregarded. Whenever Russian military units proved incapable of effective urban operations, the Russians simply fell back on massive, overwhelming firepower to reduce a strong point. It was through the use of excessive, unrestrained firepower, and a complete lack of regard for collateral damage, that finally enabled the Russians to gain control of Grozny.

(6) The Chechens negated Russian supporting fires by "hugging" Russian units. Whenever possible, Chechen fighters adopted the tactic of staying as physically close to Russian units as possible, so that the Russians could not employ supporting arms without the risk of injuring their own troops.

(7) The Russians found "bunker busting" weapons to be invaluable. The Russians came to rely heavily on "bunker busting" weapons for engaging Chechens, primarily snipers, entrenched in buildings.

(8) Russian helicopters needed stand-off weapons, that would allow them to engage targets while outside the range of Chechen weapons. The Chechens employed 23mm cannons and heavy machine guns against Russian helicopters, causing the Russian's helicopter gunships to fire from ranges of 3,000 meters or more. This required the use of wire-guided munitions and large caliber (240mm) rockets.

(9) The Russians made extensive use of precision-guided weapons. When the weather allowed, the Russians were able to use precision guided weapons, such as laser-guided bombs and missiles. They had great effect against priority targets such as bridges, major road intersections, and buildings.

e. Logistics.

(1) Deficiencies in the Russian logistics system degraded combat effectiveness. Significant inadequacies in the Russian logistics apparatus effected virtually all combat forces and operations. Some soldiers actually entered combat in Grozny without weapons or ammunition. The Russian transportation system failed. Supply officers were unprepared for the high volume demands for ammunition, hand grenades, smoke grenades, demolition charges, and one-shot antitank weapons.

(2) Russian logistics personnel were ill-prepared in basic soldier skills. Russian infantry often had to be used to conduct logistics operations into Grozny, because the logistics personnel lack the basic skills to defend themselves against the Chechens.

(3) The Russian ground forces required special equipment not normally found in unit T/Es. The T/E for the typical Russian infantry unit did not include such items as ropes, grappling hooks, and ladders, all of which proved essential for urban combat.

(4) Armored vehicle recovery is extremely difficult and dangerous. The Russians found the recovery of vehicles to be extremely difficult due to the terrain (i.e., narrow streets, rubble and debris), and Chechen opposition.

(5) Some Russian equipment was modified in the field to counter Chechen tactics and equipment. The Russians had to adopt add-on armor for their vehicles. They fabricated metal screens and wire mesh guards, which helped to defeat RPG and Molotov cocktail attacks.

(6) Medical support was of great importance. Russian casualty evacuation was difficult due to the lack of vehicular mobility, and the vulnerability of helicopters to Chechen fire. Many casualties were psychological, due in great part to the intense, violent nature of

.3

close quarter combat. The Russians were ill-prepared for the level of mental health problems they encountered. Additionally, logistics units were often unable to provide fresh drinking water, which caused Russian troops to consume contaminated water that resulted in other health problems.

f. Force Protection.

(1) Operations took much longer than expected, and the cost in terms of casualties was much greater than anticipated. The Russians committed unprepared, poorly trained and poorly equipped forces to Chechnya, based on an unrealistic operational time frame. Ultimately, this cost the Russians significant casualties and the war.

(2) Fratricide was a serious and continuing problem throughout the campaign. The situation facing Russian combat forces was ripe for fratricide. Poorly trained units, operating in a confused and uncertain urban environment, often unable to tell friend from foe, and lacking quality leadership and inter-unit coordination, were often as dangerous to themselves as they were to the Chechens .

(3) Chechen use of adhoc air defense systems eliminated Russian air support, particularly helicopters, in Grozny. The Chechens were able to improvise effective anti-air defenses using a mixture of anti-air and ground (i.e., heavy machine guns, RPGs) weapons, and ambush tactics specifically adapted to the urban environment.

(4) The Chechens made extensive and effective use of snipers. Snipers fired from well inside rooms versus near window openings, as well as, from rooftops and basements. The Russians lacked an effective sniper and counter-sniper capability of their own.

(5) The Chechens improvised crude chemical weapons. The Chechens, lacking access to military chemical weapons, Used off-the-shelf materials and equipment to make chemical mines (i.e., chlorine gas), detonated remotely by radio.

(6) The Chechens made extensive and effective use of boobytraps. It seemed that the Chechens mined and boobytrapped 7-8everything. They had a good understanding of the average Russian soldier's actions and reactions, and boobytrapped accordingly. Additionally, the Russians found it difficult to maintain boobytrap awareness among their troops.

Appendix K

Glossary

I. Acronyms

AAA .... antiaircraft artillery
AAAV .... advanced assault amphibious vehicle
AAV .... assault amphibious vehicle
ACE .... aviation combat element
AO .... area of operations
APC .... armored personnel carrier
APDS-T .... armor-piercing, discarding sabot with tracer
APFSDS .... armor-piercing, fin-stabilized, discarding sabot
API .... armor-piercing incendiary
ARVN .... Army of the Republic of Vietnam
ATF .... amphibious task force
ATGM .... antitank guided missile
BDA .... battle damage assessment
BMNT . . . . .beginning of morning nautical twilight
CA .... civil affairs
CAG .... civil affairs group
CE .... command element
CI .... counterintelligence
COC .... combat operations center
CONAD .... consolidated administration center
CP .... command post
CSS .... combat service support
CSSE .... combat service support element
DOD .... Department of Defense
DS .... direct support
EOD .... explosive ordnance disposal
EPW .... enemy prisoner of war
EW .... electronic warfare
FAC(A) .... forward air controller (airborne)
FASCAM .... family of scatterable mines
FDC .... fire direction center
FEBA .... forward edge of the battle area
FLOT .... forward line of own troops
FM .... U.S. Army field manual
FM .... Frequency modulated
FMFM .... Fleet Marine Force manual
FMFRP .... Fleet Marine Force reference publication
FO .... forward observer
FPL .... final protective line
FRAGO .... fragmentary order
FSC .... fire support coordinator
GCE .... ground combat element
GPS .... global positioning system
GS .... general support
HE .... high explosive
HEAT .... high-explosive antitank
HEAT-MP .... HEAT-multipurpose
HEDP .... high-explosive dual-purpose
HEI-T .... high-explosive incendiary with tracer
HLZ .... helicopter landing zone
HMMWV .... high-mobility, multipurpose wheeled vehicle
HSS .... health services support
HUMINT .... human intelligence
IDF .... Israeli Defense Forces
IFV .... infantry fighting vehicle
IPB . . intelligence preparation of the battlefield/battlespace
ITOW .... improved TOW
JTF .... joint task force
KOCOA .... key terrain, observation and fields of fire, cover and concealment, obstacles, and avenues of approach
LAV .... light armored vehicle
LCAC .... landing craft air cushion
LD .... line of departure
LOS .... line of sight
LZ .... landing zone
MAGTF .... Marine air-ground task force
MCDP .... Marine Corps doctrinal publication
MCRP .... Marine Corps reference publication
MCWP .... Marine Corps warfighting publication
MEDEVAC .... medical evacuation
MEF .... Marine expeditionary force
MET .... meteorological
METT-T ... mission, enemy, terrain and weather, troops and support available-time available
MEU .... Marine expeditionary unit
MOOTW .... military operations other than war
MOPP .... mission-oriented protective posture
MOUT .... military operations on urbanized terrain
MP .... military police
MPF .... maritime prepositioning force
MULE .... modular universal laser equipment
NBC .... nuclear, biological, and chemical
NCA .... National Command Authorities
NCF .... naval construction force
NEF .... naval expeditionary force
NEO .... noncombatant evacuation operation
NGF .... naval gunfire
NSFS .... naval surface fire support
NVA .... North Vietnamese Army
NWP .... naval warfare publication

Appendix J

Glossary

I. Acronyms

OP .... observation post
OPLAN .... operation plan

OPORD
operation order
OPSEC
operations security
PDDA
power-driven decontamination apparatus
PGM
precision-guided munition
PLO
Palestine Liberation Organization
PLRS
position location reporting system
POL
petroleum, oils, and lubricants
PSYOP
psychological operations
ROE
rules of engagement
ROK
Republic of Korea
RP
red phosphorus
RSOP . . . . reconnaissance, selection, and occupation of position
SASR
special application sniper rifle
SMAW
shoulder-launched multipurpose assault weapon
SOP
standing operating procedure
STANAG
NATO standardization agreement
STB
supertropical bleach
TOW
tube-launched, optically tracked, wire-guided missile
UAV
unmanned aerial vehicle
UCMJ
Uniform Code of Military Justice
U.N.
United Nations
VC
Viet Cong
VHF
very high frequency
WP . . . . white phosphorus

II. Definitions

A

assault—1. The climax of an attack, closing with the enemy in hand-to-hand fighting. 2. To make a short, violent, but well-ordered attack against a local objective, such as a gun emplacement, a fort, or a machine gun nest. (Joint Pub 1-02)

assault—The culmination of an attack which closes with the enemy. (MCRP 5-2A)

attack—An offensive action characterized by movement supported by fire with the objective of defeating or destroying the enemy. (FMFRP 0-14)

B

battlespace—All aspects of air, surface, subsurface, land, space, and electromagnetic spectrum which encompass the area of influence and area of interest. (FMFRP 0-14)

block—1. A tactical task assigned to a unit that requires it to deny the enemy access to a given area or to prevent enemy advance in a given direction or an avenue of approach. It may be for a specified time. Units assigned this mission may have to retain terrain and accept decisive engagement. 2. An obstacle effect that integrates fire planning and obstacle effort to stop an attacker on a specific avenue of approach or to prevent an enemy from exiting an engagement area. (MCRP 5-2A)

breach—A tactical task where any means available are employed to break through or secure a passage through an enemy defense, obstacle, minefield, or fortification. (MCRP 5-2A)

built-up area—A concentration of structures, facilities, and populations, such as villages, cities, and towns, that forms the economic and cultural focus for the surrounding area. (MCRP 5-2A, modified)

bypass—A tactical task that involves maneuvering around an obstacle, position, or enemy force to maintain the momentum of advance. Bypassed obstacles and enemy forces are reported to higher headquarters. (MCRP 5-2A)

C

clear—A tactical task to remove all enemy forces and eliminate organized resistance in an assigned zone, area, or location by destroying, capturing, or forcing the withdrawal of enemy forces such that they cannot interfere with the friendly unit's ability to accomplish its mission. (MCRP 5-2A)

combat arms—Units who close with and destroy enemy forces or provide firepower and destructive capabilities on the battlefield. (MCRP 5-2A, modified)

combat operations center (COC)—The primary operational agency required to control the tactical operations of a command that employs ground and aviation combat, combat support, and combat service support elements or portions thereof. The combat operations center continually monitors, records, and supervises operations in the name of the commander and includes the necessary personnel and communications to do the same. (FMFRP 0-14)

combat service support (CSS)—The essential capabilities, functions, activities, and tasks necessary to sustain all elements of operating forces in theater at all levels of war. Within the national and theater logistic systems, it includes but is not limited to that support rendered by service forces in ensuring the aspects of supply, maintenance, transportation, health services, and other services required by aviation and ground combat troops to permit those units to accomplish their missions in combat. Combat service support encompasses those activities at all levels of war that produce sustainment to all operating forces on the battlefield. (Joint Pub 1-02)

combat support (CS)—Fire support and operational assistance provided to combat elements. (Joint Pub 1-02)

combat support (CS)—Units that provide critical combat functions in conjunction with combat arms units to secure victory (MCRP 5-2A, modified).

command post (CP)—A unit's or subunit's headquarters where the commander and the staff perform their activities. In combat, a unit's or subunit's headquarters is often divided into echelons; the echelon in which the unit or subunit commander is located or from which such commander operates is called a command post. (Joint Pub 1-02)

contain—A tactical task to restrict enemy movement. (MCRP 5-2A)

cover—1. Shelter or protection from enemy observation that reduces the effects of enemy direct and indirect fire. 2. A type of security operation that protects the force from surprise, develops the situation, and gives commanders time and space in which to respond to the enemy's actions. (MCRP 5-2A)

D

defend—A combat operation designed to defeat an attacker and prevent him from achieving his objectives. It employs all means and methods available to prevent, resist, or destroy an enemy attack. Forms of defensive operations are area and mobile. Choices of defensive operations are in-depth and forward. The defensive techniques are defend in sector, defend a battle position, and defend a strong point. (MCRP 5-2A)

destroy—1. A tactical task to physically render an enemy force combat-ineffective unless it is reconstituted. 2. To render a target so damaged that it cannot function as intended nor be restored to a usable condition without being entirely rebuilt. (MCRP 5-2A)

disrupt—A tactical task or obstacle effect that integrates fire planning and obstacle effort to break apart an enemy's formation and tempo, interrupt the enemy's timetable, or cause premature commitment of enemy forces, or the piecemealing of his attack. (MCRP 5-2A)

II. Definitions

F

fire support—Assistance to elements of the Marine air-ground task force engaged with the enemy rendered by other firing units, including (but not limited to) artillery, mortars, naval surface fire support, and offensive air support. (FMFRP 0-14)

fix—1. A tactical task in which actions are taken to prevent the enemy from moving any part of his forces either from a specific location or for a specific period of time by holding or surrounding them to prevent their withdrawal for use elsewhere. 2. A tactical obstacle effect that integrates fire planning and obstacle effort to slow an attacker within a specified area—normally an engagement area. (MCRP 5-2A)

H

helicopterborne operation—A military action in which combat forces and their equipment maneuver about the battlefield by helicopters or vertical-landed aircraft. Aviation activities are under the control of the aviation combat element commander who is assigned in direct or general support of one or more combat element(s). (FMFRP 0-14)

hold—1. To maintain or retain possession of by force, as a position or an area. 2. In an attack, to exert sufficient pressure to prevent movement or redisposition of enemy forces. (Joint Pub 1-02)

I

infiltration—The movement through or into an area or territory occupied by either friendly or enemy troops or organizations. The movement is made, either by small groups or by individuals, at extended or irregular intervals. When used in connection with the enemy, it infers that contact is avoided. (Joint Pub 1-02)

isolate—A tactical task given to a unit to seal off (both physically and psychologically) an enemy from his sources of support, to deny an enemy freedom of movement, and prevent an enemy unit from having contact with other enemy forces. An enemy must not be allowed sanctuary within his present position. (MCRP 5-2A)

L

linkup—An operation wherein two friendly ground forces join together in a hostile area. (FMFRP 0-14)

M

Marine air-ground task force (MAGTF)—A task organization of Marine forces (division, aircraft wing, and service support groups) under a single command and structured to accomplish a specific mission. (Joint Pub 1-02)

military operations on urbanized terrain (MOUT)—All military actions that are planned and conducted on a topographical complex and its adjacent natural terrain where manmade construction is the dominant feature. It includes combat in cities, which is that portion of MOUT involving house-to-house and street-by-street fighting in towns and cities (MCRP 5-2A, modified)

military operations other than war (MOOTW)—Operations that encompass the use of military capabilities across the range of military operations short of war. These military actions can be applied to complement any combination of the other instruments of national power and occur before, during, and after war. (Joint Pub 1-02)

N

neutralize—1. To render enemy personnel or material incapable of interfering with a particular operation. 2. To render safe mines, bombs, missiles, and boobytraps. (MCRP 5-2A)

O

occupy—A tactical task in which a force moves onto an objective, key terrain, or other man-made or natural terrain area without opposition, and controls that entire area. (MCRP 5-2A)

P

peace enforcement—Application of military force, or the threat of its use, normally pursuant to international authorization, to compel compliance with resolutions or sanctions designed to maintain or restore peace and order. (Joint Pub 1-02)

peacekeeping—Military operations undertaken with the consent of all major parties to a dispute, designed to monitor and facilitate implementation of an agreement (ceasefire, truce, or other such agreement) and support diplomatic efforts to reach a long-term political settlement. (Joint Pub 1-02)

protect—A tactical task to prevent observation of or engagement or interference with, a force or location. (MCRP 5-2A)

pursuit—An offensive operation designed to catch or cut off a hostile force attempting to escape, with the aim of destroying it. (Joint Pub 1-02)

R

reduce—1. A tactical task to gain control over an enemy position or objective. 2. A task to create lanes through or over an obstacle sufficient to allow the attacking force to accomplish its mission. (MCRP 5-2A)

retain—A tactical task to occupy and hold a terrain feature to ensure it is free of enemy occupation or use. (MCRP 5-2A)

rules of engagement (ROE)—Directives issued by competent military authority which delineate the circumstances and limitations under which United States forces will initiate and/or continue combat engagement with other forces encountered. (Joint Pub 1-02)

S

screen—A task to maintain surveillance, provide early warning to the main body, or impede, destroy, and harass enemy reconnaissance within its capability without becoming decisively engaged. (MCRP 5-2A)

search—A systematic reconnaissance of a defined area, so that all parts of the area have passed within visibility. (Joint Pub 1-02)

secure—A tactical task to gain possession of a position or terrain feature, with or without force, and to deploy in a manner which prevents its destruction or loss to enemy action. The attacking force may or may not have to physically occupy the area. (MCRP 5-2A)

seize—To clear a designated area and obtain control of it. (FMFRP 0-14)

suppression—A tactical task to employ direct or indirect fires, electronic attack, or smoke on enemy personnel, weapons, or equipment to prevent or degrade enemy fires and observation of the friendly forces. (FMFRP 0-14)

T

terrorism—The calculated use of violence or threat of violence to inculcate fear; intended to coerce or to intimidate governments or societies in the pursuit of goals that are generally political, religious, or ideological. (Joint Pub 1-02)

Appendix L

References and Related Publications

Field Manuals (FMs)

3-11	Flame, Riot Control Agent, and Herbicide Operations
5-34	Engineer Field Data
5-102	Countermobility
5-250	Explosives and Demolitions
7-98	Operations in a Low-Intensity Conflict
9-15	Explosive Ordnance Disposal Service and Unit Operations
20-32	Mine/Countermine Operations
21-75	Combat Skills of the Soldier
3-30	Grenades and Pyrotechnic Signals
23-34	TOW Weapon System
27-10	Law of Land Warfare
33-1	Psychological Operations
34-130	Intelligence Preparation of the Battlefield
41-5	Joint Manual for Civil Affairs
41-10	Civil Affairs Operations
90-8	Counterguerrilla Operations
100-20	Military Operations In Low Intensity Conflict

Technical Manual (TM)

9-2350-264-10-2	Operator's Manual for Operation Under Usual and Unusual Conditions for Tank, Combat, Full-Tracked: 120mm Gun, M1A1 General Abrams, with Change 4

Marine Corps Concept Paper

A Concept for Future Military Operations on Urbanized Terrain

Marine Corps Doctrinal Publications (MCDPs)

1	Warfighting
1-3	Tactics

Marine Corps Reference Publications (MCRPs)

5-2A	Operational Terms and Graphics

Marine Corps Warfighting Publications (MCWPs)

3-15.3	Scout Sniping (under development)
3-37.1	Chemical Operations
3-37.4	NBC Reconnaissance

Fleet Marine Force Manuals (FMFMs)

0-9	Field Firing for the M16A2
0-25	The Law of Land Warfare
1-3	Tactics
2-6	MAGTF Rear Area Security
2-11	Antiarmor Operations
3-5	Employment of Military Police in Combat
3-30	Communications
6	Ground Combat Operations
6-7	Scouting and Patrolling for Infantry Units
6-8	Supporting Arms Observer, Spotter, and Controller
6-9	Marine Artillery Support
6-21	Tactical Fundamentals of Helicopterborne Operations
7-14	Combating Terrorism
7-34	MAGTF Civil Affairs
7-34-1	Civil Affairs Guidance for the Small-Unit Leader
7-40	Helicopter Insertion/Extraction
8-2	Counterinsurgency Operations
11-9	NBC Protection
11-10	NBC Decontamination
11-11	Treatment of Chemical Agent Casualties and Conventional Military Chemical Injuries
11-12	Chemical Hazard Assessment Guide
11-13	NBC Handbook
11-17	Chemical and Biological Contamination Avoidance
11-18	Nuclear Contamination Avoidance
11-20	NBC Reconnaissance
13-7	MAGTF Breaching Operations

Fleet Marine Force Reference Publications (FMFRPs)

0-6	Marine Troop Leader's Guide
0-14	Marine Corps Supplement to DOD Dictionary of Military and Associated Terms
4-26	Enemy Prisoners of War and Civilian Internees
7-14A	The Individual's Guide for Understanding and Surviving Terrorism
7-37	Vehicle Bomb Search
12-15	Small Wars Manual

12-18	Mao Tse-Tung on Guerrilla Warfare
12-25	The Guerrilla and How to Fight Him
13-5	Engineer Field Data

Joint Publications

1-02	DOD Dictionary of Military and Associated Terms
3-07	Joint Doctrine for Military Operations Other Than War
3-11	Joint Doctrine for Nuclear, Biological, and Chemical (NBC) Defense
3-15	Joint Doctrine for Barriers, Obstacles, and Mine Warfare
3-53	Doctrine for Joint Psychological Operations
4-06	JTTP for Mortuary Affairs in Joint Operations

Standard NATO Agreements (STANAGs)

2003	Patrol Reports by Army Forces
2014	Operations Orders
2029	Method of Describing Ground Locations, Areas, and Boundaries
2070	Emergency War Burial Procedures
2088	Battlefield Illumination
2101	Establishing Liaison
2868	Land Force Tactical Doctrine (ATP-35(B))

Miscellaneous

Fratricide: Reducing Self-Inflicted Losses, Newsletter No. 92-4, Center for Army Lessons Learned (CALL), U.S. Army Combined Arms Command, Fort Leavenworth, Kansas, April 1992.

HR, 25, Annex to Hague Convention No. IV, *Embodying Regulations Respecting the Laws and Customs of War on Land*, 18 October 1907.

Modern Experience in City Combat, U.S. Army Human Engineering Laboratory, Aberdeen Proving Ground, Maryland, 1987.

FIELD MANUAL
No. 7-85

HEADQUARTERS
DEPARTMENT OF THE ARMY
Washington, DC, 9 June 1987

FM 7-85
RANGER UNIT OPERATIONS

IMPORTANT

U.S. Army Infantry School Statement on
U.S. NATIONAL POLICY CONCERNING ANTIPERSONNEL LAND MINES

TABLE OF CONTENTS

DISTRIBUTION RESTRICTION: Approved for public release; distribution is unlimited.

PREFACE

This publication contains doctrinal guidelines for the employment and training of ranger units. The intended audience is intermediate and higher level commanders, staffs of those headquarters, and newly assigned officers and NCOs of the ranger regiment. It is to be used with all other published US doctrine concerning NBC operations and the climatic or geographic influences on combat. The fundamentals apply to the employment of ranger units on worldwide operations and across the spectrum of conflict from low to high intensity warfare. This includes all aspects of the integrated battlefield: nuclear, biological, and chemical. It is not intended to apply to the employment of long-range reconnaissance patrol (LRRP) companies, even though such units carry the designation "ranger." These LRRP companies, as well as the long-range surveillance company (LRSC) (corps level) and the long-range surveillance detachment (LRSD) (division level), are organized, trained, and equipped for long-range reconnaissance in enemy territory.

The scope of this publication addresses the mission, organization, equipment, capabilities, limitations, planning, training, operations, and logistical support of ranger units. It describes how the ranger regiment and its subordinate battalions train, receive taskings, plan missions, task-organize, deploy, and integrate supporting elements to perform the mission. Once in combat, the ranger regiment maneuvers to accomplish the mission in the manner of other light infantry units. Ranger units make full use of published light infantry doctrinal guidance. This manual will repeat procedures or techniques contained in other publications only if necessary for understanding or continuity.

This publication does not contain everything the ranger needs to know to successfully execute a unit mission. Rather, it supplements unit training and the ranger unit commander's guidance; it does not supersede it.

The proponent of this publication is HQ, TRADOC. Submit changes for improving this publication on DA Form 2028 (Recommended Changes to Publications and Blank Forms) and forward it to Commandant, US Army Infantry School, ATTN: ATSH-B-ID, Fort Benning, GA 31905.

Unless otherwise stated, whenever the masculine gender is used, both men and women are included.

CHAPTER 1
Introduction

The ranger regiment is a major component of the US Army's special operations forces. It is a unique light infantry unit tasked to conduct special military operations in support of national policies and objectives. These operations require highly trained, well-disciplined units capable of employment in any environment, either alone or in concert with other military forces. Within this publication, the term *ranger force* describes any size force consisting mainly of members of the ranger regiment and led by a member of the ranger regiment's chain of command. A ranger force may be a TOE unit or it may be a specially organized task force for a specific mission.

1-1. Mission.

a. The mission of the ranger regiment is to plan and conduct special military operations. These operations are conducted by specially trained, equipped, and organized forces against strategic or tactical targets in pursuit of national military, political, economic, or psychological objectives. They may support conventional military operations or they may be performed independently when conventional forces cannot be used. (See Figure 1-1.)

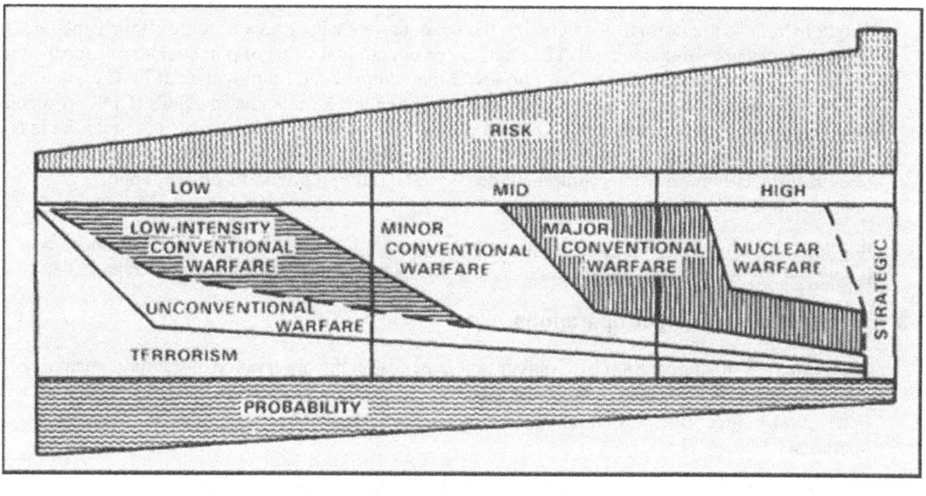

Figure 1-1. Spectrum of conflict.

b. Special military operations conducted by the ranger regiment include strike operations, usually deep penetration, and special light infantry operations. Strike operations include raids, interdiction, and recovery operations. Special light infantry operations include many of the light infantry missions assigned to airborne, air assault, or light infantry battalions and brigades. These operations are conducted in support of the AirLand Battle at all levels of intensity.

1-2. Purpose and function.

The ranger regiment provides the national command authority (NCA) the ability to move a credible military force quickly to any region in the world. The regiment uses the entire spectrum of intelligence support, from national systems to organic assets. Ranger units maintain a readiness posture that supports their immediate commitment to battle once deployed. They are often tailored for specific missions and may require augmentation from external sources. Tactical mobility may be augmented by USAF or Army special operations aviation (SOA) aircraft.

a. The ranger regiment conducts light infantry operations on the integrated battlefield as well as in low intensity conflicts. Special light infantry operations are conducted to accomplish certain contingency missions during peacetime or as part of the echelons above corps (EAC) battle.

b. The regiment conducts both strategic and operational missions. Its efforts are combined into an overall plan to destroy, delay, and disorganize the enemy, or to cause him to divert his attention and combat forces to rear area security. The limited number of ranger units and the diverse targets dictate a careful assignment of missions.

c. The mission, enemy, terrain, troops and time available (METT-T) factors govern the command and control arrangement under which the regiment operates. The ranger regiment or separate ranger battalions are normally assigned to the headquarters whose area of responsibility includes the regiment's operational area(s). The strategic or operational value of potential targets means that ranger units are normally employed at no lower than corps level. Employment at EAC is routine. The EAC headquarters employing ranger units could be the commander in chief's (CINC's) special operations command or field army. The controlling headquarters could also be that of a joint task force (JTF) or the Army component command (ARFOR) of a JTF. The ranger regiment's organization, communications equipment density, and training programs do not support operational level reconnaissance missions.

d. Ranger battalions are not oriented to a specific theater. Current force structure, contingency plans, and training needs (see Appendix G) preclude committing battalions to one region.

1-3. Fundamentals of ranger operations.

a. The success of an operation by a ranger unit depends on the observance of the fundamentals of the US Army's AirLand Battle doctrine (see FM 100-5). The ranger regiment fights the enemy as a light infantry force. It follows infantry doctrine and observes all the basic rules of infantry operations.

b. Along with initiative, depth, agility, and synchronization, ranger operations require--

(1) Detailed planning and coordination that allow the ranger unit to discern and exploit the enemy's weaknesses while avoiding his strength. Both permit the evaluation and use of information gathered from all sources. Detailed coordination integrates all supporting units and services. During planning, the ranger unit conducts pre-mission training, briefings, and rehearsals for all personnel. The use of special mission equipment is practiced and perfected.

(2) Decentralized execution IAW mission orders and the commander's intent. Special operations forces must use individual and unit initiative. Mission-type orders give the ranger force commander the flexibility to take advantage of opportunities on the battlefield.

(3) Surprise, achieved through the ranger unit's ability to move by uncommon means, along unexpected routes, and over rough terrain. Ranger units normally conduct operations during poor weather and reduced visibility, aided by night vision devices. This adds to the attainment of surprise.

(4) Survivability, achieved by using the classic infantry combat techniques of stealth and concealment. The ranger unit engages the enemy at the time and place of its own choosing. It takes full advantage of terrain, and destroys or suppresses enemy weapons. The ranger unit seeks to destroy or neutralize the enemy's command and control systems, his surveillance assets, and his mobility assets. Survivability is enhanced by rapid mission accomplishment and a prompt departure from the objective area.

(5) Mobility, speed, and violence of execution that allow the ranger unit it to close quickly on the objective area and complete the mission before the enemy can react. The speed at which events take place confuses and deceives the enemy as to the intent of the ranger unit. This forces the enemy to react rather than to take the initiative. Tactical mobility lets the ranger unit break contact and withdraw from the objective area.

(6) Shock effect, which is a psychological advantage achieved by the combining of speed and violence with the precision of the ranger attack. The ranger unit strives to apply its full combat power at the decisive time and place, and at the point of the greatest enemy weakness. The ranger unit strives to achieve maximum physical and psychological effect on the enemy by exhibiting aggressiveness and reasoned audacity.

(7) Multiple methods of insertion and attack, trying not to repeat operations thus decreasing the chance the enemy will detect a pattern. This is achieved through imaginative training and planning.

(8) Deception, practiced by the ranger unit during all phases of its operations, from deployment through the insertion phase, to the actions in the objective area and extraction. The enemy is kept unaware of the ranger unit's intentions, is fed conflicting and wrong information, and is kept from knowing the size or mission of the ranger force. The ranger force commander makes full use of ruses, feints, false insertions, electronic countermeasures, and dummy transmissions. Concurrent military operations by US or allied forces may be used as cover for ranger operations. Ranger units are careful to hide every possible aspect of an operation and to disguise those that cannot be hidden. While in the operational area, or when departing any location, ranger units ensure that no material is left that could provide the enemy with information concerning the ranger force.

(9) Audacity, achieved by a willingness to accept risk. The ranger force commander considers what the enemy expects the unit to do and the actual ranger abilities, He then chooses a course of action that may confuse the enemy, while remaining within the capability of the ranger unit. This fundamental is combined with deception and surprise to disrupt the enemy command and control.

c. Detailed planning and coordination finds the enemy. Surprise, mobility and speed, variety, deception, and audacity combine to shock and disorient the enemy, fixing him in place. The violence and precision of the ranger attack finishes the enemy while ensuring the ranger force survives.

1-4. Capabilities and limitations.

a. The ranger regiment has the following capabilities:

(1) Deploying quickly to conduct operations on all types of terrain and in all kinds of weather.

(2) Establishing a credible American presence in any part of the world to show US interest or resolve.

(3) Infiltrating and exfiltrating an area of operations, and assaulting an objective by land, sea, or air.

(4) Conducting strike operations to include raids, personnel and equipment recovery operations, and interdiction of key areas.

(5) Conducting special light infantry operations to include seizing and securing airfields, communications centers, command and control facilities, and key bridges; and other special light infantry operations.

(6) Performing short-duration reconnaissance of assigned ranger objectives for the ranger force commander.

(7) Operating for up to three days without resupply, and for longer periods when provided with accompanying or airdropped supplies.

(8) Providing liaison, communication, and coordination personnel and equipment to integrate the deployed ranger force into the logistical, intelligence, and operational system of the theater or joint task force (JTF) commander.

(9) Assuming operational control, for a limited time, of other US military forces such as engineers or infantry, airborne, or air assault battalions.

(10) Providing the focal point for all-source intelligence support to attached and assigned units of the regimental task force.

(11) Conducting limited combat operations under conditions of chemical, nuclear, or biological contamination.

b. The ranger regiment has the following limitations:

(1) Limited capability against armored or motorized units in open terrain.

(2) No organic transportation.

(3) Limited sustained combat capability due to the shortage of organic combat support and combat service support elements.

(4) Limited organic air defense weapons.

(5) Limited organic indirect fire support.

(6) No casualty evacuation capability.

(7) Reconstitution and retraining needed to replace combat losses.

1-5. Employment considerations.

a. Ranger units are characterized by the quality, motivation, training, and individual skill of their members. This produces units with superb collective abilities, able to adapt well to changing, complex situations.

b. Ranger units can conduct either deliberate or quick-response operations.

(1) Deliberate operations rely on careful planning, reconnaissance and surveillance of the target area, deception, secrecy, thorough preparation and rehearsals, and violent execution. A deliberate operation aims to complete the mission even though the enemy may have heavy forces on or near the objective area. Deliberate operations allow for detailed planning, evaluation, rehearsal, and coordination before insertion. A deliberate operation is likely to succeed against targets that the enemy has protected in depth, that have strong natural defenses, or that need a detailed and long insertion process.

(2) Quick-response operations rely on the high level of training and readiness of the ranger regiment to execute a mission before the enemy can react. These operations are conducted when there is little time for long, detailed planning. They rely on set procedures set forth by the ranger regiment and its supporting elements. A quick-response operation aims to complete the mission before an enemy can react. This type of operation may be chosen due to the time-sensitive nature of the target, political or military goals, the time frame of other operations, or the increased chance of enemy detection.

(3) Whether an operation is to be deliberate or quick response is often a difficult and time-sensitive decision by a high-level command authority. Decision-makers must consider the enemy's strength in the area, his intentions, his ability either to reinforce or to alter the target area, and the consequences of success or failure of diplomatic or military initiatives in related areas. The ranger unit commander tries to use the existing time, manpower, and resources to complete a detailed and coordinated plan. He refines that plan up to the insertion into the objective area.

c. Ranger units train to operate in any environment or weather condition. They regularly perform operations during periods of limited visibility. Ranger units maintain a high state of physical fitness and often train in close quarters combatives. In addition to completing advanced marksmanship training with standard US weapons, each member of a ranger unit trains with many foreign weapons. Ranger units are trained to operate on urbanized terrain, becoming specialists in entry and clearing techniques and quick-fire methods, especially during periods of limited visibility.

d. Because ranger units have limited vehicles, logistics operations capability, indirect fire support, and heavy weapons systems, they are not designed for continuous operations. During all phases of operations and training, ranger units need responsive external support.

e. Ranger units are normally employed against targets and under conditions that need their unique skills. Although targeting priority is set by the overall commander, ranger units are not normally assigned missions that can be done by conventional aerial bombardment or by other units.

f. Ranger units are oriented toward offensive operations. They are not normally employed as a rear area protection force. Although the ranger regiment has a small reconnaissance unit, ranger units do not normally conduct long-range reconnaissance missions. The structure, communications, and training of the ranger unit do not prepare it for LRRP missions. Ranger units engaged in strike and special light infantry operations have a secondary mission to collect and report combat information.

g. Ranger units can be deployed worldwide when US military presence or participation with a host national military activity would serve US interests. This deployment shows a readiness to commit forces into a threatened area or proves US national resolve. After the deployment, other activities include staging operations, rehearsing combat operations, securing base areas for use and deployment of other forces, and so on that provide a clear signal of US intent. Ranger units are not trained or organized to provide mobile training teams (MTT) to train indigenous forces. The US Special Forces or other special operation forces are trained to conduct such security assistance operations.

h. Ranger units can serve as an example to a host country and provide limited military advice and training. They will normally be augmented with linguists and technicians to increase their abilities. The regiment would normally still function as a unit. Most of its assistance would be through short-term, high-impact, unit-oriented operations. They would not be long-term individual efforts associated with advisory-type activities.

i. Ranger units may be deployed to engage in combined training exercises with allies. This enhances US national image by demonstrating the outstanding abilities of the American ranger. These activities may include ranger, light infantry, airborne, air assault, or amphibious operations.

j. Ranger units use standard US nuclear, biological, chemical (NBC) warning; detection; protection; and decontamination equipment and doctrine when operating on the integrated battlefield.

1-6. Command and control.

a. **External.**

(1) The ranger regiment is a key component of the US Army's special operations forces. The other elements of special operations forces are special forces (SF), psychological operations (PSYOP), civil affairs (CA), and special operations aviation.

(2) The 1st Special Operations Command (1st SOCOM) is responsible for command of all active component special operations forces elements, to include the ranger regiment, in peacetime. The 1st SOCOM has command and control of all assigned special operations forces, less operational command (OPCOM) of special operations forces units forward deployed. (See Figure 1-2.) The ranger regiment, while based in the continental United States (CONUS), is under the control of 1st SOCOM. During peacetime, the OPCOM of deployed ranger units is through channels chosen by the NCA through the Joint Chiefs of Staff (JCS) to the unified command.

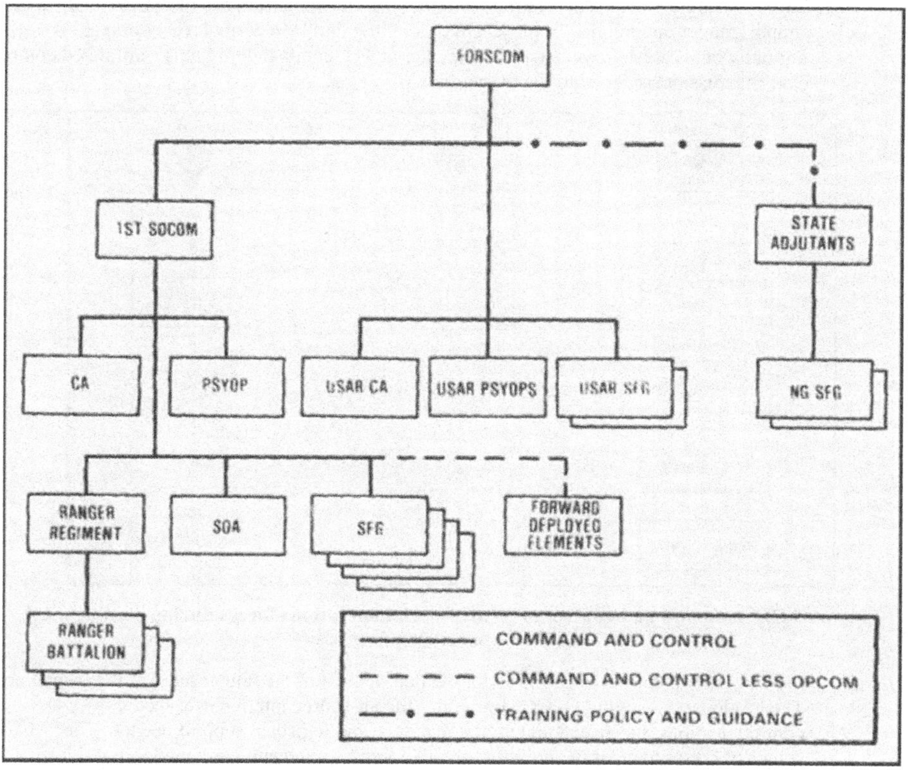

Figure 1-2. Command and control of Army Special operations forces during peacetime.

(3) In support of national policy, the NCA may direct contingency operations involving US

forces. These are politically sensitive military operations normally characterized by the short-term rapid projection or employment of forces in conditions short of conventional war. One aim of the NCA is to conclude them without going to war. The flexibility and effectiveness of the ranger regiment make it possible to use military force quickly to complete the mission and limit the spread of conflict. Command and control of special operations forces in a contingency operation is usually by a special operations task force (SOTF). (See Figure 1-3.) The SOTF and designated special operations forces are under the OPCOM of the highest level of command responsible for the operation. This command may be the unified command responsible for the contingency area (Option A), or a JTF designated by the NCA through the JCS (Option B). If no SOTF is formed, the ranger regiment could be OPCOM to the Army component commander (Option C). The ranger regiment or its battalions are not normally placed OPCOM to a division. A division's area of influence rarely contains appropriate targets. The division also lacks the assets to adequately support ranger operations. Employment of a ranger unit by a corps-level command is normally on a case-by-case basis. The theater or JTF commander retains centralized control over the mission assigned to the ranger unit.

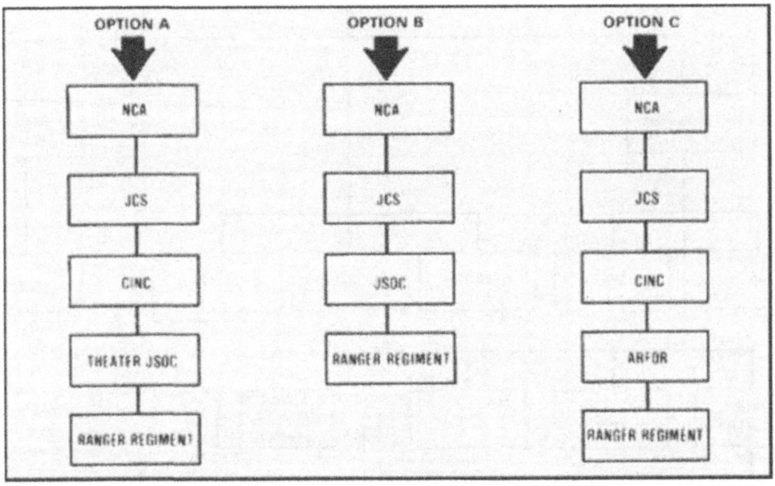

Figure 1-3. Command and control of Army Special operations forces during contingency operations.

(4) The 1st SOCOM is responsible for the deployment of the ranger force to the theater area. The headquarters having OPCOM provides the all-source intelligence, secure communications, insertion and extraction assets, and logistical support needed. The responsible commander in chief, through the Army component commander, provides support during the contingency operation.

(5) During wartime, ranger units are deployed as strategic assets to the theater of operations.

Strategic mission support is provided to the unified commander (CINC) from the NCA through the Joint Chiefs of Staff. It may involve the execution of sensitive operations. The employment of ranger units is controlled at the highest level of command directly responsible for the mission. Ranger units are assigned to a REMAB based on mission, force protection, and OPSEC considerations. The employing command and other elements provide timely and accurate intelligence, communications, logistical, and administrative support. Command and control of these units conforms to the provisions of JCS Pub 2. (See Figure 1-4.)

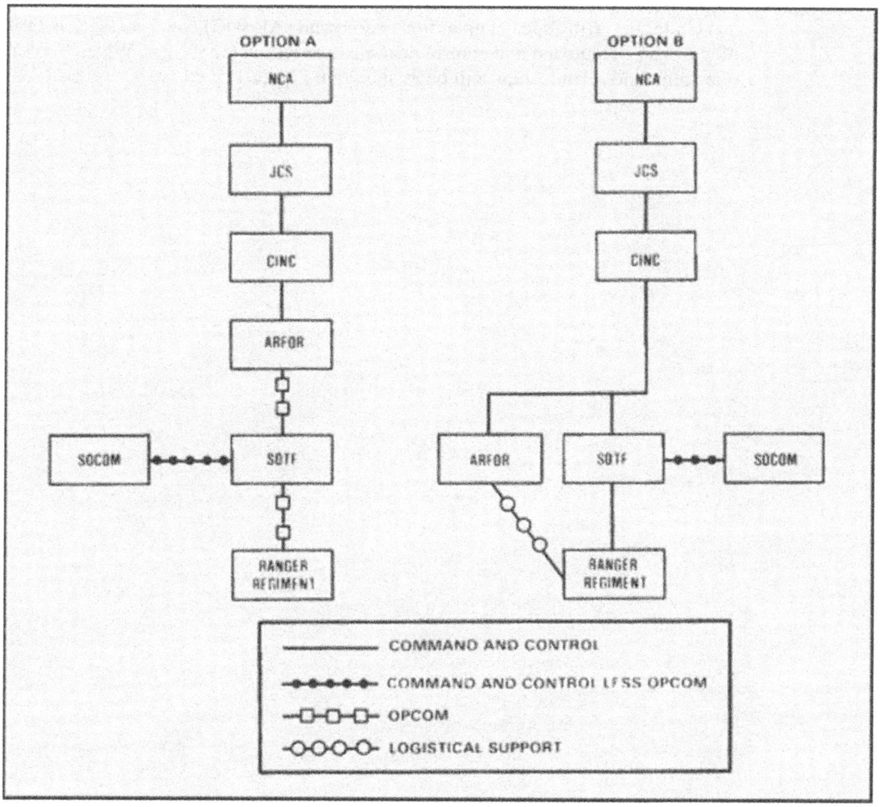

Figure 1-4. Command and control of Army Special operations forces during wartime.

(a) A theater joint special operations command (JSOC) is subordinate to the unified command to supervise the command and control, employment, and support of all US and allied special operations forces. The JSOC is a key element in the command and control of ranger units performing strike missions or special light infantry operations (Option A).

(b) The JSOC may be placed on a lateral line with other in-theater service components (Option B). This can occur when ranger units are -employed far from conventional forces and against strategic targets. It can also occur when the JCS or a CINC forms a JTF to conduct a limited-duration operation in an area where ranger units and conventional forces may operate near each other, or where operations must be coordinated on a regional level.

(c) Under the Army special operations command (ARSOC) concept, 1st SOCOM deploys as a command and control headquarters within the theater. When this occurs, the command relationships will be as shown in Figure 1-5.

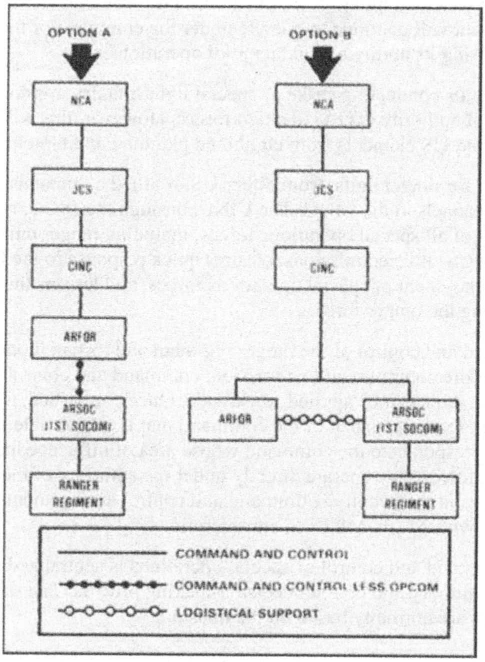

Figure 1-5. Command and control of Army Special operations forces under the ARSOC concept.

(d) Recently passed legislation will modify existing high level SOF command and control arrangements. All US Army, USN, and USAF special operations forces will be consolidated into joint commands at both national and theater level. The ranger regiment will continue to operate under the command of the 1st SOCOM, while increasing its ability to conduct joint operations.

(6) Ranger units conducting strike or special light infantry missions are not normally offered for transfer of authority (TOA) to allied forces. However, this will not preclude allied commands and US elements from combined planning and targeting for ranger missions.

(7) Requests for ranger units, from either US or allied commands, go through normal command channels to the CINC. The CINC, through the JSOC, is responsible for the employment of all special operations forces, including ranger units. This provides flexible response to JCS-directed missions, ensures quick response to the CINC's needs, enhances the joint employment of special operations forces, and lessens the chance of overcommitting and degrading the ranger force.

(8) Command and control of the ranger regiment and its battalions is IAW AirLand Battle doctrine. Before commitment to a mission, command and control is kept at a level where the unit's unique skills can be applied worldwide. Once committed, the ranger unit is placed under the operational control of the command that is responsible for the mission. This normally corresponds to the command whose area of influence includes the target area. Ranger task forces may operate directly under the authority of the senior American representative in the country. Command and control arrangements may vary according to the theater involved and METT-T considerations.

(9) The command and control of special operations is centralized and begins at the highest level of the military and civilian decision-making process. The specific command relationships are normally based on the mission.

b. **Internal.**

(1) The ranger regimental headquarters is organized to operate like that of a brigade headquarters. In addition to commanding and controlling all three ranger battalions, the regimental headquarters can assume operational control (OPCON) of conventional combat and combat support units, and other special operations forces for limited periods. It provides the regimental commander with a well-balanced responsive staff. It is structured as an operational headquarters that can deploy to the mission area and act as the ground tactical headquarters.

(a) The regimental headquarters prepares for combat by forming two small command and control groups. Each of these can deploy to an objective area and control combat operations. Other elements of the staff operate the tactical operations center (TOC) and the logistical operations center (LOC). These may be collocated depending on the situation. The TOC and LOC are normally located well behind the FEBA, at the REMAB or at the intermediate staging base (ISB). The ranger regiment also has two small liaison teams consisting of LNOS, staff representatives, and communications elements. (See Appendix B.)

(b) When a single ranger battalion is committed to an operation, the regimental commander normally exercises control from the TOC. A liaison and communications team is detached from the regimental headquarters to augment the ranger battalion. This team is attached to the deployed battalion until the end of the mission.

(c) If two or more ranger battalions are used in a single operation, the regimental commander normally deploys both command and control groups. He commands from the objective area, The use of two small, mobile control elements makes up for the possible loss of the regimental commander and the primary control element.

(d) Whenever the regimental commander is controlling an operation from the REMAB or the objective area, he normally provides a liaison team with a communications element to the higher headquarters exercising operational control. This cell can provide ranger representation to the controlling staff, operations interface, logistics interface, fire support planning and coordination, intelligence analysis and dissemination, and communications between the higher headquarters and the ranger battalion.

(2) The ranger regimental headquarters can serve as a ranger or special operations force task force headquarters. With augmentation, it can function as the Army special operations force command element. The regimental headquarters can also function as the Army component command of a JTF.

1-7. Communications.

a. Ranger missions need unique and dedicated communication support assets. Along with the ranger regiment's organic communications means, communications support assets are organized within the 1st SOCOM that-can deploy in support of the ranger regiment. This ensures that there are reliable communications to all elements of the SOTF. A special operations communications battalion (SOCB), deployed by the 1st SOCOM, provides more communications systems to support the special operations plan. The SOCB does not replace or duplicate existing in-theater communications. It augments the CINC's communications with the US Army's portion of the critical special operations force command, control, and communications link unique to the SOTF.

b. The ranger regiment's communication platoon can provide secure, long-range communication links from the CINC to a deployed ranger force. It can also enter the SOCB and the joint communications support element (JCSE) nets to provide a direct link with the NCA. Its equipment consists of four light vehicles equipped with various multiple communications systems. The two regimental liaison teams use one of these vehicles each. The TOC and LOC each use one.

c. Secure satellite communications (SATCOM) and radios with AM or FM capability are the main means of communication within the ranger regiment. These are also used for communications within the ranger battalions. The regimental commander has two light vehicles equipped with multiple communications systems. One of these is used to support each of the two mobile command and control elements of the regimental headquarters. Some of the radio systems can be dismounted and man-packed.

d. The ranger regimental headquarters has both man-packed and vehicular tactical satellite communication systems. Each of the ranger battalions has one base station and three man-packed systems.

(1) The man-packed radio is battery-operated. It provides two-way communications in both line-of-sight (mobile) and satellite (at-halt) modes. The digital message device group (DMDG) is the input and output device for data transmission. Secure voice communications is achieved using various encryption devices.

(2) The vehicular net control station uses the basic receiver/ transmitter unit installed on a vehicle. Each can serve up to 15 terminals in a communications net.

e. The reconnaissance platoon is equipped with multiple, man-packed communications systems. These let its teams operate in most of the communications nets within the regiment. Each five-man reconnaissance team has this communications ability, as does the platoon headquarters.

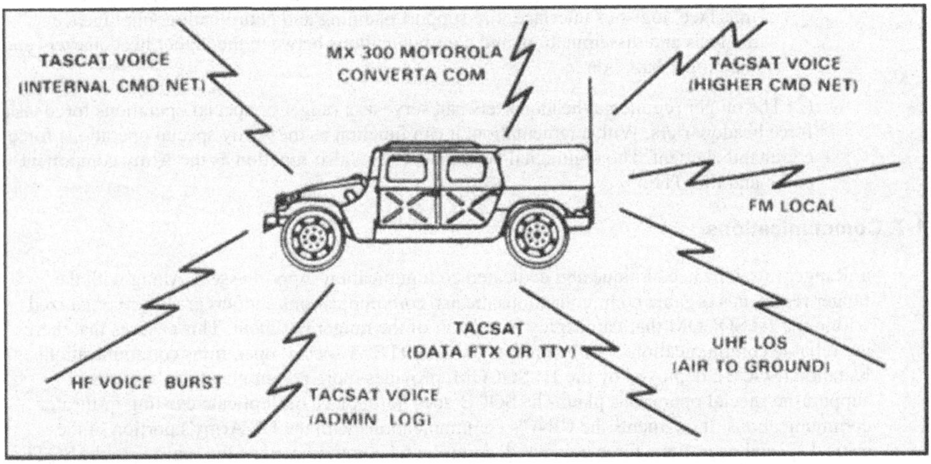

Figure 1-6. Communications capabilities within the regimental headquarters.

CHAPTER 2

Organization and Equipment

The ranger regiment is a specially organized and trained light infantry force consisting of about 2,000 personnel. It includes a regimental headquarters and headquarters company and three ranger battalions. (See Appendix A.) The regiment can operate as a single unit, separate battalions, or separate companies on specific missions. Independent missions are not normally assigned to ranger platoons.

2-1. Headquarters and headquarters company.

The HHC of the ranger regiment consists of 130 personnel, the majority being senior noncommissioned officers, commissioned officers, or soldiers with special skills to support the operations of the ranger battalions. (See Appendix A.) The HHC consists of two elements: the regimental headquarters and the headquarters company (minus).

2-2. Regimental headquarters.

a. The regimental headquarters consists of the regimental commander and his staff. The primary staff is composed of five sections: personnel (S1), intelligence (S2), operations and training (S3), supply (S4), and civil affairs (S5). These staff sections perform the standard functions and are supervised by the executive officer (XO). The regimental commander has a deputy commander and a special staff consisting of a communications officer, fire support officer, surgeon, and staff judge advocate. The S2 officer has a large intelligence analysis and counterintelligence section. The S3 section has extra liaison officers and assistant operations officers to coordinate the widespread and complex ranger missions. A United States Air Force (USAF) staff weather officer and a tactical air control party (TACP) are permanently attached.

b. The regimental headquarters:

- Analyzes assigned missions.
- Plans, coordinates, and controls training and operations.
- Coordinates the infiltration, resupply, exfiltration, and debriefing of the reconnaissance teams.
- Provides liaison support.
- Provides administrative, medical, NBC, and logistical planning and coordination support to the committed ranger battalions.
- Coordinates special support requirements (rigger, transportation, communications security [COMSEC]) for the ranger battalions.

2-3. Regimental headquarters company.

The regimental headquarters company (minus) consists of the company commander and his administrative personnel, plus the limited combat support (CS) and combat service support (CSS) elements assigned to the regiment. These are the fire support element, the communications platoon, the reconnaissance platoon, and the medical treatment team.

a. The company headquarters:

(1) Provides command and control of assigned elements.

(2) Provides administrative, personnel, logistical, and maintenance support to the assigned elements.

b. The fire support element:

(1) Plans fires to support the commander's ground tactical plan.

(2) Coordinates the integration of all available fire support means (mortars, artillery, armed helicopters, close air support, naval gunfire) to support mission execution.

c. The communications platoon (see Appendix A):

(1) Provides secure SATCOM teletype (TTY) and facsimile transmissions (FAX) abilities at the remote marshalling base or intermediate staging base (REMAB or ISB).

(2) Provides secure SATCOM, TTY, and FAX to the regimental headquarters and ranger battalions in the objective area.

(3) Provides cryptographic support to the regimental headquarters.

(4) Provides limited repair of radio and cryptographic equipment.

d. The reconnaissance platoon (see Appendix A):

(1) Trains, equips, and controls the employment of three reconnaissance teams.

(2) Infiltrates the objective area by parachute (high-altitude, low-opening [HALO]; high-altitude, high-opening [HAHO]; static line), helicopter, fixed-wing aircraft, self-contained underwater breathing apparatus (SCUBA), small boat, or other means.

(3) Remains undetected in the objective area for up to five days.

(4) Performs reconnaissance.

(5) Performs demolition target analysis.

(6) Uses small boats and inflatable rafts.

(7) Emplaces and monitors unattended ground sensors.

(8) Uses night image enhancement devices.

(9) Uses photographic equipment.

(10) Emplaces electronic target acquisition and designation devices.

(11) Collects specific combat information to satisfy priority intelligence requirements (PIR).

(12) Reports information using secure, long-range communication equipment.

Figure 2-1. Reconnaissance platoon communications.

(13) Performs drop zone (DZ) and landing zone (LZ) selection, marking, and reception duties.

(14) Escapes and evades enemy areas.

(15) Performs limited attacks or ambushes when tasked by the regimental commander.

(16) Reports weather in the objective area.

e. The medical treatment team:

(1) Provides medical support to the regimental headquarters.

(2) Coordinates required medical support and evacuation.

2-4. Ranger battalion.

a. The ranger battalion is organized and equipped to conduct both special and conventional combat operations. It contains organic combat, CS, and limited CSS elements. It consists of a battalion headquarters, a headquarters company, and three ranger rifle companies. (See Appendix A.)

b. The ranger battalion is the combat element of the ranger regiment. It can conduct independent combat operations or subdivide into its separate ranger rifle companies and their subordinate elements who can conduct semi-independent operations.

c. The ranger battalion headquarters and headquarters company (see Appendix A) includes:

(1) **Battalion headquarters.** The battalion headquarters consists of the commander; the executive officer; the command sergeant major; the S1, S2, S3, S4, and S5 officers; the chaplain; the communications officer; the fire support officer; and two radio operators.

(2) **Battalion headquarters section.** The battalion headquarters section consists of the S3 air and assistant S8, an assistant S2, an intelligence sergeant and a senior intelligence analyst, two operations sergeants and an operations assistant, a personnel staff noncommissioned officer (PSNCO), the chaplain's assistant, two clerk typists, two radio operators, a chemical officer and NCO, a personnel administration center (PAC) NCO, three PAC clerks, and a legal specialist. There is a USAF tactical air control party permanently attached.

(3) **Communications section.** The communications section consists of one telecommunications technician, one tactical communications chief, two combat signal radio team chiefs, one tactical communications systems supervisor, three radio operators, one switchboard operator, one single channel radio operator, and two equipment mechanics. The section provides communications support and associated maintenance.

(4) **Battalion medical section.** The battalion medical section consists of one general medical officer, one physician's assistant, two medical specialists, and twelve aidmen. The section provides unit-level medical support and trains selected members of the battalion in advanced medical procedures.

(5) **Battalion support platoon.** The battalion support platoon consists of a battalion support platoon leader, an ammunition NCO, and a food service section. The platoon provides limited CSS for assigned and attached units. This platoon does not have organic transportation. The battalion food service section consists of a food service sergeant (E7), a first cook (E6), and eight cooks.

(6) **Fire support team headquarters.** The fire support team (FIST) headquarters is supervised by the battalion fire support officer. It consists of the fire support sergeant and two fire support specialists. The headquarters controls three 9-man fire support teams. These teams are normally with the ranger rifle companies in a habitual support relationship. When operations are conducted near water, the fire support team headquarters may be augmented by spotter teams from the naval shore fire control party.

(7) **Company headquarters.** The HHC company headquarters consists of a commanding officer, an executive officer, a first sergeant, a supply sergeant, and an armorer.

d. The ranger rifle company (see Appendix A) includes a company headquarters, three rifle platoons, and a weapons platoon. Each company can conduct semi-independent operations for a short time when augmented by assets from the ranger battalion headquarters company.

(1) **Ranger rifle company headquarters.** The ranger rifle company headquarters consists of a commanding officer, an executive officer, a first sergeant, a supply sergeant, a tactical communications chief, two radiotelephone operators, an armorer, and an NBC NCO. It also has a two-man sniper team equipped with a sniper weapon. It normally has one of the 9-man FISTs attached from the ranger battalion headquarters company. Elements of this FIST may be further attached to the ranger rifle platoons.

(2) **Ranger rifle platoon.** The ranger rifle platoon consists of a headquarters section, three rifle squads, and a machine gun squad. The headquarters section personnel include a platoon leader, a platoon sergeant, and a radiotelephone operator. Each rifle squad consists of a squad leader and two balanced fire teams of four men each. The machine gun squad consists of a squad leader and three M60 machine gun crews of three men each. The machine gun squad is normally employed as a single element to provide quick, accurate, long-range supporting fire. One or more crews may be attached to other squads within the platoon.

(3) **Ranger weapons platoon.** The ranger weapons platoon consists of a headquarters section, a mortar section, and an antiarmor section. The headquarters section consists of a platoon leader, a platoon sergeant, and a radiotelephone operator. The mortar section has two 3-man 60-mm mortar squads. The mortar squads are normally employed as a section. They may be attached by section or squad to other platoons. The antitank section normally consists of a section sergeant and three 3-man 90-mm recoilless rifle or Dragon missile teams. They are normally employed as a section. The antiarmor section may also be attached by section or team to other platoons.

2-5. Special teams and elements.

The ranger regiment forms special teams to do critical tasks. These teams are made up of members of the unit and are used when needed. Examples of these special teams and elements are--

a. SCUBA teams whose members are graduates of the Special Forces Underwater Operations School. They conduct special insertion and reconnaissance duties.

b. HALO teams whose members are graduates of the Special Forces Freefall School and who conduct high-altitude, low-opening parachute operations.

c. HAHO teams whose members are also graduates of the Special Forces Freefall School and who conduct high-altitude, high-opening parachute operations, allowing offset parachute insertions.

d. Scout swimmer teams whose members are specially trained by the Landing Force Training Command (United States Marine Corps [USMC]) to conduct operations along coastlines, waterways, and inland waters. These teams conduct special insertion, reconnaissance, and security tasks often in support of amphibious operations.

e. Sniper teams, whose members are authorized by the table of organization and equipment (TOE), are used for both conventional and special operations. Along with the two-man sniper teams in the rifle company headquarters, each rifle platoon has two trained snipers.

f Demolition teams whose members are trained in advanced demolition and incendiary techniques and the use of special explosives,

g. Air defense teams trained in the use of man-portable air defense system (MANPADS) weapons. Such teams give the ranger battalion limited self-defense against air attack.

h. Liaison teams that are used by the ranger regimental commander to augment the existing command and control links from the employing unit headquarters to the ranger regiment. These teams are placed where they are most needed and may assist the higher headquarters staff in planning and target selection, as well as performing other liaison duties. See Appendix B for a detailed listing of the personnel and equipment included in the ranger liaison teams.

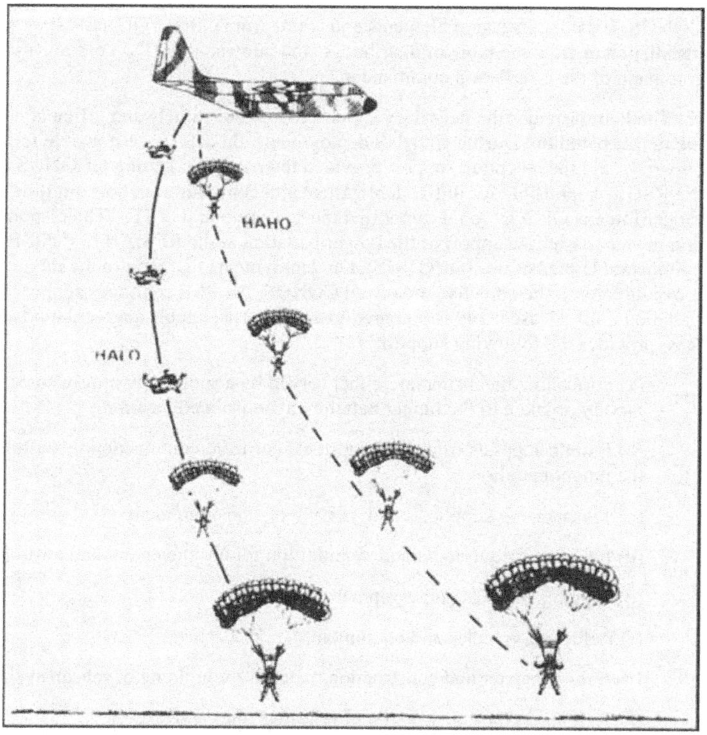

Figure 2-2. HALO and HALO parachute operations.

i. The ranger support element (RSE).

(1) The RSE is a special support element provided by host installation commanders at CONUS posts. It consists of elements and teams from either TOE units stationed at the installation or from the table of distribution and allowances (TDA) organizations under the command of the installation commander.

(2) The RSE provides the necessary support to ensure a timely and efficient deployment of the ranger battalion. During CONUS deployments, the RSE is responsible for both the departure and the reception support provided the battalion. During OCONUS deployments, the RSE is responsible for initial deployment and continuing support until the deploying ranger force is OPCON to the supported unified command or JTF. The responsible theater commander assumes support of the ranger battalion at the REMAB or the ISB if outside continental United States (OCONUS). On deployments that require the ranger battalion to move directly to the objective area from CONUS, the RSE continues support of the ranger battalion until released, and is prepared to accept considerable augmentation assets. The RSE provides the following support:

(a) Furnishing meals that are either served by a supporting unit in an established facility or taken to the ranger battalion at an isolated location.

(b) Issuing supplies from existing stocks, or from contingency items stored for use by the ranger battalion.

(c) Transporting supplies, food, personnel, and equipment.

(d) Palletizing equipment and ammunition for air movement and airdrop.

(e) Rigging vehicles and equipment for airdrop.

(f) Refueling vehicles and equipment.

(g) Issuing barrier and construction material for building of rehearsal sites.

(h) Helping in the construction of rehearsal sites and targets.

(i) Securing, storing, moving, and issuing ammunition and other Class V items.

(j) Maintaining ranger weapons and equipment, to include communications equipment. This is normally done by maintenance contact teams.

(k) Setting up and operating the RSE emergency operations center.

(l) Securing the REMAB if it is in CONUS.

CHAPTER 3

Deployment

Ranger units can quickly deploy by any means of transportation to any location in the world.

3-1. Readiness.

One ranger battalion is always in an advanced readiness condition as the Ranger Ready Force (RRF) available for immediate worldwide deployment. A second battalion is prepared to deploy later and a third battalion to follow. The regimental headquarters maintains command and control, liaison, communications, and reconnaissance elements immediately available for deployment. Higher status of readiness in response to specific world situations can be achieved.

3-2. Ranger ready force.

In order to maintain advanced readiness conditions, ranger battalions are rotated as the RRF.

a. Before taking over the duties of the RRF, battalions must do the following:

(1) Complete preparation for oversea movement (POM) qualification, to include an individual records check and required inoculations.

(2) Test fire all weapons, confirm battlesight settings, and repair or replace faulty weapons.

(3) Assemble and load on pallets the unit basic load (Classes I and V) and other supplies and equipment.

(4) Check unit-derived packing list components for accountability and serviceability.

(5) Prepare special operations troop listings.

b. During an RRF period, ranger units:

(1) Have a positive recall system for all personnel.

(2) Store enough air items for a battalion-size parachute assault.

(3) Conduct no off-post exercises, unless approved by the regimental commander.

c. Each unit keeps an alert roster and a telephone notification system for recalling personnel. The rosters ensure that the duty officer and the unit charge of quarters make a minimum number of calls. Each commander briefs personnel assigned to his unit on the recall system and tests it often.

d. Ranger units can maintain an increased readiness condition for long periods, but they should not be kept in that condition longer than needed. The effort to maintain this level of readiness will degrade the unit's training and effectiveness. If the needs of the operations dictate long periods at the highest readiness condition, provisions must be made for sustainment training. Also, added support will be required from the RSE.

e. The ranger regiment maintains a strong family support system when the unit is training or deployed on combat operations. Each member of the chain of command is responsible for family support.

3-3. N-hour sequence.

The N-hour sequence is a departure time sequence designed to help the ranger force complete all required deployment actions within a certain time. The length of the N-hour sequence depends on the readiness condition of the battalion before deployment. The N-hour sequence begins when the battalion is alerted and ends when the first elements depart. The N-hour sequence does not include travel time to the objective area.

3-4. Intelligence and security.

a. **Essential elements of friendly information.** Items of sensitive information and subject areas, if not secured, can be collectively analyzed by a hostile force. This analysis can reveal the intentions of a unit or compromise a mission. These items of information are designated as essential elements of friendly information (EEFI). They are never transmitted without encryption. The EEFI associated with a ranger unit alert normally are the mission, the use of staging bases, nature of the target, special terrain or weather conditions, and time limits. The ranger force commander must ensure that EEFI are not discussed outside of secure areas or made available to personnel without a security clearance and a need to know.

b. **Security plans.** Security plans for billets, marshalling areas, and out-loading areas must remain flexible to adapt to any situation. Certain items are always addressed in security plans. These are access control and physical security.

(1) **Access control.** Ranger units control access to their respective areas by allowing only authorized personnel to enter or leave. All entrances are secured and an armed guard with the access roster is posted. Access to telephones is restricted.

(2) **Physical security.**

(a) Equipment is open to theft and sabotage during the early phases of deployment. Units must provide guards to secure all unit and personal equipment.

(b) Weapons, ammunition, and sensitive items must be stored in a restricted area and guarded.

(c) Classified documents are open to compromise during deployment. Steps must be set up for maintenance, transportation, turn-in, and destruction of classified material. The most secure facility available must be used.

c. **Rear detachment security.** Security after deployment (maintained by the rear detachment) is as important as security before deployment. No information should be given to the news media from other than official sources. The rear detachment may continue to play a role in the unit cover and deception plan.

d. **Tactical operations center security.** Only personnel listed on TOC access rosters are allowed free entry into the TOC area. Personnel wanting to enter the TOC, who are not listed on the access rosters, are detained until their identity and purpose are proven. Such personnel are escorted at all times while within the TOC area.

e. **Marshalling and out-loading area security.** The ranger unit normally marshalls its personnel and equipment in the unit area, providing its own security force. The USAF security police provide this support at the departure airfield. On Army airfields, Army military police provide security. En route security is provided by the RSE.

f. **Personnel and unit equipment sanitization.**

(1) Before deployment, personnel are relieved of non-mission-essential material having intelligence value. This reduces the intelligence available to hostile forces if they capture a ranger force member.

(2) The ranger force commander sets the degree of sanitization IAW mission requirements.

(3) If strict sanitization is imposed, guidelines are issued as follows:

(a) The only personal items retained are identification card and tags.

(b) Unit insignia and markings are removed from uniforms and equipment.

(c) Personal items, such as wallets and photographs, are secured by the rear detachment commander.

(d) Personnel are briefed on how to compose and retain combat diaries. No personal diaries are allowed.

(4) The commander decides how much US currency is to be taken with the unit. '

g. **Area sanitization.** To avoid the compromise of classified or sensitive mission-related material, searches are conducted before and after leaving the host installation, REMAB, or ISB. These searches are held to locate and remove any material, indicators, litter, or other items that could compromise the ranger unit's mission. Unit commanders ensure that a thorough sanitization of their areas is conducted, to include dismantling rehearsal areas, destroying terrain models, erasing chalkboards, and emptying all classified and nonclassified waste containers. During their searches, commanders must pay close attention to emergency operations centers, tactical operations centers, and marshalling and loading areas. The ranger unit's intelligence officer coordinates support from local counterintelligence elements to search the area again after the ranger unit leaves.

3-5. Operations.

a. **Actions upon receipt of the OPORD.**

(1) Task-organize and equip units for special mission(s) IAW the warning order.

(2) Allocate/reallocate mission-essential equipment.

(3) Distribute augmentee personnel.

(4) Redistribute key personnel,

(5) Prepare a deployment time sequence.

(6) Prepare a movement plan.

(7) Prepare and maintain a current operation estimate.

b. **Advance elements.** A planning element is normally deployed to the REMAB/ISB or JTF headquarters in advance of the main body if the ranger force does not deploy directly to the objective area.

c. **Emergency operations center.** An EOC is setup by the ranger battalion headquarters. The EOC is responsible for relaying all reports to higher headquarters. It remains operational until ordered to cease.

d. **Air movement.** The ranger force always deploys by USAF aircraft. This is normally an administrative air movement to a REMAB or ISB. It may be followed by tactical air deployment from there. For some missions, tactical deployment begins at the unit's home station. Even if the movement is administrative, key personnel are cross-loaded. This is done under the direction of the operations section air officer based on aircraft allocations and load plans.

3-6. Logistics.

The following describes logistics readiness needs and steps for deployment.

a. **Supply.**

(1) **Class I.**

(a) Combat rations are normally palletized when the unit is RRF. These rations are to be used during the tactical operation. If the mission calls for deployment directly into the objective area, the palletized stock of rations is off-loaded in the unit area and issued as needed.

(b) Each ranger deploys with emergency combat rations.

(2) **Classes II and IV.**

(a) **Individual clothing and equipment.**

- Individual A bags are packed IAW a predetermined packing list, tagged, and marked. They are palletized, moved to the departure airfield by the RSE, and loaded IAW the air movement plan.
- Individual B bags are packed, assembled, and forwarded on order by the rear detachment.

- Individual environmental equipment is issued. Equipment shortages are replenished out of stocks maintained by the RSE.

(b) **Air items.**

- While the unit is on RRF, a contingency load of main and reserve parachutes is maintained by the RSE.
- Individual and unit air items are palletized by the RSE and shipped in bulk to the REMAB or ISB.
- For direct-deployment missions, air items and packing material are taken to the unit by the RSE.

(c) **General supplies.** The RSE maintains pallets of common contingency items and special equipment. Pallets are picked up by the out-loading support unit and moved to the unit area for issue or moved with the other pallets to the departure airfield.

(d) **Bulk supplies.** Barrier material is identified, relocated, palletized, and out-loaded IAW the air movement plan.

(3) **Class V.**

(a) The unit basic load of ammunition is secured, palletized, and loaded by the RSE IAW the air movement plan.

(b) For direct deployment missions, selected pallets are taken to the unit area and ammunition is issued. The convoy route is coordinated through the post EOC, and secured by the RSE.

(c) Designated items are rigged by the RSE for airdrop, if enough time exists before movement to the REMAB or ISB begins.

(4) **Class VI.** Personnel deploy with comfort items as needed.

(5) **Class VII.**

(a) All weapons and sensitive items not deployed are assembled and secured by the rear detachment personnel.

(b) Unit vehicles are prepared for airland or airdrop.

(c) Inoperable mission-essential equipment is repaired or replaced.

(d) Special equipment requests are forwarded to the logistics section as they are identified.

(6) **Class VIII.**

(a) Individuals deploy with one set of individual Class VIII items.

(b) Unit medics deploy with modified aidbags.

(c) Controlled medical items (less drugs) are palletized and deployed with section supplies. Controlled drugs are secured and issued to unit aidmen on order.

(d) Medical supply chests are maintained in the unit aid station.

(7) **Class IX.**

(a) Units deploy with initial prescribed load list (PLL) items.

(b) Critical PLL shortages are identified to the RSE for follow-on issue.

b. **Maintenance.**

(1) Upon alert, priority maintenance or replacement of key items of equipment is coordinated by the logistics officer. Only minimum mission-essential equipment is replaced.

(2) Items to be returned from supporting maintenance activities are identified, and pickup is coordinated by the property book officer.

c. **Out-loading coordination and support.**

(1) Upon alert, the RSE provides a liaison officer to the ranger logistics section.

(2) The support platoon leader (SPI,) attends the air movement conference and coordinates the following:

(a) Movement times (for both troops and equipment) to the departure airfield.

(b) Steps for weighing and joint inspection of the pallets and vehicles upon arrival at the departure airfield control group (DACG).

(c) Limitations on the pallet height and weight due to aircraft type, allowable cargo load (ACL), and availability.

(d) Changes in the planned number of pallets due to the mission requirements or airframe availability.

(e) Rigging and loading of the heavy drop and supply bundles.

(3) Empty pallets are placed in position for loading by the RSE in coordination with the SPL.

(a) Companies and staff sections prepare pallets.

(b) For airdrop operations, vehicles are prerigged at the unit location. On order, vehicles are driven to the heavy drop rigging site. The SPL and the officer in charge of the rigger detachment coordinate for a rigging site and the delivery of the required air items to that location. Rigging is done by the RSE, and supervised by a member of the ranger unit. Loading aboard the aircraft is IAW the air movement plan and supervised by RSE and USAF aerial port personnel.

(4) The DACG is provided by the host installation. Upon arrival at the DACG area, unit personnel and equipment are under the control of the DACG. As the deployment begins, a

member of the ranger unit is present at the DACG to ensure an orderly loading of personnel and equipment.

(5) The DACG plans for troop-holding areas in the event of delayed or aborted aircraft.

(6) Unit movement data are maintained by the air officer in the operations section.

3-7. Communications-electronics.

a. **Alert notification communications.** Alert notification of the ranger battalions by the regiment is done by secure communications means. Units in local training areas are informed by radio or through their assigned range field telephone. Units operating out of normal FM radio range during RRF periods set up a high frequency net, or other communications, to their higher headquarters.

b. **Telephone service.** Strict control is maintained over key telephones being used during unit alert. Care is taken to ensure that COMSEC and OPSEC violations do not occur. All telephones are assumed to be monitored; classified, sensitive, or official information is not discussed.

c. **Radio communications.**

(1) **Voice communications.**

(a) A secure radio net is set up early in the deployment. The ranger unit EOC, the installation EOC, and the RSE EOC operate in this net. A liaison officer from the ranger unit is placed at the installation EOC.

(b) The ranger support element liaison officer, attached to the logistics section, operates a secure voice station on a separate logistics net to the RSE EOC and the RSE commander.

(2) **Cryptographic material.** During RRF, COMSEC codes and enabling equipment are kept available to set up the secure voice net.

3-8. Remote marshalling base and intermediate staging base.

a. Ranger units can deploy from their CONUS base directly to the objective area. A more common method would be for the ranger unit to deploy first to either a REMAB or an ISB before insertion into the area of operations. (See Figure 3-1.)

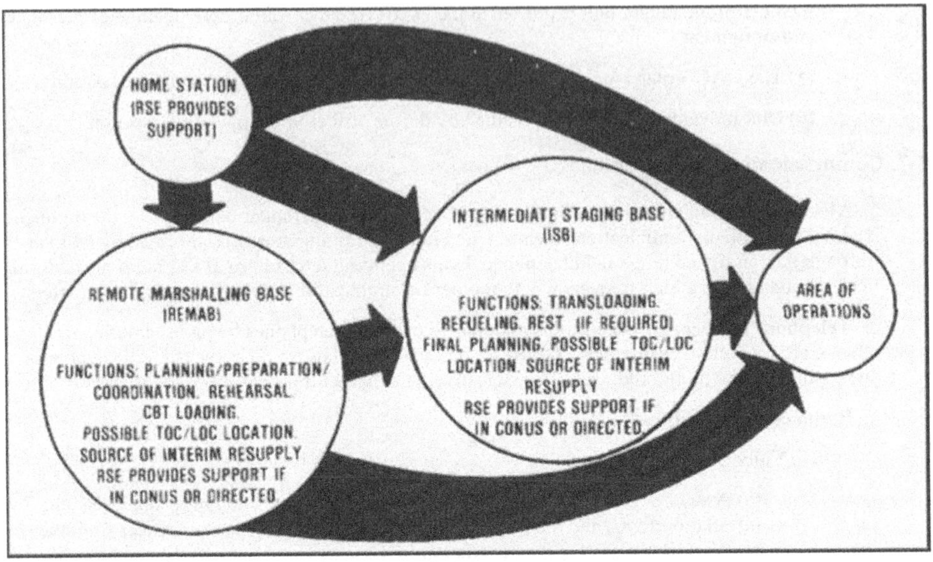

Figure 3-1. Deploying ranger units.

b. The REMAB is a secure base to which the entire ranger unit, including organic and attached support elements, deploys and where mission planning continues. The REMAB is normally within the geographical area encompassed by the command authority of the theater or JTF commander. This ensures that the CSS elements providing support to the ranger unit are operating within their normal area. It prevents or lessens out-of-sector support requirements for CSS elements.

c. The REMAB also provides a secure location for the ranger unit to conduct detailed planning and coordination with the controlling headquarters staff. In the REMAB, the ranger force commander conducts rehearsals, refines and modifies plans, determines PIR, and coordinates with the proper intelligence source to receive that information. It is in the REMAB that specialist augmentees to the ranger force are integrated into the unit, if they have not already joined. Specially trained supporting units, such as aviation and communications elements, also join the ranger force at the REMAB. The REMAB must provide--

(1) Access to the controlling headquarters staff.

(2) Physical security of billeting, planning, maintenance, and communications areas.

(3) Mess, billeting, latrine, and shower facilities for the ranger force and its supporting elements.

(4) Access to a C-141 or C-130 capable airfield, possibly with all-weather operations.

(5) Access to secure communications and processed mission intelligence.

(6) Access to rehearsal areas where sites can be built and live-fire rehearsals can be conducted.

(7) Access to the unit locations of major supporting elements, such as naval landing craft or Army aviation units.

(8) An external security force and an active counterintelligence agency.

(9) Vehicular transport for troop lift, equipment transfer, and administrative use.

(10) Access to maintenance support facilities.

(11) Medical support facilities to augment the ranger force medical personnel.

(12) Covered area to allow for packing of parachutes and rigging of airdrop loads.

d. The ISB is a more austere location. Elements of the ranger force deploy to an ISB to perform final planning, coordination, and task organization. The unit's organization and composition are finalized for movement to the objective area. The ISB is normally not occupied for long periods. Ranger units may deploy to the ISB from the REMAB or CONUS. Deployment to the ISB is more common when terrain or distance precludes insertion to the objective area directly from the REMAB or CONUS. Quick-reaction missions often involve the use of an ISB. Facilities needed at the ISB are limited to the following:

(1) Austere messing arrangements for the combat elements of the ranger force and its attachments.

(2) Austere billets or rest areas.

(3) Water point.

(4) Fuel for aircraft and vehicles.

(5) Areas for test firing of weapons.

(6) Covered and concealed areas for assembly of the ranger force and rigging of parachutes and door bundles.

(7) Austere airfield support facilities, possibly capable of all-weather operations.

(8) Remote location far away from civilians or traffic routes.

(9) Security and counterintelligence elements.

(10) Secure communications.

e. The REMAB and ISB should be located in an area similar in terrain and climate to the objective area. Time spent at the REMAB or ISB lets the ranger force begin acclimatization. Ranger units train yearly in all climates and emphasize physical conditioning. This allows acclimatization quicker than units not as well trained or conditioned. However, ranger units still need an acclimatization period if the terrain and climate of the objective area are different from that of their CONUS base. Not enough acclimatization means that the ranger force could be less efficient, especially when the objective area is hot, dry, or at a great altitude.

CHAPTER 4

Insertion, Extraction, Escape, and Evasion

Insertion is the entrance of a ranger force into a hostile area to execute a particular mission. It is the first critical phase of a ranger operation. These operations often involve deep penetrations of enemy territory by air, water, or land. This chapter emphasizes fundamentals and techniques employed by ranger units when their mission requires insertion and extraction. Extraction is conducted quickly after the mission to avoid casualties.

4-1. Intelligence.

a. Insertion plans are based on timely and accurate intelligence. The headquarters directing the operation must provide current intelligence information to the ranger force. The ranger regiment normally takes action to solicit required information from the controlling headquarters and from other agencies, if security permits. The type of intelligence needed varies depending on the specific mission, but normally includes the following:

(1) Current photographs from aerial or overhead reconnaissance.

(2) Detailed maps.

(3) Data on the population, terrain, and weather from area studies.

(4) Current enemy order of battle, enemy troop locations, and enemy capabilities.

(5) Details on the target area and infiltration routes.

b. To meet its need for current intelligence, the ranger force also relies on aerial reconnaissance and surveillance of the terrain over which it will move. Emphasis is on getting information on the enemy's ability to detect and engage the inserting forces. The location and abilities of air defense radar and weapons systems are critical. Terrain analysis is detailed and focuses on areas suitable for the insertion. Weather information must be detailed and current.

c. In getting information and intelligence data, OPSEC measures are stringent and must be enforced. Coordination with other agencies is kept within security constraints approved by the directing headquarters.

4-2. Deception.

There are always plans to deny the enemy knowledge of the ranger unit's insertion, or to deceive him as to the location or intent of the operation. Feints, false insertions, and other deception operations all add to ranger deception plans. Selection of unexpected means, time, place, and routes of insertion, as well as use of speed and mobility, give less reaction time to the enemy. Planning may include the use of diversionary fires to direct the enemy's attention away from the inserting unit. Deception techniques that may be used include--

a. Multiple airdrops, water landings, or both. This precludes the entire force from becoming engaged at the same time and place, if detected.

b. Dispersion of insertion craft (air or water) both in time and location.

c, Landing a force in an area closer to other potential targets than to the actual targets.

d. Leaks of false information and false messages.

e. False insertions or radio traffic to disguise actual insertions.

f. Diversionary actions such as airstrikes in other than the target areas.

g. Increased reconnaissance flights, either by manned aircraft or remotely piloted vehicles, over false targets.

4-3. Speed and Mobility.

a. Speed of action must be emphasized when the enemy can react quickly to the ranger force presence or when there are limited means of withdrawal. The operation must be conducted swiftly, allowing the ranger force to withdraw before the enemy can react.

b. Major problems facing ranger units are the need for speed and mobility, and the heavy loads carried by soldiers for sustained operations. Ranger units are required to be flexible and to be able to react to any problem within the objective area. However, if equipment is carried to cover every encounter, the load will become too heavy. This will reduce the speed and mobility of the ranger unit. The ranger force commander must always have a balance between the equipment carried and the mobility of his unit. He does this by making three different types of loads: fighting, mission, and existence. He decides what items will make up these loads, where they are carried, and what will be done with them upon contact with the enemy.

(1) The fighting load consists of those items of equipment, weapons, ammunition, food, and water that are common to all soldiers and are necessary for immediate use in combat. These items are normally carried and are retained at all times while in contact with the enemy.

(2) The mission load consists of those items of equipment or ammunition that are required by the mission, but are not normally carried by all rangers in a unit. The load may include spare batteries, demolitions, radios, antitank weapons, or mines. This load is normally spread out among members of the unit and is rotated often. It may be dropped upon enemy contact, but is normally carried to the objective and either used or emplaced there.

(3) The existence load consists of any items designed to provide protection from the elements: sleeping gear, changes of clothes, spare rations, or tentage. This load may or may not be brought into the objective area, depending on the tactical situation. If it is brought, it is normally dropped upon enemy contact or left in an assembly area and retrieved later.

4-4. Stealth.

Methods and steps chosen for insertion must emphasize stealth and try to avoid detection by the enemy. Multiple routes or methods of insertion may be used to preserve the ranger force if some soldiers or units are seen.

4-5. Suppression.

Effort must be made to suppress enemy detection devices, weapons systems, and command and control facilities. This may include electronic countermeasures or direct attack. This helps prevent the enemy from finding and engaging the ranger force en route to the objective area. Deception tactics add to suppression activities. If the ranger force is inserting by infiltration, suppression must not disclose friendly activities.

4-6. Security.

Emphasis on OPSEC prevents compromise of an impending operation. The operation must not be disclosed by overt rehearsals or training activities, the open use or procurement of special items of equipment, or the location of the marshalling area. Several measures can help in maintaining security:

a. Guarded and secure facilities for the headquarters and staff during planning.

b. Issue of operation orders only after units are moved to secure marshalling areas.

c. Alerts, rehearsals, and training to get the local people used to conditions that will exist during actual marshalling.

d. A secure facility set up to isolate units selected for an impending operation.

4-7. Night vision and electronic devices.

a. Rangers use night vision devices to detect and avoid the enemy's forces and his detection devices. These devices also assist in controlling and speeding movement and allow units to traverse almost impassable terrain.

b. The use of electronic homing devices permits precise navigation during reduced visibility. These devices may also be used with visual marking systems.

c. Passive night vision devices are used to assemble rapidly and to reorganize.

d. The use of active night vision devices in assembling, and the degree to which drop zones or assembly areas are "sterilized," depends on the following:

(1) Mission success may depend on not being seen during insertion. When a clandestine insertion is made, the time needed to get to the objective will normally be greater, the use of passive devices will prevail, and the sterilization of the area will be vital. Each ranger must be briefed on sterilization plans and techniques for erasing signs of the insertion.

(2) When detection is likely and the mission depends on getting to the objective before the

enemy can react, speed becomes crucial to the success of the insertion. Insertion may be near or on the objective. Rangers must get in and out quickly. They do minimum sterilization of the area and can use active RSTA devices.

4-8. Rehearsals.

Rehearsals should be close to the actual conditions of insertions and extractions. Security of rehearsals and rehearsal sites is important to the success of the operation. Rehearsals are held on terrain and structures like those in the objective area. If such structures are not available, they are built by the ranger force and the RSE with its supporting engineers.

4-9. Sand tables and terrain models.

Sand tables are useful to acquaint personnel with insertion sites and surrounding terrain. The use of sand tables and terrain models during the issuance of prejump orders and briefings enhances an orderly and speedy tactical assembly.

4-10. Communications.

Reporting during insertion and extraction missions is normally by exception, using an operations schedule.

4-11. Planning insertions.

a. **Centralized planning.** Planning for insertion of ranger units is centralized at the lowest level responsible for the overall operation. This is normally at the joint task force level, with detailed coordination required among all supporting forces. These forces may include other Army elements, USAF, United States Navy (USN), United States Marine Corps (USMC), other US agencies, and allies. Preparation must include detailed contingency planning for use of firepower, shock effect, speed to break contact (if detected), control measures, procedures, and planning. When planning takes a long time, intelligence must be continually updated through all available sources. This includes overhead photography and ground sources to ensure that the initial planning data are still current. Planning stresses the use of deception. Escape and evasion plans, external communication nets, and the location of caches must be known to leaders at all levels.

b. **Reverse planning sequence.** Insertion planning is done in the reverse planning sequence with the ground tactical plan being prepared first. Actions at the objective serve as the basis for the landing and assault plan, the plan for insertion and movement, the loading plan, and the plan for marshalling. All plans are interrelated. Some may be developed at the same time. Consideration of the factors of METT-T guides all planning. Special considerations that apply to these operations are:

(1) **Mission.** The nature of actions at the objective and requirements for special items of equipment and nonorganic personnel impacts on how the unit inserts. When the mission requires rapid deployment, the fastest method of insertion may be needed. In other operations, where success of the mission depends on maintaining secrecy, speed may be less important.

(2) **Enemy situation.** Enemy disposition, strength, and security measures along the route(s) near the objective area affect the method of insertion to be used. Planners must take into consideration whether the enemy:

(a) Uses RSTA detection devices or not.

(b) Has a strong air defense system or possesses air superiority.

(c) Has gaps between his defensive positions.

(3) **Terrain and weather.**

(a) **Topography.** Terrain such as mountains, swamps, or heavy forests favors land infiltration. Deserts, arctic regions, and open prairies favor air or water insertion. Areas not inhabited or with a friendly but scattered population favor land infiltration. A heavily populated area, or one which contains large enemy forces, favors air or water insertion. Hydrography--tides, currents, surf, reefs, and sandbars--must be considered when planning water insertion.

(b) **Weather.** Bad weather with reduced visibility and high winds favors land infiltration, and adversely affects air or water insertion. Limited visibility favors land infiltration; good visibility does not. Adverse weather aerial delivery systems (AWADS) lessens the impact of reduced visibility as a limiting factor on air infiltration. High winds may preclude airborne insertion.

(4) **Troops.** The number of men to be inserted and their level of training may be limiting factors. A need for special skills may call for use of nonranger unit personnel whose physical stamina and ability to perform many infiltration techniques may not equal that of the rangers. Whether supporting forces and special items of equipment are available must also be considered.

(5) **Time.** The main concern is whether the ranger operation is quick-response or deliberate-response. During a quick response mission, planning, preparation, and rehearsal time is short. The timing of the insertion is crucial with respect to weather, the enemy, and if an insertion craft is available.

4-12. Air insertion.

The most rapid form of insertion is by air. Rangers and equipment can be delivered by parachute (static-line or freefall technique), fixed-wing aircraft (airlanded), or helicopter (airlanded, parachute, or rappeling).

a. Suppression of enemy air defense along the insertion corridor may be necessary. The USAF will normally assume the major responsibility for suppressing enemy air defense artillery (ADA) capabilities that may interfere with the inserting forces. This is done by a variety of sophisticated electronic countermeasures applied against enemy equipment and by strikes against known enemy positions. Army fire support elements, special electronic mission aircraft (SEMA), and ground jammers may participate, as well as Navy aircraft, jammers, and gunfire.

(1) A main danger area is the forward edge of the battle area (FEBA) or frontier areas where

the enemy employs most of his sophisticated weapons systems and air defenses. Other dangerous areas are population or troop concentration areas behind enemy lines protected by point air defense systems.

(2) Since the majority of the enemy's detection devices and air defense weapons may be located near the point of entry, fire support, smoke screens (even at night), and suppressive measures may be critical. The use of special items of equipment may be needed to counter the enemy's RSTA effort.

(3) Long-range fires (artillery and naval gunfire) should be planned on known and suspected enemy antiaircraft locations and on prominent terrain features along the route(,s). Once beyond the FEBA, inserting units may be beyond the range of conventional indirect fire. More reliance will then be placed on air assets for fire support and electronic countermeasures.

b. There must be close coordination between the ranger unit and the transporting unit. What the transporting unit does during the insertion has a great impact on the success of the mission. To decrease the chance of detection, maximum use is made of periods of reduced visibility and tactical cover and deception. Drop zones and landing zones are chosen behind tree lines, in small clearings in forests, or on other types of terrain not easily seen.

c. All flight routes over enemy territory should be planned in order to avoid occupied areas and enemy air defenses. Routes should complement the cover and deception plan. They should avoid giving away final ranger objectives.

d. In-flight emergencies, mainly in deep-pedetration flights, should be considered, The ranger force receives a preflight briefing on the route to be flown and is informed of checkpoints along the route. Simple ground assembly plans for contingencies are set up before enplaning. In an emergency, the leader of the unit involved decides whether to continue the mission or abort. This is based on METT-T factors, contingency plans, and distance to the objective as compared to the distance to friendly territory. Contingency provisions for air and water rescue are made.

e. Deciding the DZ location is a joint effort. After receiving input from the air and ground commanders, the joint task force commander makes the final decision. During air assault operations, the ground commander is responsible for the overall process. After reviewing MEYF-T factors, it may be determined that surprise is more important than speed, or vice versa. This may determine the number, size, and locations of DZs and LZs.

f. Planning and preparation continue during the marshalling phase of any operation using air delivery for insertion. A review must be made of items of equipment that are crucial to the ranger force reaching its target area, accomplishing the mission, and returning safely. The need for special supplies and equipment, their transport, the distance to be traversed, the delivery means, and external support are all factors that enhance or reduce the chance for success. Once the means of insertion has been decided, logistics planning continues with emphasis on requesting and coordinating external support. The following must be considered:

(1) Detailed planning for mission logistical needs.

(2) Extra sets of key items of equipment to ensure redundancy and self-sufficiency of

inserting units.

(3) Plans for accompanying, planned, and emergency resupply. The ranger force should prepackage loads to meet specific contingencies. It should consider the use of supplies and equipment already in the operational area.

(4) In quick-response operations, only manportable items of equipment needed to accomplish the mission are carried. Plans and signals must be established for resupply of key items.

(5) Specific plans to evacuate casualties.

(6) All attached personnel and augmentation equipment must be properly prepared for movement.

(7) Provisions must be made for evacuating captured enemy equipment and personnel.

(8) All inserting units should be cross-loaded. Items to be considered in a cross-loading plan are the ground tactical plan, the assembly plan, the configuration of the DZ, the light route and formation of the aircraft, and the number of aircraft available. Cross-loading ensures that key leaders and equipment--for example, door bundles--are evenly distributed throughout the assaulting force. If one or more aircraft abort or are shot down, some key leaders and equipment will still arrive at the DZ. This permits mission accomplishment. Cross-loading aids rapid assembly on the DZ in support of the ground tactical plan. If a ranger is to assemble near the leading edge of a DZ, he is loaded on the aircraft to exit near the front of his stick.

(9) A bump plan is used if an aircraft should abort while still on the ground and there are no spare aircraft. The bump plan ensures that key leaders and equipment are not left behind when the main body takes off. If an aircraft aborts on the ground, key personnel move to another aircraft and replace nonessential personnel. This concept promotes success on the objective.

4-13. Airborne insertion.

a. During the planning phase of an airborne operation, there is no room for error in judgment. It can be planned and conducted well only if there is a constant exchange of information. All personnel involved must be informed of all changes in plans, times, and locations.

b. The reverse planning sequence is the key to the success of an airborne insertion. Most of the planning is based on the ground tactical plan, followed by the air movement plan to support it.

c. The USAF is responsible for flight planning, initial point (IP) selection, and crew procedures throughout the flight. The senior ranger in the aircraft must remain oriented and keep abreast of any last-minute changes. He must coordinate with the aircraft commander.

d. Most airborne insertions are made at either very high altitudes (HALO/HAHO) or very low. For combat operations, drop altitudes may be less than 500 feet above ground level (AGL), depending on the type of parachute used.

e. In airborne insertions, emphasis is on the use of special delivery or navigational techniques since normal insertion is during periods of limited visibility.

(1) **Adverse weather aerial delivery system.**

(a) A self-contained navigational system in some USAF C-130 aircraft is AWADS. The system allows the aircraft crew to fly a specific route to a DZ during darkness or bad weather. The system works with a Doppler radar that determines the ground speed of the aircraft, and an AWADS computer programmed with special flight route information (start point, route, checkpoints, wind speed and direction, and air speed). These two instruments show where the aircraft is in relation to the programmed checkpoints. This eliminates the need for airdrops under visual weather conditions.

(b) Only certain aircraft in formation are equipped with the actual AWADS device. Other aircraft are equipped with station-keeping equipment (SKE). This SKE consists of a computer that tells the aircrew where their aircraft is in relation to other aircraft in the formation.

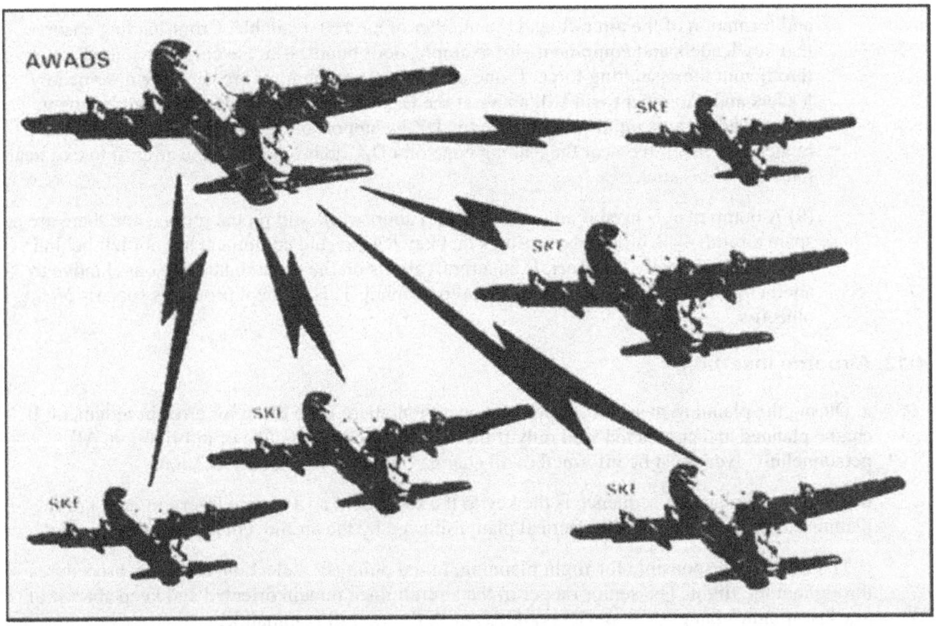

Figure 4-1. AWADS abd SKE aircraft formation.

(c) Airborne insertions using AWADS give the ranger commander the ability to airdrop personnel and equipment during low or zero visibility. This increases the safety of USAF aircraft. It protects them from visually tracked enemy antiaircraft guns and non-all-weather fighters. It also uses adverse weather as a combat force multiplier. Using AWADS, personnel and equipment may be inserted day or night without a pre-positioned USAF combat control team (CCT) or an Army assault team. Time length of the air train is shortened. The air corridor has to be cleared only once. This enhances the element of surprise. The CCTs are brought in with the assault elements. They are able to direct other low altitude parachute extraction systems (LAPES), container delivery systems (CDS), and air-land or heavy drop missions. Use of AWADS enables the ranger commander to conduct rapid, vertical reinforcement of units threatened by an enemy attack during bad weather.

(d) The use of AWADS requires detailed planning, rehearsal, and close coordination among the forces involved. Intense training of both USAF and Army personnel is required.

(e) Personnel must be briefed on terrain through use of photographs, maps, and terrain models. Limited visibility relies mainly on compasses and audible and visual assembly aids.

(f) During reduced visibility, essential heavy-drop serials may precede personnel aircraft. This reduces the time lag between personnel and heavy-drop chalks due to the need to move friendly troops off the DZ. Remaining heavy-drop serials or LAPES can be scheduled following P-hour.

(2) **Reconnaissance teams.**

(a) The operations of the main ranger force may be enhanced by use of reconnaissance teams. These teams are inserted ahead of the main body to provide early warning or to conduct reconnaissance of key objectives. If reconnaissance teams are required at a point along the flight path or near the main DZ, they may be cross-loaded on board any aircraft within the formation. They exit at planned times so as to land as close to their objective as possible. Should such teams be required at a point perpendicular to the axis of the main formation, one of the following is used:

- Another flank aircraft delivers teams to designated points before or with arrival of the main body.
- Selected aircraft deliver the main body, turn, and then deliver the reconnaissance teams.
- The teams are cross-loaded as complete elements on aircraft delivering other units whose flight path crosses the desired locations.

(b) Missions assigned to reconnaissance teams include:

- Providing local security for a USAF CCT, in which case, by joint doctrine, they form the Army assault team (AAT).
- Establishing surveillance or blocking positions on likely avenues of approach into the objective area or DZs.
- Locating and marking targets or assembly areas for the main assault force.

(c) Reconnaissance teams may be inserted by static line or HALO/HAHO operations.

(d) If the reconnaissance element is to locate a specific target or objective, it may be inserted up to 24 hours early. The longer the reconnaissance team remains near the objective area, the greater the danger of it being seen.

(e) In order for the information to be used by the ranger force commander, secure communications and a linkup plan must be prepared and rehearsed.

(3) **Assembly.**

(a) The ranger airborne force can conduct a quick assembly of combat power (weapons, personnel, and command and control elements) on the DZ. Quick assembly and reorganization is crucial because the unit is open to attack. Assembly must be fast and precise. Assembly areas and plans are set forth after considering all the factors of METT-T, mainly the location of the enemy, visibility, terrain, drop information, dispersion pattern, and cross-loading.

(b) The number of assembly areas depends on the location of the DZ, size of the assembly areas, and enemy's detection ability. Unit members move directly to the assembly areas without breaching security. Dissemination of information to all jumpers is the key to a successful assembly plan.

(c) For battalion-size or larger operations, general assembly areas are prescribed in the ranger force OPORD. Company commanders select and designate the exact location of their assembly areas. Although assembly area locations are terrain-dependent, they are generally on the DZ if the drop is at night and off the DZ if the drop is during daylight.

(d) Personnel chosen to carry assembly aids are cross-loaded in the center of their unit sticks. Strong, aggressive leaders should jump first and last in each stick. Upon landing, the leaders roll up their sticks toward the center, locate and move to the proper assembly aid. They may also carry or wear assembly aids. Individual rangers then follow these key personnel. During daylight operations, radiotelephone operators (RATELOS) can display color-coded pennants on their radio antennas to facilitate linkup with leaders. Key personnel may carry and jump with their own radios, and also display color-coded pennants on the radio antennas. Key personnel and RATELOs put their radios into operation before moving to assembly areas. For drops occurring near dawn or dusk, both daylight and night assembly plans are required.

(e) When tactical vehicles or door bundles are dropped, they are normally marked with panels, chemical lights, or strobe lights. The delivery parachutes are often color-coded.

(f) A ranger force should be able to assemble on an unfamiliar DZ, at night, within 45 minutes of the final aircraft pass. Because most ranger operations are time-sensitive, the unit normally moves from the assembly area to the objective when enough of the force has assembled to accomplish the mission. A straggler control point remains at the assembly area under the control of a designated leader. When everyone assembles, or at a specific time, the leader moves to link up with the unit. If available, a radio is left with the straggler control point.

(g) Assembly area security plans are prepared and disseminated. These plans must address hasty defense, sectors of responsibility, observation posts, fire support, and chain of command.

(4) **Contact on the DZ.** During the conduct of parachute insertion, rangers must be ready for enemy engagement at all times. Rangers are open to attack on the DZ before assembling. Just as contingency plans and chance immediate-action drills are developed, briefed, and rehearsed for other chance enemy contacts, rangers must be prepared to execute immediate-action drills for enemy contact on the DZ. The following actions are taken when there is enemy contact on the DZ:

(a) Rangers immediately assemble in fire team-size groups and fight off the DZ to designated assembly areas. Elements attempt to break contact and assemble.

(b) Assembly areas designated by commanders in the mission OPORD serve as primary rally points.

(c) Alternate assembly areas (normally 2 to 5 kilometers from the DZ) are designated in the mission operation order (OPORD). They are used by elements that cannot link up or assemble in the primary assembly area. These elements continue with the mission once assembled.

(d) During limited visibility, personnel should move directly and quickly off the DZ. They must avoid mistaking other friendly elements for enemy, which could cause firefights among friendly forces.

(5) **Reorganization.**

(a) The unit reorganizes according to plans, using designated assembly areas, aids, and identification markings for personnel and equipment. Replacements are named for key leaders or personnel who are lost or injured.

(b) Security teams assemble and move to their positions. Remaining units move quickly to their assembly areas, carrying supplies and equipment.

(c) Subordinate leaders keep the commander informed during assembly and reorganization.

(d) Runners may be used to report assembly status. This maintains radio silence. Designated personnel quickly recover supplies and equipment and move to the assembly areas.

(e) Reorganization is complete when all units have assembled or have been accounted for, and control has been established.

(6) **Pop-up technique.** This involves close terrain flying until the DZ is reached, then the aircraft quickly climbs to jump altitude. Special navigational equipment and aircrew training

are needed but may not always be available.

(7) **HALO/HAHO.** This involves high-altitude, low-opening and high-altitude, high-opening jumps with high performance parachutes designed to maneuver the jumpers to a specific point on the ground. During these operations, midair assembly procedures may be used.

(8) **Arctic.** Airborne insertion in extremely cold conditions is difficult. Times allocated for all parachute and equipment rigging must be increased. Rigging of individual equipment must be IAW FM 57-220. Ahkios can be rigged, packed inside a double A22 container, and dropped in platoon bundles. Because of the increased amount of individual equipment needed, allowable cargo loads for USAF aircraft are reduced.

(9) **Water.** Because of a lack of suitable land DZs, or to avoid detection, rangers may conduct airborne operations onto water DZs close to land. Procedures and techniques are similar to land DZ operations.

(10) **In-flight rigging.** On long-range operations involving extended flight time to the objective area, personnel may rig in flight. In-flight rigging requires detailed planning, coordination, of all jumpers. A rehearsal of these procedures should be conducted before deployment. Aircraft loads must be reduced to carry out in-flight rigging procedures. (See FM 57-220.)

4-14. Air-land insertion (fixed-wing).

a. **Planning factors.** Most of the planning factors that apply to airborne insertion operations also apply to air-land insertion operations. Most USAF aircraft cannot conduct air-land operations in total instrument weather conditions. The AWADS equipment is accurate enough for airdrop navigation, not air-land. Air-land operations are conducted, however, under conditions of extremely reduced visibility. Specially trained USAF crews, using sophisticated navigational techniques, can locate and land on small airfields in bad weather. As in airborne operations, ranger reconnaissance teams, as well as specially trained CCT elements, can jump before the air-land aircraft arrive. They emplace navigational and location markers to assist the pilots. Rangers may rely solely on air-land operations for insertion. This method does not provide for delivery of the force in the shortest time possible. Air-land insertions are normally in concert with an insertion of airborne elements.

b. **Time.** Since turnaround time is crucial, loads in aircraft should be unchained as the aircraft taxies. This also applies to personnel and seat belts. Aircraft should stop as close as possible to the assembly areas where personnel and equipment are off-loaded.

c. **Method.** Air-land insertions normally begin after a ranger Army assault team (AAT) and a USAF CCT have parachuted into the initial objective area and conducted reconnaissance, clearing, marking, and security operations. The ranger ground force commander is the arrival airfield commander. He has control over all arriving, taxiing, and departing aircraft (fixed-wing or rotary) through the CCT, which locates near him. The ranger force command and control element is usually on the first aircraft in and the last aircraft out.

d. **Security.** Air-land operations concentrate large numbers of men and equipment, which presents an open target to the enemy. The ranger ground force commander must ensure rapid off-loading, assembly, dispersion, and movement toward the objective.

e. **Communications.** The ranger ground force commander must have secure radio communications with the ground force, CCT, fire support elements, and higher headquarters. Wire should be laid to connect ground command and control locations.

f. **Operation duration.** Depending on the factors of METT-T, the aircraft may remain on the ground for extraction while the mission is being accomplished or take off at once. They may then orbit, pending recall, or return at a prearranged time.

g. **Extraction.** Extraction of the air-landed ranger force is normally planned and accomplished like an insertion, only in reverse order. A simple, yet effective, way must be found to see that no ranger personnel are left in the objective area.

4-15. Air assault insertion.

Ranger air assault (air-land, rappel, or combination) insertions normally take place during limited visibility. They quickly place the ranger force on, or close to, the objective. The ranger force (depending on the mission) normally plans on fighting when it hits the ground, completing the mission, and then being extracted. Air assault insertions need detailed planning, surprise, flexibility, speed, shock effect, concentration of combat power, and precise timing.

a. **Ground tactical plan.** The reverse planning process is important. The ground tactical plan, developed from the mission assessment, is the first planning area. All other planning begins from this point. The ranger commander and his staff decide stated and implied tasks, and request the proper helicopters for the mission.

b. **Fire support plan.** The fire support available to the ranger commander may be artillery, naval gunfire, mortars, attack helicopters, and USAF or USN tactical aircraft. Because of the timing of ranger air assault operations, preparatory fires are often not used. If they are used, enemy ADA sites normally have priority.

c. **Landing plan.** This plan puts the ranger force into the objective area at the time, place, and proper sequence to support the ground tactical plan. The ranger commander selects primary and alternate LZs and the priority for landing. The exact location of an LZ is determined after a detailed analysis of METT-T. Rehearsals are important. The ranger force must quickly assemble, reorganize, and accomplish the mission. The senior ranger on each aircraft ensures that the personnel on his aircraft offload at the correct LZ.

d. **Air movement plan.** The air movement plan provides for the coordinated movement of the ranger force into the zone of action in support of the 'landing plan. Flight routes, air movement tables, flight formation, altitude, and air speed should be considered. The ranger commander must have secure in-flight communications with his subordinate commanders.

e. **Loading plan.** This plan supports the air movement plan and provides for the pickup zone (PZ) setup and control. It sets priorities for the commander for loading personnel, equipment, supplies, and cross-loading. Air-loading table, aircraft loading formations, bump plan, and rehearsals should be considered.

f. **Staging plan.**

(1) This plan includes PZ selection and operation. The selection of PZs and LZs needs planning and coordination for best use of air assault assets. Site selection in or near the objective must be coordinated in person between the supported ranger commander and the aviation commander. The tactical situation is the key factor in PZ and LZ selection. Other factors are size of landing zones, surface conditions, ground slopes, approach and departure directions, prevailing winds, obstacles, communications, cross-loading, aircraft command and control, PZ and LZ markings, and rehearsals.

(2) The PZs are set up and secured about 15 minutes before touchdown of first aircraft. The fire support plan must support the other plans. All supporting fires must be coordinated with the air mission commander (AMC). Because of the depth of most objective locations, USAF tactical aircraft will be the main fire support method. Orbit locations for fire support aircraft must be planned.

g. **Other planning considerations.** These include escape and evasion, actions at the last LZ, assembly plan, downed aircraft procedures, control measures, forward area rearm and refuel point (FARP), weather delays, and deception plans.

4-16. Amphibious insertion.

Insertion by water may be by SCUBA, surface swimming, small boats, submarines, surface craft, helocasting, or a combination of these methods.

a. **Planning.** Planning for waterborne operations must be thorough. The plan is decided upon between the ranger commander, the transporting unit, and the joint headquarters directing the operation. Plans must be made for all types of enemy action and weather while on board the transporting craft. The transporting unit is given information on a need-to-know basis. Information that could endanger the operation may be withheld until the mission is under way. Initial planning includes:

(1) Time schedule of all events, from the beginning until the end of the joint operation, to be used as a planning guide. Timing for each event must be exact; the success of the operation depends on it. (See Figure 4-2)

Figure 4-2. Waterborne insertion.

(2) Embarkation site, where troops and equipment are loaded on the transporting craft.

(3) Drop site, where the ranger force will depart the primary craft and load into smaller boats.

(4) Handing site, where the unit will beach its boats, or land directly from amphibious craft.

(5) Loading, when loads are lashed and waterproofed IAW unit standing operating procedures (SOPs). Inspections are important. A bump plan for key personnel and equipment must be developed.

(6) Cross-loading, which will help preserve the ranger force. location of leaders, crew-served weapons, and needed equipment are stressed.

b. **Landing site.**

(1) **Beaches.** Beaches are checked to find one that allows an undetected approach. Landing sites that cannot be approached from different directions are avoided. Stretches of defended or patrolled coastline should be avoided. If sand beaches are used, tracks and other signs of

operation that may compromise the mission should be erased. Rural, isolated areas are best. The area behind the landing site should provide an assembly area and concealed routes for exit.

(2) **Barriers.** Reconnaissance may find that good landing sites are not suitable because of barriers between the site and the objective. Such barriers may be patrolled roads, large areas giving little or no concealment, enemy obstacles or defensive positions, areas populated with enemy sympathizers, or an area used for enemy troop maneuvers and training. Unless a means of avoiding such barriers is devised, landing sites should be limited to those without barriers.

(3) **Other factors.** Beach selection should also consider:

(a) The scheme of maneuver.

(b) Surprise.

(c) Enemy dispositions.

(d) Distance to the target.

(e) Landing and exit sites.

(f) Trafficability of the beach.

(g) Cover and concealment.

c. **Routes.** The route the transporting craft takes to the drop site should be planned to deceive the enemy. A major change in the route immediately after the unit's debarkation could compromise the mission. The route should be the same as a route that would be used in some other naval operation--for example, minelaying or sweeping, or a naval patrol. An alternate route must be planned.

d. **Tactical deception.** There should be other plans to deny the enemy knowledge of the insertion besides the water approach route plan. Plans may include electronic countermeasures (ECM), diversions, feints, ruses, or demonstrations.

e. **Actions at the drop site.** Some operations may permit landing directly from the transporting craft. If not, primary and alternate drop sites must be set. The drop site should be at least 1,500 meters offshore to cover any noise in the loading and launching. If the enemy has a surface radar capability, a drop site should be chosen to avoid detection. The drop site may have to be several miles offshore or may need the use of ECM. The manner in which men, small boats, and equipment are loaded must promote fast action at the drop site.

f. **Navigation.**

(1) Ship-to-shore navigation (to the landing site) may be done by dead reckoning. The course may also be maintained by compass navigation, reference to a shoreline silhouette, radar, or prearranged signals from the shore.

(2) Boat handlers, navigators, and other experts are relied on in various phases of the

transport operation.

(3) Lead reconnaissance teams or scout swimmer teams may be used to plant radar, radio, or infrared buoys, or distinct, shielded lights. Boats may be equipped with radar reflectors or beacons to permit accurate tracking.

g. **Actions on the beach.**

(1) Boats hold offshore while ranger scout swimmers move to a covered and concealed position. After reaching the beach, they make a short listening halt. They check the beach and the area immediately beyond the beach for enemy activity. Swimmers signal the force "all clear," mark the landing site, and provide local security.

(2) When the boats land, chosen personnel move directly into covered and concealed security positions. They prepare to defend the landing site.

(3) If the ranger unit is not carrying equipment, boats may be moved off the beach and buried or camouflaged. If heavy equipment makes it hard to move the boats, chosen personnel start unloading at once. The remainder of the force secures the landing site. Each boatload moves into position under the control of a designated leader. Teams are positioned with the best cover, concealment, and fields of fire available.

(4) The ranger force leader sets up a command post from which he can control the operation. He may establish secure communication with his headquarters.

(5) Boats may be deflated and buried or camouflaged near the landing site or away from it. If the boats are to be hidden near the landing site, a team may be chosen to dig holes or cut brush for camouflage along with the unloading of the boats. A small team may be chosen to sweep the beach to erase tracks and drag marks.

(6) The ranger force should be assembled and prepared to move within minutes after the last boat lands. Every ranger must be briefed on the unit's plan should enemy contact be made upon landing.

h. **Insertion by air from ship.**

(1) Helicopters from a ship can launch the ranger force from over the horizon and move at a low level to the LZ.

(2) Helicopters may be vectored from ships to a landing zone. The aircraft may return to the ship while the ranger force conducts the mission, they may orbit offshore, or they may stay at the LZ. Other aspects of landing and assembling are the same as discussed for helicopter insertions.

i. **Helocasting.** This form of insertion combines helicopters and small boats in the same operation. It is planned and conducted the same as air assault operations, except that the LZ is in the water. While helicopters are moving at low levels and low speeds, rangers drop small boats and jump into the water. They then assemble, climb into the boats, and move to shore. Ranger elements are usually assembled and loaded into the small boats within 20 minutes of the last ranger entering the water. This type of operation is normally conducted during limited visibility.

j. **Fire support.** Fire support may be ship-to-air, ship-to-shore, air-to-ground, or ground-to-ground. Specific missions include:

(1) Suppressing enemy air defense.

(2) Neutralizing enemy field artillery within range of boat or helicopter routes.

(3) Employing on-call fires against shoreline targets.

(4) Destroying enemy aircraft.

k. **Contingency planning.** The following must be planned for:

(1) Actions on enemy contact.

(a) En route.

(b) Helocast site.

(c) Flares.

(d) Aerial attack.

(e) Small arms fire.

(f) Indirect fire.

(2) Bump plan with priority to key leaders, crew-served weapons, and radios.

(3) Downed craft procedures.

(4) Escape and evasion.

(5) High surf.

(6) Adverse weather.

(7) Separation.

l. **Rehearsals.** The ranger force must rehearse all aspects of the amphibious insertion. This includes boat launching, paddling, beaching and assembly, boat commands, and capsize drills.

4-17. Land infiltration.

Land infiltration to the objective may be the best way to complete a mission. This is normal when the enemy has total air control or effective air defenses in the operational area to be penetrated, or when the objective is close to friendly territory. Rangers can infiltrate over any type of terrain, in any climate. However, thick forests, swamps, and broken or steep terrain offer the best chance for success.

a. **Planning.** There are many methods of infiltrating by land, and each has its good and bad points. Plans are designed to get the infiltrating force to the objective area without being seen. Plans should include:

(1) Selecting concealed routes for surprise and protection. Primary and alternate routes are based on detailed map and aerial photo study, ground reconnaissance, and data from agent reports.

(2) Avoiding obstacles, populated areas, silhouetting (when forced to cross obstacles or ridge lines), enemy positions, main avenues of approach, and movement along often used routes and trails.

(3) Selecting the time of infiltration to make use of reduced visibility, such as darkness, rain, snow, and fog, and periods of reduced enemy alertness, such as early morning, holidays, and times following protracted movement or combat. The time of the infiltration is important during critical phases (crossing borders or passing through enemy troop concentrations or populated areas).

(4) Using speed and stealth. All rangers must know routes, linkup points, alternate linkup points, time schedules, danger areas, and actions on enemy contact and at linkup points.

(5) Centralizing planning and decentralizing execution. Centralized coordination must ensure that units are acting IAW cover and deception plans. Control should be given to the leaders of the infiltrating groups. Within their designated zones, these leaders select their own routes and formations. Leaders control the rate of advance. When the enemy has electronic detection devices, radio communications must be kept to a minimum. Key leaders normally go with the group that is using the fastest lane or the lane least likely to be detected.

(6) Seeding infiltration routes to the assembly area with unattended ground sensors. These sensors can help the infiltrating force by reducing the chance of surprise contact during the infiltration, and by helping in controlling the forward progress of the units.

(7) Planning for fire support, which must include coverage of routes, danger areas, assembly areas, objectives, and PZs.

(8) Rehearsing actions at the objective, actions on enemy contact, actions at danger areas, and movement techniques.

b. **Movement.** Land infiltration is conducted using one of three techniques:

(1) **Movement in small groups along one axis.** All members of the force use the best route. Small groups are harder to detect, easier to control, and do not compromise the total force if detected. This technique may take too long and it needs an assembly area or linkup point. If the lead group is detected, other groups may be ambushed.

(2) **Movement in one group.** This technique has no reassembly problem. The same route is used by all, making navigation easier. A large force can fight out of a dangerous situation easier than a small one. It has no coordination problems between other infiltrating units. This form of movement is easier for the enemy to detect. If the force is detected, the mission may be endangered.

(3) **Movement in small groups along several routes at the same time.** This technique avoids putting the total force in danger, is less likely to be seen, forces the enemy to react in

many locations, and makes it harder for the enemy to determine the size of the force or its mission. Some groups travel over poorer routes and have problems of assembly, control, and support; the capture of personnel may endanger the mission. Regardless of which technique is used, ranger units normally move to reduce the chance of enemy contact.

c. **Assembly area.** Following movement into the general area of operations, infiltrating units converge on the assembly area or linkup point. Assembly may begin at once after the passage of lines, or it may take place at any time before reaching the objective. The infiltrating force must remain undetected before assembly. If an infiltrating unit is seen at this time, the enemy may learn the location of the assembly area, and the objective. It may take action to destroy the infiltrating force.

d. **Linkup point(s).**

(1) Selection characteristics include terrain that--

(a) Is free of enemy troops.

(b) Offers cover and concealment from enemy observation and patrols.

(c) Does not restrict movement of friendly units.

(d) Accommodates the unit.

(e) Is easily identified.

(f) Offers escape routes to alternate linkup points.

(g) Offers concealment from air attack.

(2) Action sequence for occupying the linkup point.

(a) Halt short of the point.

(b) Establish local security.

(c) Conduct a reconnaissance of the linkup point.

(d) Occupy the linkup point.

(e) Emplace observation posts and listening posts.

(f) Link up with other groups.

(g) Conduct final preparations.

(h) Continue the mission.

(3) Security may be enhanced with the aid of devices such as night vision goggles and night sights. Unattended ground sensors may be placed on likely avenues of approach into the linkup point. Normally, only passive observation devices are used to assist in avoiding detection.

(4) Plans must set a time schedule for actions at the linkup point. All unit members must move in quickly, consistent with security, so that the unit may proceed to the objective rally point (ORP) and on to the objective.

4-18. Stay-behind technique.

A ranger force can be positioned behind enemy lines by employing the stay-behind technique. In this method, a ranger force allows itself to be bypassed as the enemy advances. This is done to allow the ranger force to perform a specific mission. Stay-behind operations may require the concealment, or cache, of supplies before the enemy bypass. Rangers may need to use captured enemy supplies. The location of the assembly area and the plan for extraction are important parts of stay-behind plans.

4-19. Extraction.

The ranger force is normally extracted after the mission is completed. The enemy may also force the unit to abort its mission and to call for extraction. Contingency plans must be made to extract a force under enemy pressure, as well as one that is not in contact. Primary and alternate extraction sites are planned and coordinated with all ,porting forces. Ranger operations are normally conducted against deep targets. Extraction may begin with land movement but normally it includes the use of boats or aircraft. Plans may include primary I alternate means of extraction.

a. **Terrain.**

(1) The terrain is important in choosing the point of extraction - for example, whether or not the extraction site offers good cover from enemy direct-fire weapons and positions from which the extracting force can suppress the enemy. Tidal conditions and PZ size must be considered.

(2) Unlikely locations such as swamps, jungles, or mountain areas are often used for extraction. Specially trained air and naval crews, using the latest extraction devices, make such terrain useable.

b. **Enemy.**

(1) Extraction is best conducted without enemy pressure. However, such pressure could develop during the operation. Detailed plans are made for emergency or contingency extractions forced by the enemy as well as clandestine extractions not under enemy pressure. If under pressure, units will have to use planned fire support to suppress enemy weapons. At the beginning of the extraction with security posted to provide early warning, the unit may start to thin out the same as with a night withdrawal. Ground security may be relied on until the last unit withdraws under cover of suppressive fires.

(2) If enemy contact is made during an extraction, the commander must decide whether to reinsert units or break contact and move to another extraction point. Fire support and suppression are important and include suppressing enemy air defense by deception, ruses, or decoys.

(3) The ranger unit may have special weapons systems or equipment to aid in the extraction. Shoulder-fired antiaircraft missiles may be used to defend against enemy aircraft threatening

the extraction.

c. **Extraction by air or water.** Extraction by air or water is preferred when the resources are available and their use will not endanger the mission. Other factors that favor their use are when--

(1) Long distances must be covered.

(2) Time of return is essential.

(3) The enemy does not have air and naval superiority.

(4) Heavily populated hostile areas block land movement.

(5) Ranger forces are burdened with prisoners of war,casualties, or critical items of equipment.

d. **Extraction by land exfiltration.** Stealth and evasiveness are key elements of exfiltration. This method of extraction is favored when--

(1) The enemy has air superiority.

(2) The enemy is able to prohibit air or water extraction.

(3) The distance to friendly lines is short.

(4) The terrain provides cover and concealment (for movement on foot) and limits enemy mobile units.

(a) Multiple routes of exfiltration may be used to prevent destruction of the force if soldiers or units are detected. Captured enemy vehicles and equipment may be used to assist in the exfiltration.

(b) The ranger force may exfiltrate in one body or in small groups. Exfiltrating in small groups avoids the time delay of assembly.

(5) The exfiltrating force is lightly equipped and not burdened with captured personnel or material.

(6) The exfiltration route passes through an area occupied by friendly civilians or guerrilla forces who can assist the movement.

(7) Areas along exfiltration routes are not inhabited.

(8) The enemy force is dispersed or is not able to concentrate against the exfiltrating force.

4-20. Escape and evasion.

a. A vital part of all premission planning is the development of en-route and post mission E&E plans. Such plans enhance survival of the ranger force, and the transporting air or boat crews. The ranger force commander is responsible for--

(1) Helping in the development of an E&E plan in coordination with all supporting agencies.

(2) Ensuring that all members of the ranger force and the supporting boat or air crew are briefed on the E&E plan.

b. Each E&E plan is unique. Each situation has unique problems associated with E&E. The E&E plan devised by the ranger force commander must address these problems, while profiting from the individual abilities and training of the rangers and their supporting air and boat crews. The following considerations apply to E&E plans devised during ranger operations:

(1) The plans enhance survival of personnel who can no longer accomplish their assigned mission. The senior ranger in a group must decide if there are any remaining missions that the group can accomplish. If not, then he must try to evade and escape enemy capture if unable to link up and be extracted with the rest of the ranger force.

(2) Because of the depth of penetration behind enemy lines, most successful E&E plans may involve either air or water movement away from enemy-held territory. The air and boat crews supporting ranger operations are highly skilled and dedicated. If there is any chance of a successful rescue, these forces continue to try to retrieve rangers escaping from an enemy-held area.

(3) Rangers possess stamina and endurance. They can move a great distance (especially at night) and over rugged terrain to reach an area where they can rendezvous with search-and-rescue (SAR) aircraft or boats. Rangers are also resourceful and highly trained in communication, stealth, and camouflage. This, coupled with their self-discipline, means that successful E&E can be conducted over a great distance, through heavy concentrations of enemy, and over a long period.

(4) Escape and evasion plans for ranger elements should include avoiding contact with the local civilians. However, the aid of friendly insurgent forces may be enlisted. The ranger E&E plan may also include the use of other special operations forces E&E networks that are in place behind enemy lines. These nets must not be compromised by the volume of evaders.

c. Escape and evasion plans should be based on three phases:

PHASE ONE: That portion of the insertion following entry into the hostile zone where personnel survival and recovery is the only thing considered. Should an aircraft or boat carrying rangers be disabled during phase one, the ranking survivor assumes command. He must consider E&E planning and the experience and skills of other survivors. He should be concerned with the security of the mission, caring for wounded, and moving away from enemy activity while contacting SAR elements for recovery.

PHASE TWO: That portion of the insertion that is near the objective and could permit the ranger element to successfully pursue its mission. Although part of the E&E plan, actions by the ranger force during this phase become a forced land infiltration rather than an E&E. The senior surviving ranger must decide if enough personnel and equipment can be moved to the objective area in time to accomplish the mission. If so, he must proceed, taking along any surviving air or boat crew members. If not, he may proceed directly to the extraction point and wait, or he may contact SAR elements for recovery.

PHASE THREE: That portion of the operation after mission accomplishment. It is a hard time to escape and evade since the enemy is alert and reacting to the ranger forces. The confusion caused by the violence of the attack, coupled with the rangers' ability to move quickly cross-country away from the objective area, offers the best chance of a successful escape and evasion. The surviving ranger elements that cannot be extracted can do one of the following:

(1) Move over land at once to a PZ or beach for extraction by SAR elements.

(2) Move to an area where reentry into friendly lines can be made.

(3) Move to a neutral or nonaligned country where either internment or return to friendly control can be effected.

d. Any plan for E&E should provide for the issue to each ranger of an area map, a covert means of recognition, and a small compass. An emergency ration may also be included and any items needed for the climate in that objective area. The ranger force commander must balance the need for other survival equipment against the added weight and bulk involved.

CHAPTER 5

Strike Operations

The ranger regiment conducts strike operations in support of the Air Land Battle. Ranger strike operations are integrated into an overall plan designed to destroy, delay, and disorganize the enemy. They may also cause him to divert a major part of his combat forces and power to rear area security. Strike operations also help create an environment in which other special operations forces can operate. Strike operations can complement air power in disrupting enemy forces in depth. They allow the theater commander to regain the initiative and lead to decisive action.

5-1. Application.

a. The ranger regiment can conduct strike operations either as deliberate-response or quick-reaction missions. It can be inserted into the area of operations by land, sea, or air.

b. Strike operations can be conducted throughout the depth of the battlefield in support of conventional or unconventional warfare. Multiple strike operations can have a synergistic effect and can create situations that help friendly forces seize and maintain the initiative.

c. Strike operations include raids, personnel and equipment recovery, and interdiction of lines of communication. They are best accomplished by ranger units who use conventional techniques and special skills. Strike operations vary in size. When well augmented and supported, the ranger regiment can conduct strike operations deep in enemy territory using up to three battalions. Strike operations normally are limited in time and may be conducted unilaterally or in conjunction with other special operations forces or allied forces. They are normally ended by the extraction or exfiltration of the ranger force.

5-2. Raid operations.

a. A raid is a strike operation conducted behind enemy lines against strategic objectives, targets of high tactical value, time-sensitive targets, or key personnel and facilities in enemy rear areas. It normally involves a swift penetration of hostile territory to confuse the enemy or destroy his installations. Ranger units can conduct deep penetration raids when given the appropriate assets for insertion.

b. A strategic raid by the ranger regiment would be directed by the NCA and conducted under the OPCOM of a unified or JTF commander. An operational raid would be directed by the unified or JTF commander and conducted by that command directly or under the OPCOM of a corps commander. Typical targets would include:

(1) Command, control, communications, and intelligence centers at front level or combined arms army level.

(2) Nuclear and chemical weapons storage sites and delivery means.

(3) Key logistical centers, warehouses, ammunition complexes, or fuel pumping centers supplying logistical support to a specific front.

(4) Air defense and air traffic control integrating centers and air defense weapons sites.

(5) Key power generating and transmitting facilities, hydroelectric dams, and irrigation systems.

(6) Key ports or rail complexes.

(7) Key installations or facilities, such as airstrips, buildings, bridges, dams, or tunnels.

c. The size of the ranger raid force is determined by a detailed analysis of METT-T. The unit committed to the ground phase of the raid mission is kept as small as possible. It may be as small as a ranger company or it could include all the battalions of the ranger regiment. The force must be tailored to complete the mission quickly, violently, and with few casualties. The size of the force may also be affected by the augmentation of specially trained personnel needed for that mission.

d. Raids are normally conducted in five phases. (See Figure 5-1).

Figure 5-1. Five phases of a raid.

PHASE 1. The ranger force inserts or infiltrates into the objective area.

PHASE 2. The objective area is then sealed off from outside support or reinforcement, to include the enemy air threat.

PHASE 3. Any enemy force at or near the objective is overcome by surprise and violent attack, using all available firepower for shock effect.

PHASE 4. The mission is accomplished quickly before any surviving enemy can recover or be reinforced.

PHASE 5. The ranger force quickly withdraws from the objective area and is extracted.

e. This paragraph discusses the raid after insertion/infiltration. All considerations discussed in Chapter 4 must be applied in order for the unit to get deep behind enemy lines. Upon receipt of a raid mission, the following basics are emphasized:

(1) **Maximum use of intelligence information.** The gathering and dissemination of information must be continuous and provided to the raid force even while en route to the target area. To ensure mission accomplishment, the ranger unit must be kept informed of the latest enemy developments in the objective area to prevent being surprised. Only then can the full combat power of the rangers be concentrated at the decisive time and place.

(2) **Plan development.** The reverse planning sequence and the planning steps set forth in FM 101-5 are followed in the plan development.

(3) **Coordination.** Coordination is through the operational headquarters with friendly units (such as higher and supporting Army or joint headquarters, intelligence agencies) for fire, RSTA, special equipment, and personnel and logistical support.

(4) **Rehearsals.** Rehearsals help validate all aspects of planning for the raid and ensure precision in execution-They allow changes to be made in the plan before it is carried out.

f. A successful raid is ensured by--

(1) Launching the raid at an unexpected time or place by taking advantage of darkness and other periods of limited visibility, and moving over terrain that the enemy may consider impassable.

(2) Avoiding detection through proper movement techniques and skillful camouflage and concealment to include taking advantage of natural cover of the terrain, and using sophisticated equipment to detect and avoid enemy forces.

(3) Timing the operation as close as possible.

(4) Using all available support, both organic and nonorganic, to include use of special weapons such as Air Force "smart bombs" and artillery cannon-launched guided projectiles, with the ranger unit using laser target designators.

(5) Performing quick, violent, precise, and audacious actions that focus full combat power at the decisive time and place.

(6) Disengaging quickly upon mission completion.

(7) Withdrawing swiftly using planned routes and including a deception plan.

g. Four functions are normally performed by ranger tactical formations when conducting a raid. Each subelement is organized and equipped to do a specific part of the overall mission. Depending upon the specific mission, nature of the target, enemy situation, and terrain, these four functions are as follows:

(1) The command group controls movement to and actions at the objective. This unit normally consists of the ground commander, other subordinate leaders in the raid organization, and communications to support these leaders.

(2) The security element, whose organization is determined by the mission of the raid force, size and type of enemy force and its mobility and state of alert, terrain and avenues of approach into the area, and the time needed to seal off the objective area. The security element may perform the following:

(a) Secure the ORP.

(b) Give early warning of enemy approach.

(c) Block avenues of approach into the objective areas.

(d) Prevent enemy escape from the objective.

(e) Provide overmatch for the units at the objective and suppressive fires for their withdrawal.

(f) Provide short-range air defense.

(3) The support element provides the heavy volume of fire needed to neutralize the objective. Because fires from this unit are violent and devastating, they must be closely controlled to ensure the precision needed. On order or as planned, fires are lifted and shifted to cover the maneuver of the assault element by suppressing enemy fire from the objective or aerial fires. The support element may also be given specific locations to cover by fire in support of the security element if an enemy quick-reaction force moves toward the objective area. These may include routes into and out of the objective site, key terrain features, or installations adjacent to the main objective. Once the assault has been completed, or on order from the raid force commander, the support element displaces to the next planned position. Organization of the support element is determined by the following:

(a) Size of the objective, the geography of surrounding area, and the enemy threat (to include air) in the area. This element should be able to neutralize the objective (when supported by air or naval gunfire) and to lift or shift fires either when the assault is launched or when ordered by the raid force commander.

(b) Mission of the assault unit.

(e) Suitable firing positions.

(d) Size and nature of the enemy force in the objective area and those enemy forces capable of reinforcement at the objective.

(e) Fire support from other units (air strikes, naval gunfire, surface-to-surface missiles, and artillery fire).

(4) The assault element seizes and secures the objective and protects demolition teams, search teams, prisoner-of-way teams, or other teams.

(a) The organization of the assault element is always tailored to the mission. Each target area must be examined carefully. To assault, seize, and destroy an installation, position, or equipment, the assault element could be organized into one or more assault teams. The element's mission is to overcome resistance and secure the objective and to destroy the installation or equipment. Other specialized teams may also be needed. For example, sniper teams could be needed to remove key sentries. To capture prisoners, liberate personnel, and seize or destroy equipment, the assault element could be organized into assault teams, prisoner teams, search teams, medical teams, demolition teams, or breach teams.

(b) To destroy a point target or installation in a heavily defended area where the USAF cannot get close en to be effective, the assault element might be organized with one small team equipped with laser target designators. From covered and concealed positions, members of the team could then provide guidance for USAF delivery of laser-guided munitions from a safe distance.

h. The site chosen for the raid LZ or DZ must support the planned actions at the objective. There are two options when choosing sites:

(1) The ranger force can land on or near the objective and seize it before the enemy can react. (See Figure 5-2.) This avoids forced marches over land carrying heavy combat loads. If there is no suitable landing area near the objective, or the enemy has a strong reaction force nearby, this option is not favored.

Figure 5-2. Airborne landing on the objective.

(2) The ranger force can land unseen far from the objective. It then assembles, reorganizes, and moves into an ORP near the objective. (See Figure 5-3.) The objective is seized after security and support elements are in place. This option makes coordinated action easier by setting up control of small units before engaging the enemy.

Figure 5-3. Airborne landing distant from the objective.

5-3. Interdiction operations.

a. AirLand Battle doctrine set forth in FM 100-5 is a shift in emphasis toward keeping the initiative in battle. The deep battle seeks out enemy elements separated from the main battle area and attacks them before they can be deployed. Interdiction of selected enemy forces by attacking them or blocking chokepoints hinders, delays, or interrupts the enemy's use of his lines of communication; it denies him the use of certain areas; it destroys industrial facilities, military installations, and equipment; and it stops or slows his advance along a particular route of approach. The ranger regiment's ability to conduct interdiction is vital to execution of the AirLand Battle doctrine.

b. Interdiction by ranger forces includes blocking or channeling approaches, holding key terrain for a limited time. and destroying facilities.

c. Ranger forces conduct interdiction missions that supplement those conducted by long-range Army and USAF weapon systems. Large-size ranger forces are normally used only when other systems fail or are not available, and the target is of such priority that it must be attacked. Small ranger laser target designator teams may work in concert with USAF or USN strike aircraft to direct terminally guided weapons systems. Rangers should be directed against crucial interdiction targets that disrupt, confuse, or delay the enemy and allow friendly forces to gain the initiative at the forward line of own troops (FLOT).

d. Typical interdiction operations are--

(1) Destroying rail lines, switches, rail yards, and rolling stock.

(2) Destroying bridges, overpasses, tunnels, and truck parks.

(3) Closing mountain passes or routes in restricted terrain.

(4) Cutting oil or natural gas pipelines and destroying pumping and distribution systems.

(5) Destroying dams, locks, spillways, and flood control gates.

(6) Destroying rail transload points, switches, and yard engines.

(7) Destroying shipyards, dry docks, wharfs, lighters, cranes, and stevedore facilities.

(8) Destroying electrical generation facilities, power lines, transformers, and grid monitoring centers.

(9) Destroying water pumping stations and filtration systems.

(10) Destroying radio and TV stations, phone lines, microwave terminals, satellite receiving stations, and telephone exchanges.

(11) Ambushing enemy forces and personnel moving along lines of communication.

e. Considerations when using rangers for interdiction are:

(1) Insertion and extraction.

(2) Support resources available.

(3) Ranger force vulnerability.

(4) Target identification.

(5) Threat environment.

(6) Other or supporting means of interdiction.

(7) Command and control.

(8) Time and magnitude of the effect on the enemy forces at the forward edge of the battle area.

(9) Collateral effects of target destruction.

f. Interdiction targets are not identified and attacked at random. They are part of an overall deep attack plan to destroy or cripple an entire system. Interdiction is based on the assigned deep campaign plan that directs which targets are to be attacked, by whom, the results desired, and the priorities of attack for specific systems. Based on his plan, the ranger force commander selects those ranger elements best suited to conduct the attack. Target selection, which requires detailed intelligence and planning, is based on:

(1) **Target criticality.** A target is critical when its destruction or damage would have a major influence on the enemy's ability to conduct or support operations. Each target is considered in relation to other elements of the particular target system designated for interdiction. The criticality of a target changes with the situation. For example, when the enemy has few locomotives, a railroad bridge is less critical.

(2) **Target accessibility.** The ranger force must be able to insert into the target area. Terrain, security controls, location, and means of insertion are considered when studying a target for accessibility. systems. Based on his plan, the ranger force commander selects those ranger elements best suited to

(3) **Target restorability.** The enemy's ability to restore a damaged facility to normal operating capacity affects target selection. The enemy's repair and maintenance ability and his reserve stocks must be considered.

(4) **Target vulnerability.** A target is vulnerable when it is open to attack by means available to ranger forces. Vulnerability is affected by the nature of the target such as type, size, disposition, and security. It is also affected by the means available to attack it such as explosives, incendiaries, and special devices.

g. Key to the interdiction process is the ability to locate and evaluate target systems for attack and the timing of the attack. A target system is a series of related functions that serve a common purpose.

(1) A target is one element, installation, or activity identified for attack such as a locomotive, train, bridge, or prison.

(2) A target complex is several targets in the same general area, such as a railway marshalling yard, an airfield, or port and dock facilities.

(3) A target system consists of an industrial system and its sources of raw materials; the rail, highway, waterway, or airway systems over which these materials are transported; the source of power and method of transmission; the factory complex; and the means by which the finished product is transported to the user. The targeting process is usually accomplished in the special operations force cell at EAC and the battle coordination element (BCE) at corps level.

h. Interdiction techniques include ambush, emplacement of conventional demolitions, target designation for laser-guided munitions, and emplacement of beacons or sensors.

NOTE: The enemy surface-to-air threat normally precludes the use of an airborne platform for direct observation and designation. This may be accomplished by the on-ground ranger force.

i. Through interdiction, ranger units are an ideal force to contribute to the Army's AirLand Battle operational concept. This includes taking the initiative, moving quickly to strike deep in the rear of the enemy, and defeating the enemy force beyond the FEBA. Planners and commanders must ensure that the ranger units are used on missions that are synchronized to contribute to the overall deep battle campaign. Strikes against critical targets must combine with efforts of other assets to take the initiative from the enemy, limiting his flexibility, initiative, and momentum.

5-4. Personnel and equipment recovery operations.

Recovery operations are specialized raids to liberate imprisoned or detained personnel or to return certain items of equipment to friendly control. These operations include recovering and extracting downed or hijacked aircrews and political or military leaders. Rangers use surprise and combat power to overwhelm resistance before harm can come to the detainees or prisoners.

a. **Planning.** The planning of personnel and equipment recovery operations is the same as for the raid. The main difference is that the ranger force commander must plan for the extraction of the recovered personnel and the loading and extraction of sensitive equipment. Personnel and equipment recovery raids are often executed as quick-response missions. They normally require the ranger force commander to plan and execute the recovery quickly.

b. **Augmentations.** This type of strike operation often requires that the ranger force be augmented with nonranger personnel with special skills or abilities. Examples are medical personnel, technical experts, mechanics or crew members trained in repair and retrieval operations, linguists or translators, and other special operations force elements.

c. **Organization.** The organization of the recovery force is the same as for a raid. Sometimes special teams are needed to perform certain missions involved in the recovery. This mission is conducted aggressively, usually ending in close combat during darkness or other limited visibility conditions. Reconnaissance teams may be inserted ahead of the ranger force to reconnoiter the objective and to locate the detained personnel.

(1) All planning and execution takes place as described for the raid. The emphasis must be on detailed, timely intelligence.

(2) Medical teams must be available to care for the detainees.

(3) The ranger personnel must be adept at searching and clearing of buildings.

(4) The ranger force must--

(a) Locate the specific building or area where the friendly detainees are being held.

(b) Gain entry into buildings, use stun grenades, and avoid obstacles and booby traps.

(c) Engage all enemy personnel using selective-kill techniques.

(d) Locate, identify, and secure detained personnel.

(e) Move personnel to a secured extraction site.

(f) Leave no detainees or rangers behind in the objective area.

d. **Execution.** The key to success is surprise and speed. The enemy must be assaulted when least expected. The ranger force must ensure that friendly detainees are not harmed during the assault - either by rangers or enemy personnel.

e. **Extraction.** The ranger force commander should plan for the extraction point to be as close as possible to the point of recovery. This prevents the detainees from having to move cross-country or for the ranger force to transport bulky equipment long distances. For small groups of detainees or small items of equipment, the extraction point can be further away. Recovery operations may use any method or combination of methods of insertion and extraction (see Chapter 4). Close planning and coordination is normally required with US Army, USAF, or USN aviation for both delivery and evacuation of the target area.

CHAPTER 6

Special Light Infantry Operations

Ranger units are normally employed on strike operations. However, broad US strategic needs demand a wide range of other military abilities. The ranger regiment may conduct special light infantry operations when larger conventional forces could not be used. The regiment conducts these operations using light infantry doctrine, methods, and techniques.

These operations include, but are not limited to, movement on foot, military operations on urbanized terrain (MOUT), passage of lines, ambush, and reconnaissance. Ranger units do not normally conduct planned operations against enemy armored forces. Battlefield surprises or METT-T considerations may cause ranger units to engage armored forces for short periods. The ranger regiment can conduct a conventional light infantry defense for short periods. A defense is not normally conducted adjacent to other friendly forces. It is normally a perimeter defense from which the unit then breaks out from encirclement and links up with friendly forces.

6-1. Movement.

a. Most missions involve movement on foot in tactical operations. Ranger units infiltrating behind enemy lines seek to avoid chance enemy contact. They choose the movement that allows them to retain security and control. To avoid loss of surprise and initiative, casualties, and mission failure, ranger units normally--

(1) Avoid chance enemy contact, if possible.

(2) Move on covered and concealed routes.

(3) Avoid likely ambush sites and other danger areas.

(4) Practice camouflage, noise, and light discipline.

(5) Maintain all-round security.

(6) Make unavoidable enemy contact with the smallest element possible.

(7) Retain the initiative to attack at the time and place of the ranger unit's choice.

b. In planning the movement of his unit, the ranger commander must consider all the elements of METT-T and determine how they will affect the movement. He must determine the route, formations, security measures, methods of command and control, location of key leaders and weapons, and action to be taken on chance enemy contact.

(1) **Routes.** The ranger force commander conducts a careful route analysis, using the factors of observation, concealment, obstacles, key terrain, and avenues of approach (OCOKA). He chooses a route that maintains the security of the force while ensuring surprise. He tries to choose a route that will avoid enemy contact and speed movement. The commander makes a detailed route reconnaissance using maps, aerial photographs, and reports from persons familiar with the area. If possible, he reconnoiters the route himself or uses a reconnaissance team. He may have that team act as guides or mark the route using covert marking devices.

(2) **Formations.** Ranger units use the movement techniques found in current light infantry doctrinal publications. The traveling, traveling overmatch, or bounding overmatch technique is chosen based on the likelihood of enemy contact. The ranger force normally tries to avoid detection and contact with the enemy while moving. The commander chooses a movement technique that ensures that if unexpected contact were made, it would be with the smallest element possible and not the entire formation. The distance between moving ranger elements depends on the terrain, visibility, and enemy situation. March interval is normally great enough to allow each succeeding element either to deploy or change direction if enemy contact is made. Distances are close enough that each element could quickly assist the element to its front.

(3) **Security.** Security is to the front, rear, and flanks when enemy contact is likely. A front security element is placed well forward with adequate communications with the main body. The security element is used to detect the enemy and warn the main body. The main body then either changes direction or hides until the enemy is past. If the enemy does detect the main body, the security element may be used to attack the enemy from the flanks or rear. Flank security elements are placed next to the route of march. They move forward either by alternate or successive bounds. An alternative is to have the security element moving adjacent to the column along routes paralleling the direction of march. Rear security is handled the same as front security. Plans are made for the rear guard to support the main body during chance enemy contact. This is done either by maneuver or by furnishing supporting fire.

(4) **Command, control, and communications.** All available communication is used, consistent with OPSEC and movement security, to assist in maintaining control during movement. March objectives, checkpoints, and phase lines may be used to aid in control. The number of reports is reduced, normally only exception reports are needed. The ranger force commander should be well forward in the formation but may move throughout as the situation demands. Communications with security elements are mandatory. Operations security often precludes the use of radios. Therefore, connecting files, runners, and visual signals are normally used. Detailed planning, briefing, rehearsals, and control are valuable if there is enemy contact. Alternate plans are made to cover all possible situations.

(5) **Weapons.** Rangers are placed within the formation to allow a mix of their individual weapons to cover the front, flanks, and rear of the formation. Rifles, grenade launchers, and light automatic weapons cover all sectors. Crew-served weapons such as machine guns, antitank weapons, and mortars should be given designated locations within the formation. This allows the element leader to employ them quickly. They should be placed near the lead element, but not so close as to be engaged along with it. Crew-served weapons are used to

provide overmatch, especially when employing the bounding overmatch technique of movement.

(6) **Rehearsals.** The ranger force, expecting movement as part of an operation, conducts the following combat drill rehearsals:

(a) Conduct movement.

(b) React to enemy contact.

(c) Maneuver under fire (fire and movement).

(d) Cross danger areas.

(e) Break contact.

(f) Conduct a hasty ambush.

(g) Make a temporary halt.

(h) Establish a perimeter.

(i) Establish a rally point.

(j) Take actions upon consolidation.

(k) Use arm-and-hand signals.

(l) React to flares and indirect fire.

c. After a ranger force is inside enemy territory, it must be alert to avoid detection before reaching the objective.

(1) Should the ranger force security element become aware of enemy presence, it alerts the main force. The main force tries to move away without alerting the enemy and continues the mission.

(2) If the enemy does make contact, the security element then engages the enemy, trying to deceive him as to the size of the ranger force and its direction of movement. After breaking contact, the security force divides into small groups and moves to a rally point to rejoin the main force.

(3) The ranger force does not fight en route to the objective unless there is no alternative, then it quickly breaks contact and moves away.

(4) Ranger units may need to set up a temporary defensive position for resupply, evacuation of wounded, planning, or extraction. One technique that can be used is the temporary perimeter defense.

(5) If the moving ranger force comes under indirect fire, it moves quickly out of the impact area and continues the mission.

(6) The most effective means of combating an ambush is through the immediate return of a

high volume of fire by all weapons. This requires alertness, discipline, dynamic leadership, and rehearsed plans.

(a) Regardless of the method of movement, all weapons should be positioned for immediate use. A well-executed ambush is hard to counteract. However, the effectiveness of an ambush can be reduced by the immediate return of fire. Enemy weapons may be silenced, thus creating weak points in the ambush position. Smoke and high-explosive (HE) grenades are effective in executing counterambush actions. They create confusion, provide a screen for movement, and disrupt the ambush plan.

(b) When all elements of a ranger force are not trapped within the ambush, the elements that are free to maneuver should initiate an immediate flank or rear assault against the ambush force. A flank assault permits better coordination with personnel in the ambush. Supporting fires are provided by those weapons that cannot be hand carried and fired in the assault.

(c) Specific actions against a near and far ambush are as follows:

- **Near ambush.** Elements caught in the kill zone return fire at once and move out of the kill zone. Elements not in the kill zone help by maneuvering and conducting a flanking attack. The unit assaults the ambush position.
- **Far ambush.** Those ranger elements not in the kill zone set up an overmatch position and lay down a base of fire for the elements in contact. The elements in contact move out of the kill zone and prepare to assault the enemy, using individual movement techniques and fire and movement.

(7) The reorganization after enemy contact involves the use of rally points, plans for local security, reorganization of the force, evacuation of dead and wounded, and movement based on the unit mission.

6-2. Ambush operations.

a. **Definition.** An ambush is a surprise attack, from a concealed position. It is used on a moving or temporarily halted enemy to destroy or capture him and his equipment. An ambush is a useful tactic because--

(1) Small, well-trained, disciplined forces, with limited weapons and equipment, can destroy much larger enemy forces.

(2) It reduces the enemy's overall combat effectiveness by destruction and harassment of his forces.

(3) Enemy morale and effectiveness suffer heavily at little cost to the force executing the ambush.

b. **Execution.** A successful ambush must be executed with precision, violence, speed, and audacity of execution. For success, ambush operations must emphasize the following:

(1) **Surprise.** Surprise, more than any other single aspect, enhances the value of an ambush. Surprise increases the potential for inflicting damage on the enemy with less danger to the

ranger force.

(2) **Coordinated firepower and shock effect.**

(a) Coordinated firepower is used for maximum shock effect. Massive volumes of accurate fire, explosives, and mines, coupled with an aggressive attack, breaks the enemy's spirit to fight back. Surprise increases shock effect and the chances for success. Shock effect can cover unexpected defects in an ambush--for example, ambushing a much larger force than expected.

(b) All weapons must be sited with interlocking fires in the kill zone and along likely avenues of entrance or exit. Mortars should be used if the terrain permits. Tripods and traversing and elevating mechanisms are normally used with machine guns to lock in fires. All riflemen use firing stakes to mark left and right limits, and elevation stakes. There is a tendency to shoot high in an ambush--especially at night. The M203 grenade launchers are sited to cover the dead space and routes of escape. When the rules of engagement permit, plastic bags containing CS crystals may be taped to the front side of Claymore mines.

(3) **Control.** Control is essential.

(a) Leaders must have contact with all members of their unit to alert them to the oncoming enemy. Leaders should not move around the ambush site during this crucial period. A method used to alert members can be to tie strings or vines to rangers' legs or arms. By a series of light tugs, all members of the ambush can be alerted to enemy presence.

(b) The leader must initiate the ambush with a casualty-producing device. A bank of Claymore mines on a double-ring main is an excellent device to spring an ambush. Other good techniques are to use a 90-mm recoilless rifle firing antipersonnel (APERS) rounds or a machine gun. All must be under the direct control of the ranger leader. Whistles or pyrotechnics must not be used. They will allow the enemy time to react.

- As soon as the enemy is hit, he reacts. The ambush force has only a few seconds to destroy the enemy in the kill zone. He will quickly recover from the initial shock and leave the kill zone--either with a direct counterattack or withdrawal, Subsequent fires and other banks of Claymore mines must be planned.
- The leader always executes the ambush, except when a member of the ambush knows he has been discovered. He then has the authority to execute--with killing fire, not by yelling.
- The cease-fire must be controlled by the ranger leader. A whistle or other device may be used to get attention, and then cease-fire is signaled.

(4) **Security.** The flanks and rear of an ambush site are open to counterattack. Flank and rear security may be enhanced by -

(a) Echeloning in depth.

(b) Designating sectors of observation.

(c) Positioning of RSTA devices.

(d) Enforcing noise and light discipline.

(e) Having a good withdrawal plan.

(f) Securing routes of withdrawal.

(g) Executing with speed and violence.

(h) Positioning security forces to sea] off the ambush area.

(i) Having good camouflage.

(5) **Simplicity.** A simple, direct plan improves the chance of success. The ambush plan must be clear yet concise to offer the greatest likelihood of success. For example:

(a) Mission statements for security, support, and assault elements must be clear, concise, and direct.

(b) Tasks to be performed by the ambush elements should be easy to understand.

(c) Contingency plans should be simple.

(d) Routes into positions and withdrawal routes should not cross. They should be the shortest, most secure routes.

(6) **Training and self-discipline.** Precise and violent execution is how ranger units win against enemy forces. All advantages must be exploited. Discipline must be strict. There must be no sleeping, talking, eating, or smoking in the ambush site. If an ambush is to be set up for long periods, then the elements of the ambush must be pulled back at set times to the ORP for rest. Extended ambushes of 24, 36, or 48 hours require six- or eight-hour shifts. It may take a company to man an extended platoon ambush position. Tired troops cannot man an ambush well. Other vigorous operations cannot be performed all day and have an alert ambush all night.

(7) **Maximum use of specialized equipment.** This includes the most sophisticated weapons, communications equipment, and RSTA devices. The effectiveness of any mission is enhanced by the use of this equipment.

(8) **METT-T.** Every ambush mission must be evaluated by the ranger commander IAW METT-T to design an ambush that works.

(a) **Mission.** Clearly state the mission--for example, stop enemy resupply, ambush enemy armor columns, or capture personnel or equipment.

(b) **Enemy.** State the type of enemy units--for example, guerrillas, main force, logistics carrying parties, or armor units. What is the size of the force, its habits when attacked, and type of support? What is the enemy night vision ability and expected time of movement?

(c) **Terrain.** Is it wooded, mountain, jungle, or urban? Some good ambushes can be set in a city or suburb. Are the routes available to the enemy suitable to his activity?

(d) **Troops.** Squads and platoons are the basic ranger ambush forces. An ambushing force should be able to engage a force two to three times its size--but evaluate enemy and terrain, A platoon ambushing an enemy tank company in daylight in open terrain is risky. But engaging that same column at night or in a forest may give the ranger platoon an advantage.

(e) **Time.** How long will it take to emplace the ambush? A complex, lethal ambush with plenty of "dirty tricks" takes hours to set up; a hasty ambush on a trail takes 15 to 30 minutes. How long is the ambush to be in place? What hours of the day, or at nighttime only? How soon must it be set up? Do not set up an ambush in darkness--even if it is to be executed in darkness. Try to allow every ranger to '6see" and "lay" in his weapon in daylight. Time is important.

c. **Organization of the ambush force.** A ranger unit conducting an ambush must be task-organized to perform the following functions: assault, security, and support. The ambush forces should be task-organized according to the TOE--by platoons, squads, and fire teams. The TOE should not be changed to create smaller elements for an ambush. The TOE formations may be reinforced with machine gun, recoilless rifle teams, or a 60-mm mortar squad.

(1) **Assault.** The elements assigned the assault mission either move directly into their positions or through a release point. The mission may include any combination of the following:

(a) Conduct the main assault.

(b) Halt an enemy's motorized column or any moving target,

(e) Kill or capture personnel.

(d) Recover supplies and equipment.

(e) Destroy vehicles and supplies.

Search teams are not always used. The ranger leader must decide how and when to use search teams. When rangers leave the security of their well-chosen, concealed ambush position, they are subject to the fires of the enemy who may also be hidden and ready. If it is at night, do not throw tripflares or shoot illumination to light the search area as this will illuminate and expose rangers to the enemy. Always assume there is hidden enemy--the ambush will not kill them all. Night observation devices (NODs) or a red filtered flashlight taped to M16s should be used to quickly search in ranger buddy teams. If the return fire from the enemy is great or if the ambush missed the main body, then the leader may choose to break contact and leave without searching the kill zone.

(2) **Security.** The elements assigned the mission of security may move to their positions

directly or by way of a release point. Their missions may include any or all of the following:

(a) Secure flanks, rear, or ORP.

(b) Provide early warning.

(c) Seal off the kill zone to prevent the enemy from escaping or reinforcing.

(d) Assist in executing the ambush.

(e) Cover withdrawal of main ambush force.

(3) **Support.** The units assigned a support mission provide fires that may include employment of--

- Heavy automatic weapon fires.
- Antitank fires.
- Mortar fires.
- Mines.
- Flame munitions.

d. **Ambush site.**

(1) **Choose an ambush site.** When choosing an ambush site, all sources of information must be used to enhance surprise, exploit the enemy's weak points, and take advantage of the terrain. An analysis of the terrain is also based on the other factors of METT-T to include maps, aerial photos, and, if possible, personal reconnaissance. Emphasis is on--

(a) Natural cover and concealment for the ambush force.

(b) Routes of entry and withdrawal (at least two) that should be direct and easy to reach.

(c) Good observation and fields of fire.

(d) Harmless-looking terrain.

(e) Few enemy escape routes.

(f) Limited enemy reinforcement ability.

(g) Nearby assembly or rendezvous area.

(h) Terrain that will canalize enemy into kill zones, and natural obstacles to keep him there.

NOTE: Try to select a site covered by friendly supporting indirect fires.

(2) **Take advantage of the terrain.** Emphasis must be on exploiting all natural cover and concealment afforded by the terrain. Site the ambush and individual positions based on the terrain rather than trying to adapt the terrain to a fixed geometric design.

(3) **Restrict enemy movement.** Restricting enemy movement by natural or man-made

obstacles should also be planned.

e. **Types of ambushes.**

(1) Ambushes have two categories for ranger purposes--area ambush and point ambush.

(a) Area ambushes may be set up by ranger platoons, companies, or battalions. They are used to interdict enemy movement in a given area or inflict casualties on his forces. Area ambushes consist of a series of point ambushes. The size and location of the ambushes are dictated by the METT-T analysis.

(b) Point ambushes are set at the most ideal location to inflict damage on the enemy. Such ambushes must be able to accept the enemy force from more than one direction. The ranger force levels may enable it to execute an ambush in two or three main directions. The other directions must be covered by security that gives early warning of enemy attack.

(2) The basic ambush is linear (Figure 6-1). An ambush is set on an expected avenue of approach. This ambush can accept contact from three basic directions--left, right, and front. The rear is secured by a security team.

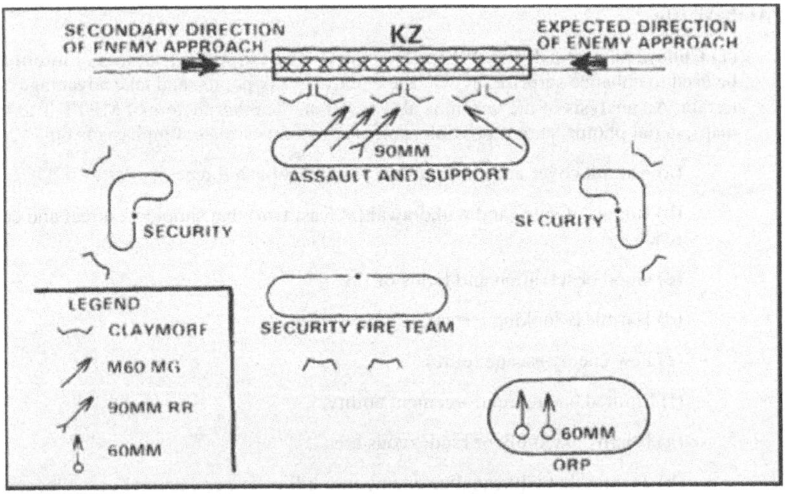

Figure 6-1. The linear ambush.

(3) A variation of the linear ambush is the L-shaped ambush (Figure 6-2). The L-shaped ambush is formed with the base (bottom) of the L perpendicular to the expected enemy direction of advance. This is a good ambush for a road, jungle trail, or an area where the enemy is canalized and his approach route is known. The L-shaped ambush can handle an enemy coming from the expected direction (toward the base of the L) from the front (stem

of the L). It is less effective against an enemy formation that comes from the opposite direction. The ambush must be executed when the enemy main body has cleared the base of the L. The flank security must protect the rear of the base of the L. In the L-shaped ambush, weapons must be sited to avoid direct or ricochet fire into friendly forces. The security elements must be considered as fire fans are being planned.

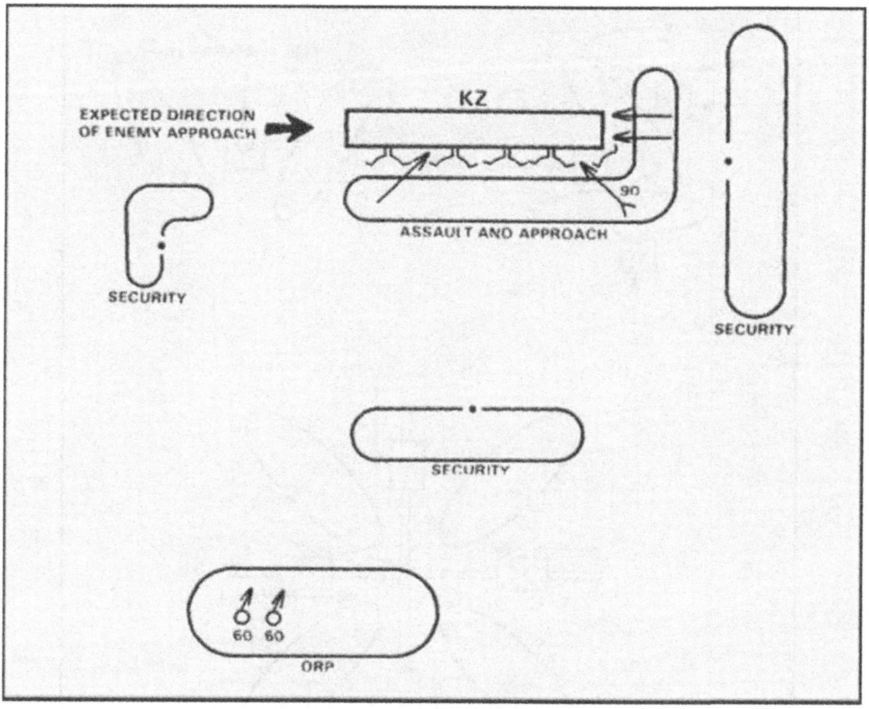

Figure 6-2. The L-shaped ambush.

(4) Other ambushes (the Z, the X, the V, and so on) are all variations of the linear and L formations (Figure 6-3). The more advanced ambush formations are difficult. Units must be expert in the basics before they are ready to employ such techniques.

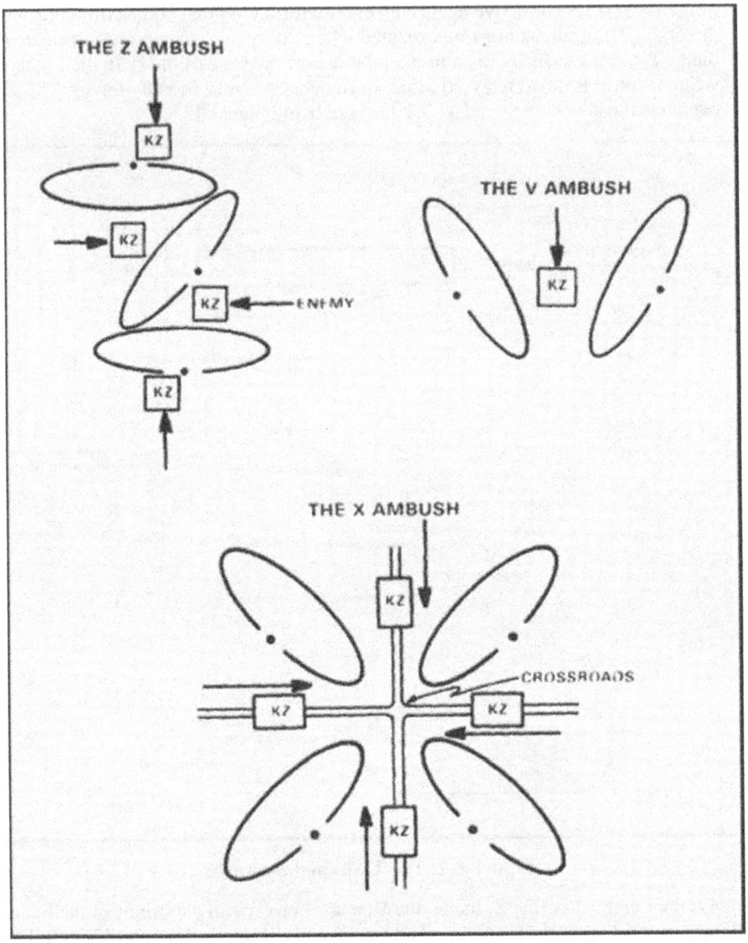

Figure 6-3. Other ambush variations.

(5) Another type of point ambush is the mechanical ambush (Figure 6-4). It consists of Claymore mines set in series with a double-ring main. It is command detonated or detonated by a triggering device activated by the enemy. Mechanical ambushes are normally manned. Rangers prepare to engage the enemy with direct fire if the mechanical ambush does not detonate or if it is wholly or partially ineffective. Mechanical ambushes are an effective way to interdict a large area using a small ranger force. If the mechanical ambush is effective and

rangers do not reveal their presence, the enemy is confused. This has a devastating effect on his morale and effectiveness.

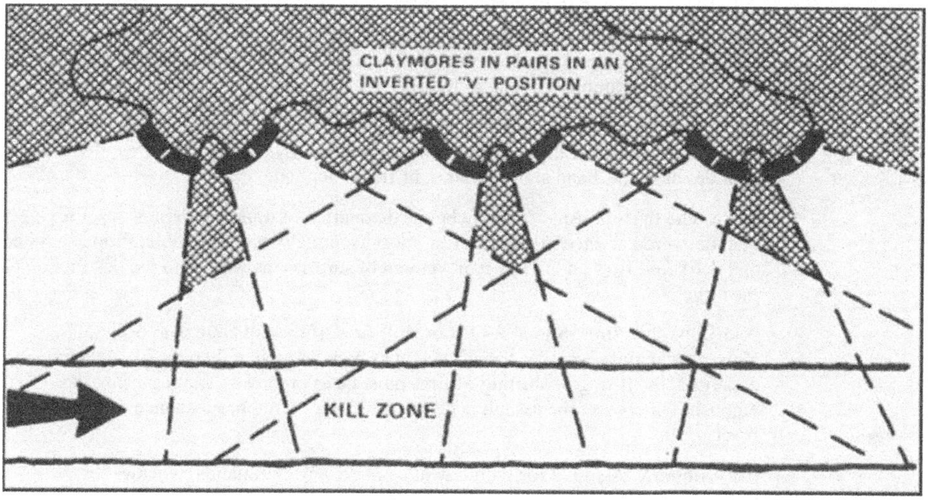

Figure 6-4. The mechanical ambush.

f. **Execution of the ambush.**

(1) **Occupation.** Stealth and security are important factors.

(a) Position security teams and early warning detection devices first.

(b) Use the rest route to main ambush position consistent with security.

(c) Quickly occupy the ambush position and set up communications and signaling devices.

(d) Position key weapons (automatic and antiarmor).

(e) Rig Claymore mines, tripflares, and booby traps.

(f) Ensure that all weapons are correctly positioned. Assign sectors of fire to provide mutual support and cover dead space.

(2) **Camouflage.** Each man must be hidden from the enemy, During preparation for the mission, each man camouflages himself and his equipment, and secures his equipment to prevent noise. At the ambush site, positions are prepared with minimum change in the natural appearance of the site. All debris resulting from preparation of positions is concealed to prevent evidence of occupation.

(3) **Movement.** Movement is kept to a minimum. The number of men moving at a time is closely controlled. Every man is as quiet as possible, especially at night. Light discipline is rigidly enforced at night. Smoking is forbidden.

(4) **Signals.** Audible and visual signals, such as whistles or pyrotechnics, must be changed often to avoid setting patterns and alerting the enemy. Three or four simple signals are needed to execute the ambush. Signals are used to--

(a) Provide early warning of an enemy approach. A signal by the security force to alert the patrol leader to the correct direction of enemy approach may be given. This includes arm-and-band signals, radio, or field telephone.

(b) Initiate the ambush. This may be the detonation of mines or explosives, Fire is then delivered at once in the heaviest, most accurate volume possible. Properly timed and delivered fires add to the achievement of surprise, as well as to the destruction of the target.

(c) Lift or shift fires, Signal for lift or shift fires, if the kill zone is to be assaulted. Voice commands, whistles, or pyrotechnics may be used. When the kill zone is assaulted, the lifting or shifting of fires must be as precise as when starting the ambush. Otherwise, the assault is delayed and the enemy has a chance to recover and react.

(d) Withdraw. A signal for withdrawal can be voice commands, whistles, or pyrotechnics.

(5) **Objective rally point.** The ORP is located far enough from the ambush site so that it will not be overrun if the enemy manages to attack the ambush unit(s). Existence loads are normally left in the ORP. Routes of withdrawal to the ORP are scouted (when possible by each man), Withdrawal routes should provide cover and concealment for the unit and hinder enemy pursuit. Withdrawal routes are a main consideration in the selection of the ambush site. They may be the key to survival after execution of the ambush. On signal, the ambush force quickly (but quietly) withdraws to the ORP. If pursued, the withdrawal may be by bounds, with mines or hasty ambushes used to delay pursuing forces.

(6) **Ambush variety.** If one method of ambush is used predominantly, the enemy will develop an effective defense against it and will be affected less by the shock of the ambush since he knows what to expect. More than one method of ambush should be used. No single method will fit all combinations of terrain, equipment, weather, and enemy capabilities. Variety should also be used when signaling - for example, use multiple signals, both audible and visual. Use weapons fire, mines, and RSTA when possible and vary signals to avoid compromise.

(7) **Swift action.** Speed in the execution of the ambush and the withdrawal should prevent enemy reaction forces from engaging the ambush force. Speed is often a shield against casualties and failure. When there is contact with reaction forces, speed may enhance quick disengagement. Ranger units must not only perform better than the enemy but also move faster than the enemy can react.

g. **Successful ambush.** Emphasize the following:

(1) Intelligence to ensure the enemy is ambushed at a time and place when he least expects or is least prepared to fight. This is preferred during reduced visibility to achieve surprise.

(2) Detailed planning and thorough training and rehearsing of all elements in all phases of the ambush. This ensures maximum shock effect through swift, precise execution.

(3) All available RSTA devices to permit daytime effectiveness at night when moving, shooting, or detecting enemy movement. To avoid detection, active RSTA devices should not be used until after the ambush has been triggered.

(4) All available firepower with emphasis on antiarmor, area and automatic weapons, and precision-guided munitions.

(5) Speed to achieve surprise and enhance security of the force.

(6) Cover, concealment, and overall protection afforded by the terrain when moving or when occupying ambush positions.

6-3. Antiarmor operations.

a. Antiarmor operations are not primary missions for ranger forces. They lack antiarmor firepower and have limited antiarmor ability. Ranger units can, however, perform limited antiarmor missions on close terrain by interdiction or ambush. The antiarmor weapons they do have are man-portable, but the ammunition is too heavy to carry many rounds cross-country. A ranger battalion has ten Dragon missile launchers and nine 90-mm recoilless rifles. It is also equipped with light antitank weapons (LAWs) and laser target designators (LTDs) that can guide Copperhead rounds.

b. Every ranger must be skilled with the limited antiarmor systems available. Proper selection, fortification, and camouflage of antiarmor ambush sites, route selections, withdrawal procedures, and initiative are important. Armor kill zones (AKZs) are normally set up after infiltrating the ranger force behind enemy lines under limited visibility.

c. A ranger antiarmor force is organized into four elements: command and control, support, security, and armor-killer teams (Dragons, LAWs, 90-mm recoilless rifles). These elements function the same as a raid force. The exception is that the assault force is the armor-killer team with the mission of destroying enemy armor, and it will probably not enter the AKZ. After the ambush occurs, the ranger force withdraws to its ORP. Indirect fires (if available) are then brought to bear on the AKZ.

d. Ranger units may conduct antiarmor ambushes and interdiction operation by use of laser-guided Copperhead rounds or air-delivered precision-guided munitions. Air-delivered family of scatterable mines (FASCAM) can be used to provide the ranger force security against attack by armored forces.

6-4. Military operations on urbanized terrain.

a. The ranger regiment normally tries to avoid the manpower-intensive, high-casualty battles associated with MOUT. However, due to the nature of strike operations and the types of targets selected, the regiment may conduct MOUT for limited periods. As in all ranger operations, the ranger force commander tries to surprise the enemy, attack him from unexpected directions, and destroy him with a violent, precise attack. The maze of cellars, rooms, sewer systems, rooftops, and alleys common to urban areas presents many opportunities for the ranger force.

b. Offensive operations by a ranger force exploit the centralized planning and decentralized execution aspects of ranger unit training. Individual action and initiative are always channeled to accomplish the commander's intent. The ranger unit's offensive tactics are based on quick action and boldness. Attacks by the ranger force are short and sharp. Execution is marked by swiftness, daring, timing, and surprise. The enemy's sleeping habits, eating habits, and relief times are watched. launched almost always at night or under smoke screens, attacks are often made without mortars or artillery, relying on surprise for success. Snipers and sharpshooters are used to shift the balance of battle.

c. Deliberate defensive operations are not good ranger unit missions. However, if a ranger unit must defend for a short period, the close terrain of urban areas allows it to function well against a stronger force. The buildings in a city break up advancing enemy formations and force them to move along streets. The ranger force executes an active defense anchored on strongpoints, such as a building or a group of buildings. It launches surprise counterattacks against the enemy's flank and rear. The aim of ranger defense is to give the enemy no respite and to make him feel threatened by a sudden, deadly attack. Ranger units should be withdrawn from defensive positions in urbanized areas as soon as possible. Prolonged MOUT and the associated casualties quickly degrade a ranger unit's combat effectiveness.

6-5. Perimeter defense.

The Perimeter defense may be conducted to control key terrain, gain time or economize forces.

a. **Planning.**

(1) **Warning order.** The order must be issued soon after mission receipt. It should contain the mission, time schedule (including rehearsal and inspection times, mission time, and place for OPORD), uniform and equipment, and special instructions.

(2) **Operation order.** The leader normally gives the OPORD on the ground to be defended. Locations and sectors of fire for squads and crew-served weapons should be designated. If time permits, leaders walk the terrain with subordinates as the OPORD is issued. The OPORD covers security, communications, signals, target reference points (TRPs), armor kill zones, and priority of engagement rules for antiarmor weapons. The OPORD also states the priority of work.

(a) Set up security and communications.

(b) Position crew-served weapons and personnel.

(c) Clear fields of fire.

(d) Prepare fighting positions.

b. **Organizing.**

(1) Squads are positioned with overlapping and interlocking sectors of fire and observation.

(2) The M60 machine guns are positioned either with an assigned final protective line (FPL) of fire or with a primary direction of fire (PDF). The PI)Fs are employed on covered avenues of approach such as gullies. The M60 FPLs should interlock across the unit front.

(3) The 90-mm recoilless rifles and Dragons are positioned with frontal cover to engage attacking armored vehicles from the flank.

(4) Riflemen are positioned to support and protect crew-served weapons. Riflemen are assigned FPLs that interlock at a point just beyond hand-grenade range.

(5) Dead space is covered with Claymore mines, antipersonnel mines, and M203 fires.

(6) Antitank mines, positioned to provide close-in protection from armored vehicles, are covered by LAW fires.

(7) Leaders plan indirect fire for long-range suppression of armored vehicles and short-range protection from dismounted assault. Mortar or artillery final protective fires (FPF) are placed on the most dangerous dismounted avenue of approach.

(8) Range cards are prepared for all crew-served weapons. Squad and platoon leaders prepare sector sketches and fire plans.

(9) The ranger force camouflages everything from ground and aerial observation.

(10) Leaders designate alternate and supplementary positions.

c. **Executing.** Leaders control the distribution and concentration of fires in their sectors. They constantly consider the overlap of fires with units to their left and right.

(1) Squads and crew-served weapons engage targets in sectors assigned by platoon leaders.

(2) Each weapon is used in its best role:

(a) Rifle, against enemy infantry within 300 meters.

(b) Machine gun, against enemy infantry within 1,000 meters.

(c) Grenade launcher, against infantry within 350 meters and armored vehicles within 200 meters.

(d) Light antitank weapon, against armored vehicles within 200 meters.

(e) Recoilless rifle, against armored vehicles within 400 meters.

(f) Dragon, against armored vehicles within 1,000 meters.

(3) Most dangerous targets are engaged first.

(4) Squads shift fires as needed or as directed by platoon leaders.

(5) The platoon masses fire on selected targets while maintaining coverage of the entire platoon sector.

(6) Platoon leaders control fires.

(7) After repelling an enemy assault, the platoon consolidates and reorganizes the defense.

6-6. Breakout from encirclement.

A breakout from encirclement is conducted when units, operating behind enemy lines, find themselves cut off from friendly forces and surrounded by superior forces. Given ranger unit missions, the chance of operating as an isolated force behind enemy lines is great. The breakout is characterized by determining enemy weak points, deception, massing of combat power, and a direct attack for a violent and timely breakout.

a. **Characteristics and fundamentals.** The breakout from encirclement involves the following characteristics and fundamentals:

(1) Set up a hasty defense. Adjust the perimeter as needed and make contact between subunits. Assign each a specific sector.

(2) Plan fire support. Plan FPFs on likely dismounted avenues of approach. Mass antiarmor weapons on the most dangerous mounted avenue of approach.

(3) Conduct reconnaissance. Identify enemy weak points. Locate enemy armored forces and the routes they must use. Locate enemy automatic weapons.

b. **Planning.**

(1) Organize all support personnel into maneuver elements--for example, company headquarters, mortars, and antitank. Organize the ranger force into a rupture force, reserve, support force, and a rear guard.

(2) Issue a fragmentary order. It should contain current local enemy situation and cover all pertinent areas of the five-paragraph field order.

(3) Select rally points.

c. **Execution.**

(1) Reduce the defense perimeter. This is done slowly until the unit is massed for the breakout attempt.

(2) Create a diversion. This is done either by fire from the rear guard force or by a limited attack in a direction other than that of the main breakout attack.

(3) Break through the defense with the rupture force. Hold the shoulders of the breakthrough.

(4) Move the reserve through to exploit the breakthrough.

(5) Move the support troops through with casualties. Have them collect casualties en route.

(6) Have the rear guard follow the support troops. It may have to fight a delaying action and use fire and movement to break contact with the enemy.

(7) Consolidate and reorganize following the breakout.

(8) Have the rupture force break contact with the enemy once the rear guard joins.

6-7. Linkup operations.

During operations behind enemy lines, ranger units may be required to link up with other friendly ground forces. Linkup may occur in airborne, air assault, amphibious, or shore-to-shore operations. It may occur during the breakout of an encircled force or during the convergence of separate forces. It may involve conventional or unconventional US forces, allied forces, or friendly guerrillas. Rangers may be required to link up with other rangers, partisans, or friendly ground forces that are conducting offensive operations.

a. **Planning.** Ranger commanders plan detailed linkup operations to avoid friendly forces firing on each other. Coordination and planning of communications, recognition signals, fires, command relationships and responsibilities, and control measures are essential. Coordination should be done before the start of the operation.

(1) Site. The linkup site should be easy to recognize, have cover and concealment, be located away from prominent terrain the enemy might use, be defensible, and provide multiple access and escape routes. An alternate site should be planned in case of compromise.

(2) **Communications and recognition signals.** Frequencies, call signs, codes, visual signals, and alternates for each of these should be planned before departure of friendly lines. Radios may be used to report the location of each unit to the site, and occupation and securing of the site. A system of mutual recognition must be devised to keep friendly patrols from firing on each other. This may include visual signals such as arm bands, panels, colored lights, or RSTA devices.

(3) **Fires.** The headquarters directing the linkup sets up fire control measures and priority of fires. The most commonly used control measure is the restrictive fire line (RFL), beyond which one force may not fire without coordination with another force. The RFL may be adjusted as two units draw closer together. Such an adjustment needs close coordination and should be planned before the operation. Successive phase lines between the two units can serve as on-order RFLS, provided that neither unit comes too close to the RFL before it is shifted.

(4) **Command relationships and responsibilities.** These are set up by the headquarters directing the linkup. Ideally, liaison personnel are exchanged before the operation. The key to success is liaison and guides.

b. **Execution.** Each member of the ranger force must understand the plan. Rehearsals are necessary in order to avoid confusion during the linkup. One uninformed ranger can start firing and cause enough confusion that friendly forces could be identified as enemy. Actions must be quick since units are open to enemy attack while they try to link up.

(1) **Moving units.** A linkup between moving units is hard. As the units move closer to each other, the chance of them engaging one another increases. Therefore, the linkup units must adjust their movements to each other and continuously coordinate on a designated secure radio net. If possible, one or both units should come to a short halt before linkup.

(2) **Stationary units.** Moving units must know the positions of the stationary units. They must keep stationary units advised of locations. Stationary units guide the moving units to the linkup point by radio. The stationary units must be ready to accept and position the moving units smoothly and quickly.

6-8. Passage of lines.

A passage of lines is an operation in which one unit moves through another unit that is stationary and disposed in a tactical formation on a FEBA. Ranger forces may conduct a passage of lines to get behind the enemy. Movement in forward unit areas must be controlled, coordinated, and kept to a minimum. This avoids conflict with friendly troops or the activation of RSTA devices. Ranger units treat the positions of forward units as danger areas. They must be assumed to be under enemy surveillance in all weather or visibility. The ranger force is open to attack during the passage. Personnel and units may be more concentrated than is desired. Fires of the stationary unit may be masked for a limited time. The passing ranger unit may not be in a good position to react to enemy contact. Detailed reconnaissance and coordination are crucial to ensure that the passage is conducted quickly and smoothly.

a. **Principles.**

(1) **Coordination.** The ranger force commander, or his representative, must coordinate departure and reentry of friendly front lines (FFL). This is done either directly with the units through which the rangers pass or through designated staff agencies. Items to be coordinated include fire support, passage lanes, control measures, guides, communications, intelligence, CSS, and actions on enemy contact.

(2) **Reconnaissance.** The ranger force commander should make a ground reconnaissance of the area through which the unit will pass. He should pick the initial rally point (IRP) at this time. The reconnaissance should include observing the area just forward of the friendly unit for routes, danger areas, and obstacles. During reconnaissance or coordination, rangers should wear the same uniform as the troops in the forward unit. This will avoid attracting the enemy's attention.

(3) **Security.** Maintain security to avoid contact with the enemy while departing or reentering. It is important that the ranger force make no enemy contact at this time. Control is hard if a fire fight starts just forward of an FFL.

b. **Departure of friendly front lines.**

(1) Designate an IRP.

(2) Maintain security. Use an appropriate movement formation when departing the FFL.

(3) Do not move within the friendly unit's area without a guide.

(4) Make final coordination with the friendly unit commander.

(5) Count the ranger force members as they depart.

(6) Make a security and listening halt.

c. **Departure of friendly units.**

(1) The ranger force arrives at the forward unit and is met by a guide. The guide leads the rangers to an assembly area chosen by the ranger commander during initial coordination with the forward unit commander.

(2) No one should move, either singly or as a unit, anywhere in the FFL without a guide.

(3) The ranger commander makes final coordination with the forward unit commander. He determines if any changes have taken place since initial coordination and learns of any recent enemy activity.

(4) Before leaving to coordinate, the ranger commander gives instructions (called a contingency plan) for what should be done while he is gone. These instructions state where he is going, who is going with him, how long he will be gone, what to do if he does not return, and actions to be taken if there is enemy contact.

(5) On the commander's return from final coordination, he may issue a fragmentary order (FRAGO) to cover any changes.

(6) Three common threats and techniques for countering them during departure from the FFL are:

(a) **Ambush and chance contact.**

- The ranger force commander must avoid enemy contact while departing.
- The ranger force goes to the friendly side of the departure point where a point team is dispatched to see that the area forward of the barriers is clear.
- The point element should check the area out to the first covered position or an area large enough to allow the unit room to maneuver if it is hit while departing.
- The distance varies with the size of the ranger force and the terrain.
- The point team notifies the ranger commander when the area is clear. The rest of the unit is brought forward.

(b) **Indirect fire.** If enemy indirect fire is falling, the ranger commander should not halt his unit after final coordination at the forward command post. Instead, he should move it quickly through the FFL. The point element should be far enough ahead of the main body to provide security from ambush or chance contact.

(c) **RSTA.** If the enemy has radar, sensors, or night vision devices, they may be offset by the following countermeasures:

- Use a well-hidden departure point, such as a reverse slope or a dense woods.

- Infiltrate the unit through the departure area and have the rangers rendezvous in a defilade.
- Pass through during rain, fog, snow, or darkness to help conceal the passage.
- Employ electronic countermeasures.

(7) Have a security and listening halt after the ranger force has moved out of sight and sound of the FFL. This is a short halt to see if enemy is in the area, and to accustom the unit to the sights and sounds of the battlefield. Halt in a position that covers the unit from chance friendly small-arms fire. If the security and listening halt must be in an exposed position, have each man lie flat on the ground.

(8) During movement, have frequent halts to see that the unit is not being followed and that no enemy is in the area.

d. **Reentry of friendly front lines.**

(1) Set up and occupy a reentry rally point (RRP).

(2) Maintain security at the RRP and at the reentry point.

(3) Use as few personnel as possible to locate the reentry point while the bulk of the unit remains in the RRP.

(4) Meet a guide at the reentry point. Normally, a forward-of FFL password is used since it may be overheard by enemy.

(5) Count the ranger force members to ensure no infiltrators follow the unit, especially during reduced visibility.

(6) Give the friendly unit commander a spot report, providing combat information of immediate tactical value.

e. **Reentry of friendly units.**

(1) The unit is moved into a rally point near the reentry point. This rally point should be on a prominent terrain feature where the ranger commander can pinpoint his location, with respect to the reentry point.

(2) The forward unit is alerted by radio that the ranger unit is ready to reenter. A code word is used for security and brevity. The code word must be acknowledged by the forward unit before the ranger force reconnoiters the reentry point. This shows that a guide has been sent to the reentry point and is waiting for the unit.

(3) If the ranger commander is certain of the reentry point location, he moves the entire unit directly to the reentry point. He has the point element coordinate the reentry.

(4) An effective method of locating the reentry point is through the use of a thermal imagery device. It detects the body beat of the reentry guide, even though he may be well camouflaged or his location masked by smoke. If ground surveillance radar is available, it can be used to vector the ranger unit to the reentry point.

(5) The reconnaissance party should never move parallel to friendly barriers or move around the wire. If the reentry point cannot be found by the initial reconnaissance or if the reconnaissance party finds only the barrier wire, the unit should notify higher headquarters and move to another rally point to wait until daylight. The unit should not stay in a rally point from which a radio transmission has been made.

(6) When the reentry point is found, the ranger commander can bring the unit forward. If the reentry point is easy to find, he can call by radio for the unit to come forward.

(7) The guide leads the unit through the barriers to the security position coordinated with the forward unit commander. The unit halts in the security position, remaining in movement formation. The ranger commander then gives a spot report to the forward unit commander. The ranger commander tells him anything of intelligence value or of immediate tactical use to the forward unit.

6-9. Reconnaissance platoon operations.

a. The reconnaissance platoon organic to the ranger regiment consists of a platoon leader, a senior reconnaissance sergeant, and three five-man reconnaissance teams that include a communications specialist in each team (see Appendix A). This platoon is responsible for short-duration reconnaissance of objectives for the ranger force commander. This reconnaissance is routinely in support of future strike or special light infantry operations. The teams can also be used by the ranger force commander to conduct deep target acquisition and designation, or to confirm or disprove information reported by other sources. Rarely will teams be used to attack enemy targets.

b. The reconnaissance teams of the ranger regiment do not replace (or perform the same reconnaissance tasks associated with) long-range reconnaissance patrol (LRRP) units or long-range surveillance units (LRSUs) that may be organic to the corps or division. These units are organized, trained, and equipped for long-term passive surveillance and reconnaissance in enemy territory. The ranger reconnaissance teams are trained and organized to support the immediate intelligence-gathering needs of the ranger force commander. They are not to act as a human intelligence (HUMINT) collection asset for other operational headquarters.

c. The reconnaissance platoon headquarters trains, equips, and controls the employment of the three reconnaissance teams. It operates the communications stations needed to receive reports from deployed teams. In the absence of the S2, it debriefs team members upon extraction or exfiltration. It can provide one or more teams to support each ranger battalion on independent operations. The reconnaissance teams can--

(1) Infiltrate the objective area by parachute (HALO, HAHO, or static lines), helicopter, fixed-wing aircraft, SCUBA, small boat, foot, or other means.

(2) Remain undetected in the vicinity of the objective area up to five days.

(3) Perform reconnaissance operations employing a full range of night observation devices, infrared detection devices, unattended sensors, and photographic equipment.

(4) Perform demolition target analysis.

(5) Operate small watercraft and inflatable boats.

(6) Emplace unattended ground sensors, omni-directional navigational beacons, hand-emplaced expendable jammers, and electronic target designation devices.

(7) Collect combat information to satisfy priority information requirements and mission-essential elements of information. Teams report that information by use of long-range, secure, burst-transmission communications equipment.

(8) Perform DZ selection, marking and reception duties.

(9) Report objective area weather conditions.

(10) Perform highly selective, limited attacks or ambushes when so tasked.

(11) Link up with the main body of the ranger force in the objective area, or escape and evade the enemy in order to return to friendly lines.

(12) Act as part of the AAT during airborne operations.

d. There are many limits of the reconnaissance teams.

(1) Mobility is normally restricted to foot movement in the area of operations.

(2) Use of radio and other active electronic or optical devices makes the teams open to enemy detection.

(3) All supplies and equipment needed by the team must be carried on the initial insertion, since resupply increases the probability of detection by the enemy.

(4) Conduct of insertion and extraction operations deep in enemy-held territory is difficult due to the need for secrecy, timeliness, security, and accuracy of location.

(5) Medical treatment of team casualties is limited to individual first aid. Casualty evacuation is difficult unless linkup with the main ranger force has been achieved.

e. The training of reconnaissance team members parallels that given to all members of ranger units with emphasis on infiltration, communications, and reconnaissance skills. Extensive training is also given in survival, evasion, resistance, and escape (SERE) and advanced first aid. Training emphasis is placed on actions at the reconnaissance site. The reconnaissance teams must conduct their mission and remain unseen. Every team member is trained to consider the need for information balanced by the likelihood of detection. Detection by the enemy compromises future operations.

6-10. Reconnaissance planning.

a. Detailed planning ensures that the reconnaissance mission is successful. It is also vital to the survival of the reconnaissance team. This planning is conducted by the ranger force planning group assisted by elements of the JTF staff. The supporting air or naval commander, navigator, and electronic warfare officer may aid the ranger force staff in the planning.

b. Upon receipt of the mission, the team selected begins an intense preparatory phase at the REMAB or the ISB. There the team receives its initial mission briefing. The items covered allow the team to complete its detailed planning. The reconnaissance platoon leader reviews the team plan. He ensures compatibility and integration into the ranger force commander's plan for employment of the main force. Staff specialists at ranger regiment and battalion--such as those concerned with intelligence, weather, or communications--provide help.

c. Planning for the use of reconnaissance teams is conducted within the framework of the ranger planning sequence for the entire mission. There are five planning considerations for the employment of reconnaissance teams:

(1) Operations security must be considered in all staff efforts to provide maximum protection for the reconnaissance teams. This includes intelligence, CE, logistics, administration, and maintenance. Operations security must be integrated throughout every ranger force operation. It consists of four main categories of security measures: signal security, physical security, information security, and deception. All are related and must be considered at the same time for each operation.

(2) Some missions may require rapid deployment to the operational area, thereby dictating the quickest method of insertion. However, others may depend on maintaining total secrecy, regardless of time.

(3) The transportation means selected for the delivery and recovery of the teams depends on the mission. This selection is also based on the abilities, limits, and availability of mission support platforms.

(4) The distance to and from the objective area must be considered, as well as the distance from the insertion point to the actual target. Reconnaissance teams can move great distances over rugged terrain. However, this movement does require time and limits the team's ability to carry bulky or sensitive equipment.

(5) Consideration must also be given to intelligence information available concerning the objective area which includes, but is not limited to--

(a) **Enemy situation.** The enemy threat--his abilities, disposition, security measures, detection, and air defense systems - affects the selection of the means for delivery or recovery.

(b) **Terrain.** Land formations must be considered in selecting the method of infiltration and exfiltration. Terrain affects the selection of altitudes, approach and exit routes, landing areas, drop zones, and beach landing sites. Air infiltration routes that provide terrain masking are desirable in static line parachute operations.

(c) **Weather.** Seasonal weather conditions affect team delivery and recovery. Factors to be considered are temperature, precipitation, visibility, clouds, and wind. If para-SCUBA techniques are used, high surface winds and their effect upon surf conditions or period of reduced visibility may preclude the use of parachutes, inflatable boats, or swimming as entry or recovery techniques. These same conditions favor land infiltration and exfiltration, The AWADS reduces the impact of visibility as a limiting factor for air infiltrations.

(d) **Astronomical conditions.** Periods of sunrise and sunset, moon phase, moonrise and moonset, and twilight must be considered.

d. A reconnaissance team mission must be specific and support the mission. Information concerning the target may not be gathered because the risk of detection by the enemy is so great. Detection may preclude the conduct of the planned operation. If more than one mission is assigned, priorities are set, All team missions must be coordinated to avoid other friendly forces operating near the target. Ranger forces conduct operations at such depth behind enemy lines that no other US or allied forces are normally present. There is a chance that a friendly indigenous guerrilla force may be operating near the objective, but ranger reconnaissance teams do not normally make contact with them.

e. The reconnaissance platoon leader joins with the intelligence and operations sections of the ranger force headquarters in the initial planning for reconnaissance team operations. The method of operation while moving, communication procedures, reporting, surveillance procedures, and actions on enemy contact should be included in the platoon's SOPS.

f. The ranger force intelligence officer normally prepares the reconnaissance plan IAW the guidance provided by the controlling headquarters. He is helped by the operations officer and the reconnaissance platoon leader. Selected team members and a member of the unit providing the insertion and extraction means are briefed on the mission early in the planning phase. They should participate in the detailed planning to follow. The essential details of a reconnaissance plan normally include--

(1) Areas to be observed and recommended areas or positions from which this can be done. These positions are determined in advance of the insertions of teams. They are based on a study of terrain, road and rail nets, enemy order of battle, delivery means, operations plans of the controlling headquarters, and the desires of the ranger force commander.

(2) Load plans and procedures. This includes both insertion and extraction. Alternate sites are chosen for possible use. Primary and alternate DZs are selected, if applicable.

(3) The flight plan. This includes insertion and extraction flight routes and alternate routes.

(4) The movement plan to and from a patrol base. This includes the times, routes,. and destinations when individuals or small groups are required to operate away from the patrol base.

(5) The fire support plan. This includes the location and times of planned air strikes in support of the main body, and all planned fires by the main assault force. The reconnaissance teams should know the frequencies and call signs for tactical air support.

(6) The timing and ground tactical plan for execution of the main force operation. This includes disposition of the main force, special mission teams from the ranger force, or any other special operations force elements.

(7) The communications plan. This includes frequencies, reporting schedule, emergency reporting procedures, and alternate communications plans. It also includes the ground-to-air

communications plan and the friendly recognition signal to be used during linkup with the main assault force or other friendly forces.

(8) Plans for use of technical specialists or special equipment.

(9) The plan for the treatment and evacuation of sick or wounded team members from the objective area.

(10) The E&E plan. This includes routes out of the objective area, passwords, and alternate extraction zones or methods far from the objective area.

(11) Mission-specific elements of information desired by the ranger force commander, and any special or on-order missions the team must conduct.

g. The team leader uses specific steps in planning, preparing, and executing missions. These procedures are comprehensive, yet adaptable to any situation. Three of the most important steps are:

(1) **Issuance of warning order.** This action gives the team maximum time to prepare for the mission. It should be issued as soon as the team leader has been notified of a mission and updated after he has made his tentative plan.

(2) **Issuance of OPORD.** The team leader conveys the information and instructions needed to accomplish the mission. He requires a briefback from each team member.

(3) **Rehearsals.**

h. The specific steps used when planning missions are as follows:

Step 1. Study the mission.

Step 2. Plan use of time.

Step 3. Issue initial warning order.

Step 4. Study terrain and situation.

Step 5. Make tentative plan.

Step 6. Organize.

Step 7. Update warning order.

Step 8. Coordinate (continuous throughout).

Step 9. Consult aerial photographs and map updates.

Step 10. Complete detailed plan. (To be combined with paragraphs 3, 4, and 5 of the operation order.)

Step 11. Issue operation order.

Step 12. Supervise (throughout), inspect, rehearse.

Step 13. Execute mission.

i. A workable E&E plan must be developed. Such a plan enhances the survival of the reconnaissance team members. It ensures that they are psychologically prepared to conduct the mission, by knowing there is a system and plan for returning them to friendly lines. The E&E plan must include a section describing the actions of the reconnaissance team members should their insertion craft be disabled en route to the objective area. It should also include what to do if the team is unable to link up with the main ranger force, or if the main ranger force has not been inserted into the objective area.

j. Rehearsals and briefbacks are best for finding flaws in procedures or planning. Thorough coordination of all procedures is essential. The exact type of aircraft or other means that will be used for insertion should be used for training. Rehearsals should be conducted under terrain, astronomical, hydrographical, and meteorological conditions close to those to be met on the operation. The team leader conducts a briefback with the platoon leader. This lets the platoon leader decide if the team is ready for the mission. It also allows him to make changes.

6-11. Execution of insertion.

a. Reconnaissance teams normally conduct blind insertions. Blind insertions are amphibious, heliborne, or parachute drops onto unmarked DZs without help from a DZ-marking party or a CCT. Blind insertions often depend on favorable light and weather conditions in the objective area. Favorable conditions for the reconnaissance team often means low-lying cloud cover, fog, and darkness. The teams are able to operate during reduced visibility by the extensive use of night observation devices. The team's extensive training and high level of land navigation skills allow it to quickly traverse rugged terrain and not be detected.

b. Team insertion into the objective area often needs a cover and deception plan. Team members can conduct HALO and HAHO operations. Teams can be inserted into areas not normally acceptable as a DZ for conventionally trained parachutists. By the use of HAHO techniques, reconnaissance teams can assemble in the air and glide to a landing at a location far from the drop point.

c. If reconnaissance teams need special or bulky equipment that cannot be carried on the individual team member, it can be airdropped, hidden near the DZ, and retrieved later. Such container drops are often used during low-level static line insertions. The high-altitude airdrop resupply system (HAARS) can be used to deliver such items at the same time as reconnaissance teams are inserted by HALO.

CHAPTER 7

Combat Support

Combat support is used to enhance the combat power of the maneuver companies. Combat support elements may be organic or nonorganic. They can conduct a variety of missions in providing indirect and direct fire support, security, intelligence, and maneuver and fire control.

7-1. Intelligence support.

a. The success or failure of a ranger mission often depends on the accuracy, detail, and timeliness of intelligence information. Both deliberate and quick-response missions need specific information that can be obtained only by a multidiscipline collection and analysis agency. An active interface with all levels of the intelligence system helps get the needed information quickly.

b. The theater or JTF commander is responsible for providing intelligence support. The 1st SOCOM also provides information and analyzed intelligence to the ranger force commander. The main source of processed intelligence is the Echelons Above Corps Intelligence Center (EACIC), similar to the division level military intelligence (MI) all-source production center (ASPC). The EACIC provides intelligence, security, and electronic warfare (EW) interfaces (see Figure 7-1). It is through the EACIC that ranger commanders can request support from national, joint, combined, or CONUS intelligence analysis systems.

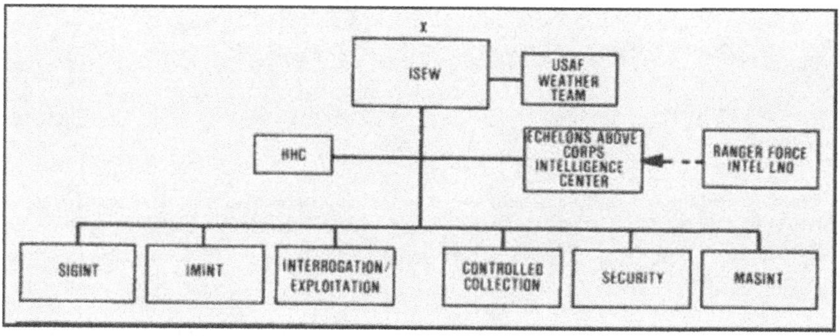

Figure 7-1. Intelligence at echelons above corps.

c. Intelligence requirements for ranger operations are unique, highly sensitive, and compartmented. The intelligence officer of the 1st SOCOM and the appropriate 1st SOCOM regional planning cell coordinate with theater intelligence agencies to provide needed information.

d. The ranger regimental staff and the ranger battalion staff have personnel in their intelligence sections to correlate and analyze information from many sources.

e. The reconnaissance platoon, organic to the ranger regiment, maintains surveillance of the objective area before the insertion of the main ranger force. It reports any last-minute developments that would affect the ground tactical plan. The platoon can also be tasked to determine any mission-specific items of essential information that the ranger force commander needs to complete his plan. This platoon is a highly specialized unit that is reserved for use by the ranger force commander. It is not a HUMINT collection agency for use by the theater or JTF commander. Reconnaissance teams normally link up with the ranger force in the objective area and are extracted with the main body. If this linkup cannot be made, they will exfiltrate enemy territory. They may use escape and evasion nets set up by other special operations forces or Department of Defense (DOD).

f. The ranger regiment normally places a liaison team at the headquarters of the command having OPCOM of a ranger force. This team provides command and control communications and liaison. It puts an intelligence liaison officer (LNO) at the EACIC or the appropriate ASPC (see Figure 7-2). This provides coordination to ensure that the intelligence needs of the ranger force are relayed to the intelligence processing center, and that the analysis is on the specific needs of the ranger force commander. This LNO coordinates with another located at the corps or EAC targeting center. Close coordination between the regiment, the intelligence analysis center, and the targeting center is essential to effective employment.

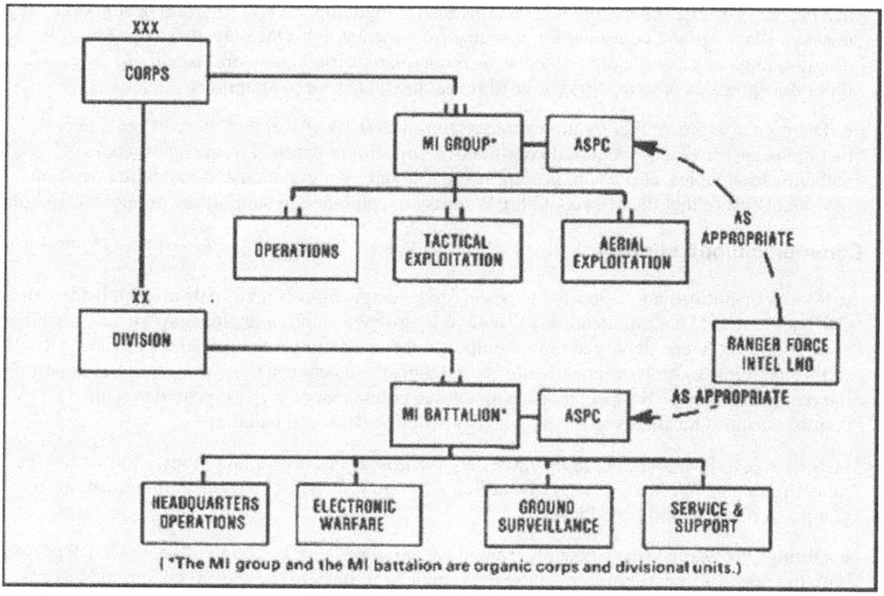

Figure 7-2. Intelligence source at corps and division.

7-2. Electronic warfare support.

a. Depending on the nature of the target and enemy abilities, a ranger mission may need support from EW units. This is mainly true during insertion. Electronic warfare operations disrupt or destroy the enemy's command and control of his forces and weapons systems, and retain friendly use of the electromagnetic spectrum. Electronic warfare is also used to support deception operations. They mislead the enemy by manipulation, distortion, or falsification of indicators, and persuade him to react in a manner prejudicial to his own interests.

b. Active jamming and chaff dispersal can prevent enemy early-warning radar from detecting the airborne force or determining its route. False transmissions can aid deception plans or feints used to help the ranger force's entry behind enemy lines. Selective jamming and imitative deception can disrupt the enemy command and control nets. This prevents the enemy from reacting to the presence of the ranger force in time to prevent mission accomplishment, or slows and interrupts his deployment of reaction forces.

c. Ranger elements, or other military forces, may conduct closely coordinated direct attacks against enemy EW sites during a ranger force's insertion or extraction. Not only does this reduce the enemy's ability to direct EW against the ranger force command, control, and communication system, but it misleads and confuses the enemy as to the true ranger objective.

d. The JTF or theater commander is responsible for planning the EW program to support a ranger mission. Planning and coordinating elements from the 1st SOCOM may also help. The EW program must be closely coordinated so as not to disrupt friendly air-to-ground or ground-to-ground communications, or to reveal the ranger force intentions.

e. The ranger regimental communications-electronics (CE) officer and those of the ranger battalions ensure that the required coordination concerning jammed frequencies, codes, authentication tables, and so on, is completed. The ranger regimental and battalion operations personnel ensure that the efforts of all EW units are combined to support the ground tactical plan.

7-3. Communications support.

a. Ranger operations are supported by secure long-range, lightweight, real-time, high-frequency (HF), and satellite communications. Effective long-range communications provide command and control links between deployed ranger units and the controlling headquarters. The JTF, ARFOR, or theater commander is responsible for communications between the controlling headquarters and the ranger regiment. The special operations force commander may also provide secure communications terminals to the ranger regiment or a deployed battalion.

b. Secure AM, FM, and SATCOM radios are the primary means of communication within the ranger regiment. Within the ranger battalion, AM and FM radios provide communications to company, platoon, and squad level.

c. During certain missions, specially trained and equipped quickreaction elements (QRE) deploy with the ranger force to provide secure communication links to the special operations force commander. These JCSE elements operate on either SATCOM or tactical satellite terminal (TACSAT) channels or through an airborne communications relay platform (see Figure 7-3). Depending on the mission, an airborne battlefield command and control center (ABCCC) aircraft or a joint airborne communications center/command post (JACC/CP) may be used. Their communications systems can operate at all levels of the national chain of command to permit a quick response to the tasking authority.

Figure 7-3. Airborne communications relay platform.

d. When only one ranger battalion is employed and the regimental headquarters is not the controlling headquarters, the ranger regiment provides a liaison cell to the controlling higher headquarters. This liaison cell includes a communications element from the regimental communications platoon. It can provide secure SATCOM, teletype, and facsimile transmission support to the ranger force in the objective area.

e. If two or more ranger battalions are employed, the ranger regimental headquarters deploys and acts as the command and control headquarters. The regimental communications platoon would then provide another communications link to the special operations force commander.

f. The ranger regimental CE officer ensures that the necessary communications links are set up and coordinated. The many communication means and channels available provide for effective control of a deployed ranger force. However, communication means must be closely coordinated at all levels of command to control the complex operations of a ranger mission. Planning and coordination with supporting aviation, transport, fire support, medical, and logistical elements before an operation are vital to efficient communications. The regimental CE officer must consider the communication systems linking the ranger force and other services. Air-to-ground and ship-to-shore communications are vital and must be set up early in ranger operations.

7-4. Fire support.

a. **Planning.**

(1) The ranger regiment has limited organic fire support assets. The six 60-mm mortars in each ranger battalion (two in each rifle company) provide the only organic indirect fire

available to the battalion commander. Each rifle company also has three 90-mm recoilless rifles and three Dragon missile systems that provide a small measure of organic direct fire support.

(2) The missions assigned to the ranger regiment and its subelements normally require fire support from organizations outside of the regiment. To plan, coordinate, and control these fires, the ranger regiment is organized with personnel dedicated to these tasks.

(3) The ranger regimental headquarters has a fire support element consisting of one fire support officer (FSO), two noncommissioned officers, and two fire support specialists. They prepare the fire support annex to the regiment's operation order. They ensure that all available fire support is planned to best support the regimental commander's ground tactical plan. They act as the primary coordinator for all external fire support, such as artillery, armed helicopter, close air support, naval gunfire, and short-range air defense. They also coordinate with each ranger battalion FSO to ensure unity of effort. The fire support element coordinates the efforts of air force, naval, and marine liaison officers when they are attached.

(4) The regimental commander may assign part of the fire support element to a committed ranger battalion as an attachment to help in the integration and planning of fires. The ranger battalions request through the fire support element that more or different fire support means be integrated into the ground tactical plan.

(5) The ranger battalion headquarters has a fire support team headquarters consisting of the battalion's FSO, two noncommissioned officers, and two fire support specialists. The FSO is the prime fire support coordinator for the ranger battalion commander. He maintains coordination with the regimental FSO and any attached fire support coordination elements, such as the USAF tactical air control party or naval shore fire control party. The FSO performs the following:

(a) Advises the ranger battalion commander on all fire support matters.

(b) Recommends allocation of fire support.

(c) Prepares fire support plans.

(d) Assigns target numbers.

(e) Processes target lists from the ranger company fire support teams to eliminate duplication.

(f) Monitors and functions as net control on the ranger battalion fire support coordination (FSCOORD) net.

(g) Reports changes in the status of fire support units to the ranger battalion commander, staff, and FISTS.

(6) When the ranger battalion is operating a single TOC, the FSO and his element are normally located there. When the ranger battalion is operating two TOCS, one TOC is normally manned by the FSO, the fire support sergeant, and a fire support specialist. The

senior fire support sergeant and the other fire support specialist would then monitor the FSCOORD net at the other TOC. The FSO ensures that any attached fire support elements also divide when needed to operate in a two-TOC configuration.

(7) The ranger battalion fire support team headquarters has three fire support teams assigned to it. These are normally allocated one to each of the ranger rifle companies. The company FSO and his team--

(a) Locate targets and request and adjust surface-to-surface fire support (mortar, field artillery, and naval gunfire).

(b) Plan fires to support the company ground tactical plan and prepare target lists.

(c) Coordinate fire support requests through the FSO for surface-to-surface and air-to-surface fires.

(d) Report battlefield information.

(e) Control air-to-surface fires of helicopters and prepare to assume control of fires from fixed-wing aircraft.

(f) Coordinate airspace use within the ranger rifle company operational area.

(g) Advise the ranger company commander on all fire support capabilities, limitations, and methods of employment.

(h) Inform all fire support units of target priority changes.

(i) Designate targets for laser-guided munitions.

(j) Assume operational control of, and administrative responsibility for, all augmentation fire support personnel (TACP, shore fire control party, and attack helicopter liaison teams).

(8) The company FSO is normally with the ranger rifle company commander. The FOs and their RATELOs are normally with the ranger rifle platoon leaders.

b. **Coordination.**

(1) Coordination of fires is a continuous process. It involves implementing the fire support plans of the ranger force commander and reacting to changing tactical situations. The success of a mission often hinges on how well fires are coordinated during the operation.

(2) The fire support planners within the many echelons of the ranger force become fire support coordinators during the actual execution of the operation. They use standard fire control coordination measures.

c. **Organic fire support.**

(1) **Indirect fires.**

(a) The 60-mortars constitute the only organic indirect fire support available within the ranger regiment. The mortar squads in each ranger rifle company normally operate as a section under the control of the mortar section sergeant. However, they can operate alone for short periods.

(b) The 60-mm mortar is used to destroy enemy equipment and light vehicles and to attack enemy personnel. Its main advantage is the ease with which it can be brought into operation against the enemy. This ease of employment allows the ranger force commander to concentrate combat power at the critical time and place. This helps to surprise and destroy the enemy.

(c) Mortar fires are employed by the ranger rifle company commander and coordinated by the company FSO. They are normally requested and adjusted by the FIST observers. The mortars are normally emplaced in a central position, and their fires are controlled by the fire direction center (FDC) using the indirect-fire method. The mortars may also use the direct-alignment or the direct-lay method.

(2) **Direct fires.**

(a) Each ranger battalion is authorized 90-mm recoilless rifles and Dragon missile launchers. These weapons constitute the heaviest organic direct-fire ability in the regiment. The antitank section in each ranger rifle company may operate as a complete section, but normally individual weapons are assigned throughout the company sector wherever their fires can best support the mission. These weapons can destroy most enemy armored vehicles and can penetrate the flank and rear armor of all enemy main battle tanks. The 90-mm recoilless rifle has a multiple projectile round that is effective against exposed enemy personnel. The HE round is highly effective against buildings and bunkers.

(b) The main disadvantage of direct-fire antiarmor weapons is their weight. These weapons, and their ammunition, are man-portable but place a strain on the unit when moving across rugged terrain.

(c) The ranger battalion is trained in the use of light antiarmor weapons, such as the US M72A2 LAW and the Warsaw Pact rocket-propelled grenade (RPG). The following provide protection for the ranger force against attack by enemy armored forces:

- The employment of light antiarmor weapons at all organizational levels.
- The placement of the 90-mm recoilless rifles or guided-missile systems on the most likely avenues of enemy-armor approach.
- An aggressive, offensively oriented employment.
- The ability to move into and across terrain impassable to armored vehicles.

d. **Field artillery fire support.**

(1) Many ranger unit operations are out of the range of supporting field artillery fires. Whenever such fires are available, they are planned for and integrated into the ranger force ground tactical plan.

(2) Coordination and exchange of call signs, frequencies, and target lists occur before insertion of the ranger force. Unless required by the tactical situation, ranger units do not normally receive augmentation forward observers from supporting field artillery units. The degree of mental and physical training necessary to execute certain ranger techniques requires more preparation than field artillery units are able to provide. The FIST from the ranger battalion requests and adjusts field artillery fires for the ranger force.

(3) Field artillery fires can be used to support the ranger force even if the objective area is out of range. Field artillery cannon fire and multiple rocket launch system (MRLS) fire can be used to suppress enemy air defenses. This helps the ranger force as it crosses the FEBA during airborne or air assault operations.

(4) Field artillery fires can be used to support the exflltration of ranger elements as they approach friendly lines. These fires can contribute to the deception plan and add combat power to feints used to support ranger operations.

(5) When in range of the objective area, field artillery units can be used to emplace field artillery delivered FASCAM to enhance the security of the ranger force.

(6) Copperhead rounds fired by 155-mm field artillery units can be terminally guided by the ranger FIST forward observers (FOs). They can attack hardened point targets or enemy armored vehicles by using a man-portable laser target designator (see Figure 7-4).

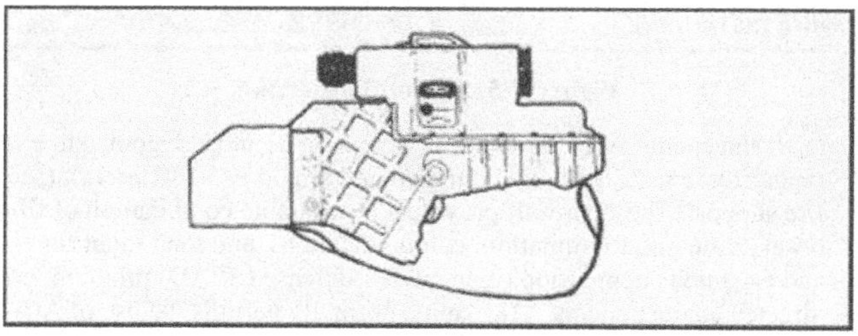

Figure 7-4. AN/PAQ-1 laser target designator.

e. **Aerial fires.**

(1) Aerial fire support is usually the prime means supporting the ranger force due to the distance behind enemy lines at which most ranger operations take place. Aerial fire support can be provided by either fixed-wing aircraft or helicopters.

(a) Fixed-wing aerial fire support may come from USAF, USN, or USMC units. The type of unit providing support, the aircraft, and the mix of ordnance carried all affect the fire support planning and coordination process. Some aircraft have a night and all-weather strike ability enabling them to support the ranger force during any level of visibility. Operations during poor weather that limits visibility to less than 3 nautical miles are still somewhat restricted. The ranger fire support coordinators must ensure that the correct aircraft are requested and employed effectively on the enemy. The

TACP directs and adjusts aerial fires in the objective area.

(b) The ranger force can use ground laser target designators to pinpoint targets for air strikes, as well as electronic navigation aids to permit nonvisual air strikes (beacon bombing). The ranger rifle company FIST or the TACP can control a laser-designated standoff air strike (see Figure 7-5).

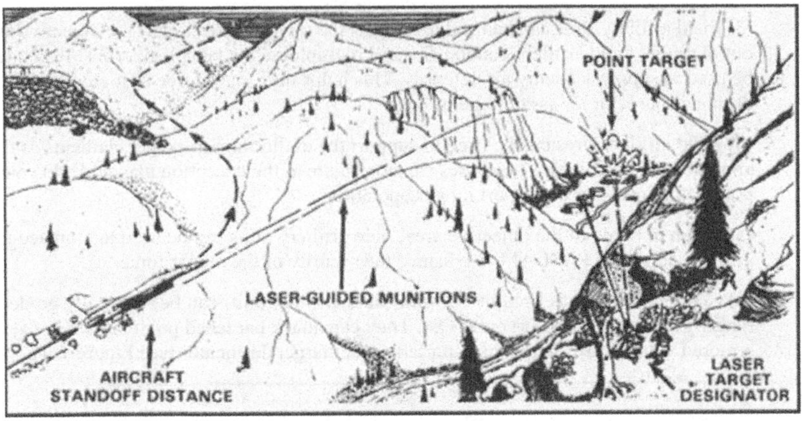

Figure 7-5. Standoff air strike.

(c) If the enemy ADA ability is not great or it can be degraded to a low level, the ranger force uses specially equipped and armed AC-130 aircraft (see Figure 7-6) for fire support. These aircraft provide an invaluable combination of firepower, night observation and illumination, communications, and long loiter time. A well-planned and executed suppression of enemy air defense (SEAD) program, coupled with ECM directed against enemy ADA units, normally permits the use of AC-130 aircraft.

Figure 7-6. AC-130 aircraft.

(2) The attack helicopter, armed with a mix of antitank guided missiles (ATGM), 2.75-inch rockets, 20-mm cannon, and 40mm grenade launchers, is accurate and responsive aerial fire support. The ranger force commander plans to use all sources of aerial fire to help accomplish his mission. However, the nature of ranger operations may preclude the extensive use of armed helicopters due to their limitations. Some attack helicopters are limited in range and lack an all-weather ability. They are restricted during operations at night by a lack of sophisticated night vision devices. They may lack the speed to go with the ranger force on some heliborne insertions. Attack helicopters may be used to escort and assist the ranger force as it crosses the FEBA. They may also be used to conduct feints and demonstrations to cover the insertion of the ranger force. As the MV-22 Osprey and the AH-64 Apache enter service, more use will be made of these aircraft to support ranger operations beyond the FERA, (See Figures 7-7 and 7-8).

Figure 7-7. MV-22 Osprey.

Figure 7-8. AH-64 Apache attack helicopter armed with Hellfire missiles.

(a) If attack helicopters are used to support a ranger operation, planned indirect fires are normally delivered along entry and exit corridors. Attack helicopters approach and depart the objective area using nap-of-the-earth flight profiles.

(b) Fires from armed helicopters are normally requested and controlled by the company FSO or one of his FOs, operating on a special ground-to-air net. The laser target designator may be used to precisely identify targets for the AH-64 Apache. Friendly unit locations may be marked by smoke, panels, lights, mirrors, or infrared sources.

f. **Naval gunfire.**

(1) During amphibious assaults or operations near a coastline, the ranger force may receive indirect fire support from naval gunfire. Normally, a ranger battalion would be supported by either a destroyer or a cruiser in a direct support role. However, the type and importance of the mission, the type of targets, the ships available, the hydrographic conditions, and the enemy capability determine how many and which type of ships are provided to support the ranger force.

(2) Naval gunfire is characterized by large volumes of highly destructive, flat-trajectory fire. Planned strikes in support of the ranger force may also include surface-to-surface missile fire. Some naval guns can fire a laser-guided projectile much like the Copperhead. This projectile, called a Deadeye round, is terminally guided the same as the Copperhead. Deadeyes can be guided either by an attached spotting party from the ANGLICO or by ranger LTD teams.

(3) When ranger units are being supported by naval gunfire, a shore fire control party is attached to the ranger force, This party normally consists of a liaison team and several spotter teams (forward observers). The liaison team is integrated into the operations of the fire support element (FSE) at the ranger battalion TOC. The spotter teams are attached to the ranger rifle companies. The shore fire control party LNO is the ship's representative to the ranger force commander, through the FSO. The spotter teams request and adjust fires from surface vessels. They can also request and control air strikes by carrier-based aircraft.

(4) The LNO and the spotter teams operate in the ground spot net, communicating with the ship by high frequency (HF) radio to request and adjust naval gunfire. The spotter teams communicate with the LNO using very high frequency (VHF) radios. The LNO also can communicate with aircraft using ultrahigh frequency (UHF) radios.

(5) Coordination and control measures that apply to naval gunfire are the same as for field artillery, with the addition of two terms, which are:

(a) Fire support area (FSA)--a sea area within which a ship may position or cruise while firing in support. It is labeled as FSA and numbered by a Roman numeral - for example, FSA VII.

(b) Fire support station (FSS)--a specified position at sea from which a ship must fire. This is very restrictive positioning guidance. It is labeled as FSS with a Roman numeral--for example, FSS VII.

g. **Air defense artillery.**

(1) Army air defense artillery units are rarely used to directly support the ranger force since

ranger operations normally take place deep behind enemy lines. However, it is possible that during the insertion or extraction phase of an operation, ADA units may support the ranger force as it crosses the FEBA.

(2) The ranger regiment's fire coordination element and the FSO in the ranger battalion are mainly concerned with adequate coordination of ADA fires. This prevents friendly units from being engaged. It allows the integration of ADA fires into the scheme of maneuver. The fire support planners ensure that enough coordination has been made so that ADA units are aware of the ranger force and know enough about its flight paths, routes, and altitudes to prevent engagement by friendly units. When conducting this coordination, care must be taken to pass only needed information to the ADA unit. To aid coordination, an ADA officer may be attached to the ranger force.

(3) The plan for a heliborne or airborne insertion and extraction of the ranger force must include the correct identification-friend-or-foe (IFF) codes, the flight profiles, the times of FEBA crossings, and the estimated number of aircraft making the crossing. There must also be a plan for crossing by straggler aircraft and battle-damaged aircraft that cannot reply to IFF interrogation. If possible, the routes selected should avoid concentrations of friendly ADA units.

(4) During an aerial crossing of the FEBA by a ranger force, the regimental headquarters may assign a liaison officer to the headquarters controlling the ADA fires in the crossing area. The LNO is then responsible for close and continual coordination. This ensures that the crossing is made without friendly ADA units engaging the ranger force while it is exiting or entering friendly lines.

(5) During ranger operations, the primary source of air defense fires is shoulder-fired antiaircraft missiles carried by specially trained individual rangers. These weapons provide the ranger force commander a limited defense against attack by enemy aircraft. The decision to take these weapons into the objective area depends on the factors of METT-T. Ranger operations should be completed before the enemy can react with ground forces. However, enemy aircraft may be able to react in time to threaten the ranger force. These lightweight missiles offer a highly effective, short-range counter to enemy air attack.

(6) The ranger force commander must define the weapons control status and rules of engagement for all short-range air defense (SHORAD) teams under his control. Priorities of target engagement must also be determined to prevent multiple engagements of a single enemy aircraft. The ranger battalion FSO and the attached ADA officer are responsible for the coordination of ADA fires within the objective area. Standard ADA weapons-control status and rules of engagement may be modified by the ranger force command.

7-5. Aviation support.

a. **Planning.**

(1) The ranger regiment may need extensive aviation support on some missions. The ranger battalion is organized and trained to move quickly to the objective area by air and to be resupplied by airdrop for short periods.

(2) Often the decision to execute a mission depends on the amount and type of available insertion and extraction aviation support. Normally, the theater commander or JTF commander allocates the type and amount of aviation assets to support a ranger operation. The theater or JTF commander's staff is responsible for the planning and coordination concerning aviation support. The ranger force commander must play an active part to ensure mission success.

(3) Due to the extensive training requirements and specialized operational techniques needed to support special operations, special air operations units in the Army and the USAF have been equipped and trained for the mission. These forces train with the ranger regiment and jointly develop tactics, techniques, and procedures of employment. These special units operate diverse aircraft that have unique capabilities. These help the ranger force commander to effectively employ his units. Command of these special units is maintained at a very high level in the chain of command. They are assigned to support a specific operation. Then they return to their parent unit's control.

b. **Coordination.** In addition to the staff of the theater commander or JTF commander, the coordination of aviation support occurs at all levels in the ranger force. These operations often are conducted in phases. A separate air mission commander is then designated. The AMC is responsible for the completion of each phase of the aviation mission, as well as the close and continuous coordination with the ranger force, commander. The ranger regimental air operations officer and the movements control officer are responsible for the planning, integrating, and controlling of aviation support.

c. **US Air Force.**

(1) The USAF provides special aviation support to the ranger regiment and other elements of the special operations force. The USAF has elements that are equipped with MC-130 Combat Talon aircraft (see Figure 7-9) and AC-130 Spectre gunships, as well as HH-53 Pave Low, UH-60, and UH-1 helicopters. These aircraft are specially equipped to conduct long-range infiltration, exfiltration, attack, and search-and rescue operations. The USAF also provides C-141B and C-130 aircraft for strategic and tactical airlift of the ranger force. There are specially trained USAF crews that can operate using AWADS and SKE. This gives them an excellent means of inserting the ranger force during reduced visibility.

Figure 7-9. MC-130 Combat Talon aircraft.

(2) The AFFOR commander provides the airlift control element (ALCE) support at the departure airfield. The ranger air operations officer ensures that coordination of aircraft load limits, air movement times, specific load plans, and manifests is conducted in an effective and timely manner. The air operations officer and the movements control officer ensure that a departure airfield control officer (DACO) from the deploying unit is designated to provide face-to-face coordination at the departure airfield.

(3) The USAF normally provides an airborne battle command and control craft. This is a specially equipped C-130-type aircraft that provides communications and data system down-links to ground stations.

(4) The terminal guidance needed for assault landings and air traffic control in the objective area is provided by a specially trained and equipped CCT. This element has habitual training and operations relationships with the ranger units. The USAF is also responsible for the specially trained and equipped weather teams that can be sent into the objective area with the CCT.

(5) The USAF Tactical Air Command (TAC) includes fighter, interceptor, ground attack, electronic countermeasure, early warning, reconnaissance, and airborne command and control squadrons, any of which may be used to support the ranger force. The Tactical Air Command provides the results of aerial reconnaissance to the AFFOR, who then processes it for use by the theater or JTF commander's staff. This reconnaissance may be by low-level photo, imaging infrared, side-looking airborne radar (SLAR), low-light-level television, or overhead photography.

(6) The Military Airlift Command (MAC) equips and trains special long-range SAR aircraft and crews. These aircraft, mainly helicopters, are equipped with sophisticated navigation, observation, and communications equipment, as well as extended-range fuel tanks and in-flight refueling capabilities (see Figure 7-10). The ranger regiment uses these aircraft for certain missions.

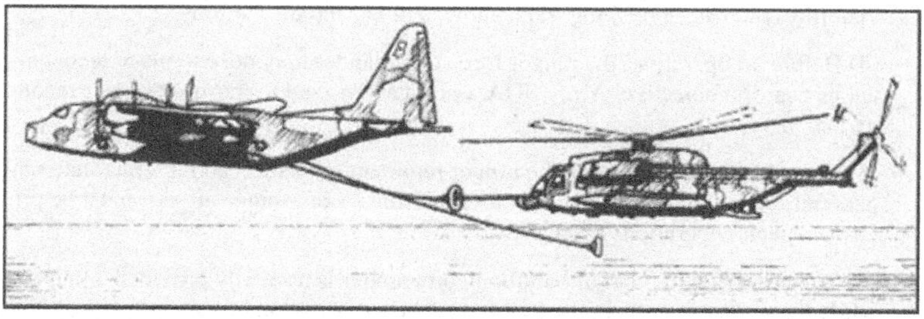

Figure 7-10. In-flight refueling.

d. **US Navy and US Marine Corps.** Both the USN and the USMC have units equipped and trained to support ground forces. The type of aviation support normally provided to the ranger regiment is close air support against targets in the objective area, or as suppression missions against enemy air defense installations. Long-range helicopters from the USMC may be used to support an amphibious operation or to insert special teams during ranger operations. The USN also has a search-and-rescue ability that may be used to support the ranger regiment.

e. **Special operations aviation units.**

(1) The US Army has specially equipped and trained SOA units tasked to support the ranger regiment, as well as other elements of the special operations force. These SOA units have a habitual training association with the ranger regiment and use dedicated aviation assets to -

(a) Insert, extract, and resupply the special operations force.

(b) Conduct armed escort, reconnaissance, surveillance, and electronic warfare in support of the special operations force missions.

(c) Provide airborne command, control, and communications for special operations force elements.

(d) Provide general support aviation during peacetime and contingency operations.

(2) Penetration of enemy airspace is normally done during limited visibility at very low altitudes. Penetrations are planned to maximize the use of existing terrain and exploit blind spots in enemy air defenses. On-board navigational aids are crucial to mission success. There normally is not any terminal guidance to the selected LZ or DZ. Deception operations of a tactical or operational nature help SOA units penetrate to the objective and reenter friendly lines. Advanced helicopters and short-takeoff-and-landing aircraft provide the help needed to do these missions.

(3) To survive enemy air defense systems and prevent mission compromise, SOA units avoid detection. Detected aircraft may reveal the location of supported special operations force ground forces or may be destroyed by enemy ADA fires. The SOA normally avoids detection by operating at night without illumination. This condition requires flight using sophisticated night vision devices. The SOA also operates in other conditions of reduced

visibility, such as clouds, fog, rain, snow, smoke, and dust storms.

(4) Before an operation, the ranger force commander may need to place reconnaissance teams near the objective. Army SOA assets can be used to transport ranger reconnaissance teams.

(5) Elements of SOA provide the ranger regiment with short and intermediate range penetration, resupply, and extraction capabilities. Long-range aircraft must be provided from USAF special operations assets.

(6) The ranger regimental or battalion commander is normally provided a liaison aircraft while the ranger force is in the REMAB. This ensures that he can travel quickly to the dispersed headquarters of the agencies providing support for the mission. This aircraft and crew are provided by the theater or JTF commander from EAC general support aviation assets. It may be a helicopter or fixed-wing aircraft, depending on the location of the REMAB to a suitable landing field.

(7) Army helicopters and fixed-wing aircraft may be used to provide SEMA aerial reconnaissance; radar, infrared and photographic imagery; and electronic intelligence support to the ranger force commander. This support is normally provided by the aerial exploitation battalion of the theater Army military intelligence group.

7-6. Engineer support.

a. The ranger regimental and battalion operations officers plan and request engineer support for ranger operations.

b. The most common engineer-type tasks needed during a ranger operation are the destruction of enemy facilities, equipment, and material by use of conventional explosives; and the interdiction of major lines of communication. To reduce the need for external support, the ranger regiment trains selected individuals in advanced demolition techniques. All members of the ranger regiment are well trained in the basic employment of conventional demolitions. Therefore, there is no need to attach combat engineers to conduct normal demolition missions. If specialized skill or knowledge is needed for a specific mission, qualified personnel can be attached.

c. During the occupation of the REMAB and the rehearsal for an operation, the ranger regiment needs construction engineer support to build simulated target areas. The simulated target areas should be camouflaged to prevent the enemy from recognizing the target. The construction engineers augment the ranger unit's self-help construction efforts. Limited-facilities construction and improvement may also be needed at the REMAB to provide security and isolation. The theater commander or JTF commander provides the construction engineer support.

d. Certain missions, such as those of interdiction, may need engineer special weapons teams. The teams are provided by the theater commander or JTF commander, or they may come from other special operations force assets. They are attached to the ranger force for a specific mission and released at the completion of the extraction and debriefing phase.

e. Some operations may need the ranger force to use captured enemy heavy construction equipment. Selected members of the ranger regiment are trained to operate such equipment.

However, construction engineer personnel may be attached to the ranger force for this mission.

CHAPTER 8

Combat Service Support

Combat service support consists of the logistical and administrative effort to maintain the battalion's ability to fight. The commander forecasts his needs for CSS. He employs CSS units to ensure accomplishment of the mission.

8-1. Ranger support element.

a. The ranger support element is a non-TOE organization formed from elements of CSS units collocated with CONUS-based ranger units augmented by selected personnel from the host installation. The mission of the RSE is to provide all the CSS needed to outload and deploy a ranger unit from CONUS. The personnel designated as part of the RSE have the same alert notification, reaction, and security requirements as members of the ranger unit they support. The RSE is normally built around a command and control group provided by a major CSS unit on the ranger unit's host installation. The remaining CSS elements and teams are then designated. An RSE consists of--

(1) An emergency operations center liaison group that provides timely coordination between the ranger unit and the functional elements of the RSE.

(2) A command and control group that coordinates and directs the actions of the functional elements.

(3) A supply section responsible for the timely issue of all Class II, III, IV, and VII items. This section maintains a stockage of contingency items marked for use by the resident ranger unit. It also issues items out of common stocks.

(4) A transportation section moves equipment and personnel from the unit area to the departure airfield and picks up equipment from local storage sites. It may also operate material-handling equipment and trailers to support loading of aircraft.

(5) An ammunition section responsible for the requisition, storage, issue, loading, and security of the ranger unit's basic load(s) of conventional munitions.

(6) A mess section provides all Class I support to the ranger unit. This includes operating a dining facility, providing food to remote sites, and drawing and issuing operational rations.

(7) A load team loads and unloads all equipment and supplies for the ranger unit. This team prepares pallets for loading, documents hazardous or dangerous cargo, and loads equipment pallets on USAF aircraft. It can also provide limited security for loads.

(8) A communications and electronics section provides secure communications among the RSE subelements, ranger unit, EOC, and DACG.

(9) An engineer platoon that is responsible for the construction of targets and rehearsal areas. This element of the RSE may or may not be activated, depending on the mission given the ranger unit. If the ranger unit deploys from CONUS to the objective area, this platoon has a key role to play in the premission preparation.

(10) A maintenance team whose primary mission is to ensure that the vehicles used by the RSE remain operational. It also provides predeployment maintenance and Class IX support for ranger unit vehicles and equipment.

(11) A medical team that provides unit-level medical care while the organic ranger medical section completes its outloading preparations. This team also provides any Class VIII supplies that the ranger unit may need.

(12) A security team responsible for security at the EOC, ranger unit location, and departure airfield. This team enforces stringent physical security precautions and maintains a constant counterintelligence (CI) surveillance of the outloading procedure. The security team provides armed guards for the protection of ammunition convoys, equipment movement, and vehicles.

(13) A rigger detachment that maintains, packs, rotates, and issues the personnel and equipment parachutes required by the ranger unit. It maintains enough stock of personnel parachutes to conduct a battalion-level parachute assault. The riggers lend technical skill to the rigging and aerial delivery of bulky, unusual, or delicate equipment.

b. The primary mission of the RSE is to provide the support necessary to quickly load and deploy the ranger units. The theater commander provides CSS to the ranger force once it has arrived at the OCONUS REMAB or ISB. The transition of support from the RSE to the CSS assets of the theater commander may be delayed for certain subelements in order to ensure a smooth deployment and enhance mission accomplishment. These subelements are released when the in-theater CSS assets can provide the support. Retained subelements of the RSE may be released from support requirements all at once, released in increments, directed to continue to provide support until mission completion, or augmented by in-theater CSS assets. The maintenance, medical, security, and rigger elements are those most likely to continue providing support OCONUS. This is likely if deployment is to an austere ISB pending immediate employment.

c. Once a ranger unit has been deployed OCONUS and has closed to a REMAB or ISB, the theater commander's staff coordinates CSS. If 1st SOCOM elements have deployed to the theater or operational area, the special operations support battalion (SOSB) of the 1st SOCOM coordinates CSS. Ranger logisticians at regiment and battalion level are mainly planners with limited logistics operations ability. Ranger logistical and personnel officers coordinate with CSS elements who provide such services.

d. The Theater Army Materiel Management Center (TAMMC) is the point of contact for information concerning in-theater logistical assets. In a developed theater of operations, a series of theater Army support groups will have been set up. One of these groups is assigned the mission of supporting the ranger force. The 1st SOCOM may also deploy certain assets available to provide CSS to the ranger force. The many CSS sources, the rapidly changing nature of SOF operations, and the unique CSS requirements of ranger units combine to make providing logistical and personnel support a challenge. The ranger unit's logistics officer must be included in the early phases of planning and deployment.

8-2. Logistics and personnel.

a. The echelons-above-corps CSS structure provides support to the ranger regiment as directed by the theater Army (TA) commander. In a low-intensity contingency involving a corps or less, there may be no echelons-above-corps logistics and personnel support elements. The corps support command (COSCOM) involved may be able to meet all needs from its own sources. However, in a combined operation in a high-intensity environment, the theater commander assembles a large CSS organization to provide support to the TA. This CSS base normally provides for all the ranger unit's requirements. During contingency operations in an area where there is no TA, the JTF commander provides the CSS elements. Until the buildup of these elements is complete, the ranger regiment must be prepared to operate from austere facilities with only the supplies it brings from CONUS.

b. The ranger regiment is normally assigned to the theater command for CSS. The TA commander, as directed by the theater commander, provides CSS to the ranger regiment and sets up support priorities. The EAC CSS organizations are under the command of the TA commander and consist of subordinate area and functional commands. The TA plans, coordinates, and executes its CSS functions through area-oriented support commands and specialized subordinate organizations. During the early stages of conflict, the TA may be organized as shown in Figure 8-1.

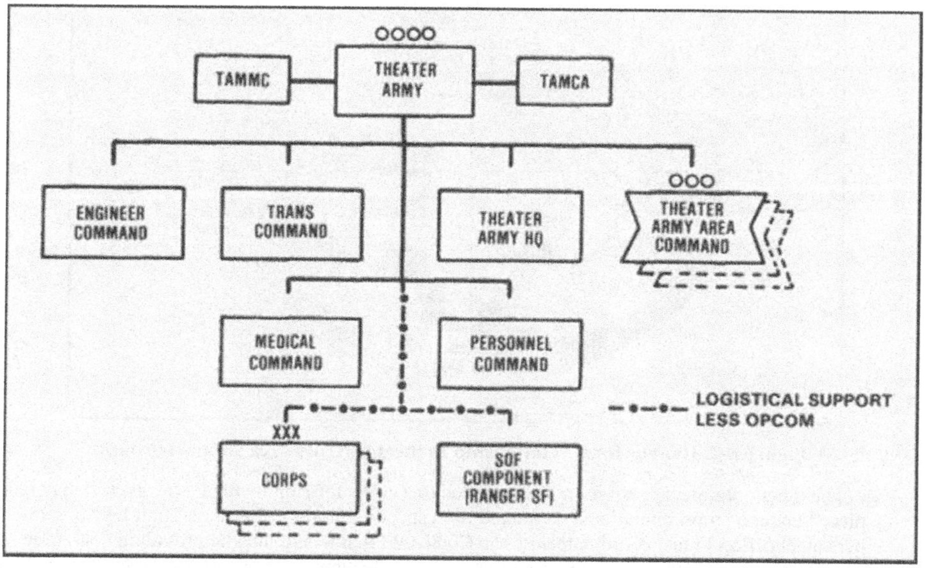

Figure 8-1. Ranger force relationship to theater Army.

c. The Theater Army Area Command (TAACOM) is the key CSS operator in the communications zone. It functions as a major subordinate command under the theater Army. The primary mission of the TAACOM is to provide direct support CSS to units such as the ranger regiment located in or passing through its assigned area. While located in an OCONUS REMAB, the ranger regiment receives most of its logistical support from the TAACOM.

d. The TAACOM area support group (ASG), which is subordinate to the TAACOM, is structured with operating units to provide direct support CSS to units located in or passing through its assigned geographical area of responsibility (see Figure 8-2). Area support groups are usually located along the lines of communication. The ASG is the prime source of CSS for the ranger regiment. It contains the units that actually provide services requested by the ranger unit logistics officer.

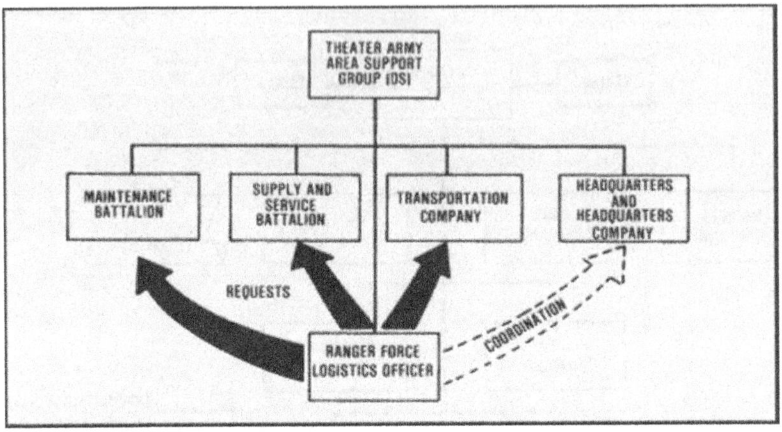

Figure 8-2. Ranger force relationship to theater Army area support group.

e. The ranger regiment is not normally assigned to a corps for employment. However, it may be placed under a corps' operational command for a specific mission. The ranger unit logistics and personnel officers must coordinate with the COSCOM that would then be providing CSS. (See Figure 8-3.) Corps support groups are major subordinate elements of the COSCOM. They provide command and control of COSCOM companies and battalions, which provide direct support (DS) and general support (GS) supply and field services to nondivisional units, such as the ranger battalion. (See Figure 8-4.)

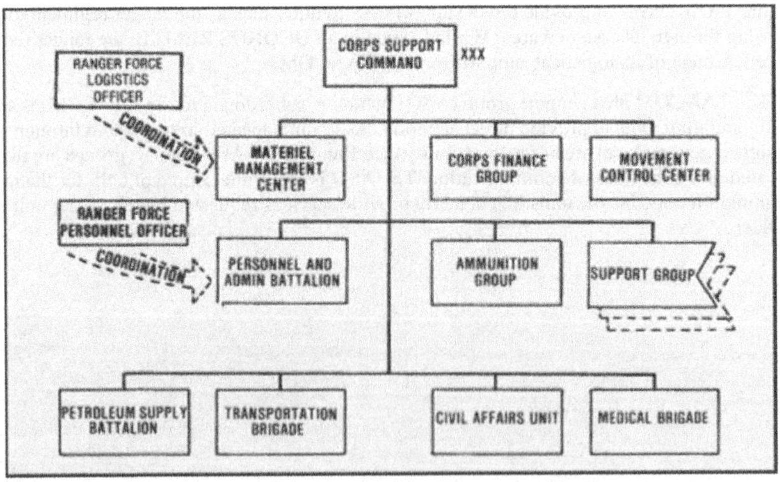

Figure 8-3. Ranger force relationship to corps support command.

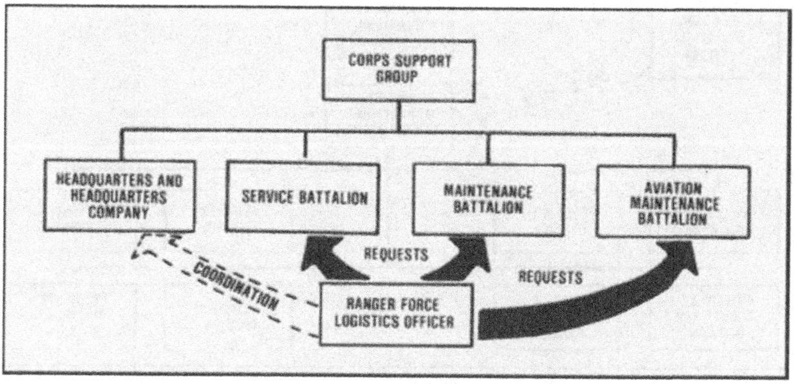

Figure 8-4. Ranger force relationship to corps support group.

f. Normally, the ranger regiment receives its logistical support from the corps support group responsible for the geographical area in which the REMAB or ISB is located. The corps support group provides all supplies except Classes V and VIII. The medical brigade provides Class VIII supply and medical maintenance support. The ammunition group provides Class V support. Financial and personnel services are provided by the corps finance group and the personnel and administration battalion, respectively.

g. Transportation needs are met by the transportation brigade. The ranger unit places ground transportation requests with either the Movement Control Center (MCC) or a designated movement control team. All airlift supply requirements are passed IAW FM 100-27. Standard and nonstandard supplies are normally requested from the supporting direct support unit (DSU). If authorized by the corps commander, the ranger regiment can place requirements for nonstandard items on the appropriate division within the corps MMC.

h. The COSCOM MMC (see Figure 8-5) performs supply and maintenance management for all classes of supply. The ranger unit logistics officer may coordinate with the MMC division chiefs or operations section.

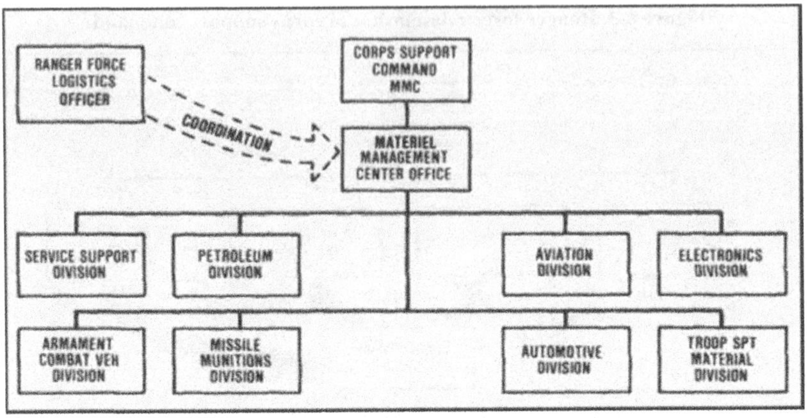

Figure 8-5. Ranger force relationship to COSCOM MMC.

i. Critical logistics operations, such as those supporting ranger and other SOF elements, need intensive management. They may be assigned to special staff sections that act as special operations force logistics managers. If designated, a logistics manager for special operations force elements would be a major point of contact for logistical support for the ranger regiment.

j. The ranger regiment may also receive logistical support from the SOSB. Normally, the SOSB provides administrative and logistical support to the Headquarters, Army Special Operations Forces (ARSOF). It complements the support provided by the Army component in the theater of operations by coordinating, augmenting, or assisting CSS forces. The SOSB provides unique quick-reaction support to the ARSOF or deploying SOF elements.

k. The ranger regiment does not make inordinate demands on the logistical system. However, a ranger unit may need certain nonstandard or low-density equipment. The concurrent need for nonstandard Class IX items requires management by logisticians to ensure support.

(1) Ranger battalion mess sections are responsible for Class I operations at the REMAB. The ranger regimental headquarters, which has no mess section, either collocates with one of the battalions or is supported by a mess facility operated by a nearby unit. It may share this facility on an area basis. The ranger unit personnel officer coordinates with the supporting DSU, which initiates the issue of Class I supplies.

(2) Class II, IV, and IX support to the ranger force is normally not of a great volume. It may consist of some nonstandard items. Depending on the need, the item, and the security, the SOSB may get some nonstandard items directly.

(3) Class III support needed by the ranger regiment is small and limited to standard items. The normal source of supply for Class III is the supply and service company providing area support.

(4) Class V support needed by the ranger regiment is small. It may include requests for nonstandard items of ammunition and explosives. The ranger logistics officer passes requests for standard items to the ordnance company (conventional ammunition) of the ammunition battalion assigned to the TAACOM ammunition group. Requests for nonstocked items are passed through the TAACOM Materiel Management Center or 1st SOCOM channels. The ranger unit must contact the ammunition company early to ensure a continuous supply of ammunition.

(5) Class VII supplies needed for the ranger regiment are often nonstandard or in excess of TOE. (See Appendix C for an equipment density listing.) All nonstandard or excess requests must be coordinated through the TAMMC or 1st SOCOM channels.

(6) Class VIII supplies for the ranger regiment are normally standard items supplied by the area support Class VIII point. The ranger regiment personnel officer sets up the Class VIII supply points of contact through the ranger battalion's medical officer.

(7) Aerial delivery equipment is provided by the airdrop equipment repair and supply company of the theater Army support group (TASG) (GS). There is normally one such company for each TAACOM. This unit supplies and repairs aerial delivery equipment. It does not pack and rig supplies for aerial delivery. The ranger regiment's requirements for aerial delivery supplies normally consist of standard items. Certain low-density items, such as HALO or HAHO parachute equipment, may also be provided through 1st SOCOM channels.

(8) The supporting motor transportation group of the transportation command is responsible for unit movement of the ranger regiment. The ranger regiment does not have any organic transportation assets. It depends on outside support whenever motor movement is needed.

8-3. Resupply.

a. Resupply operations for ranger units are normally covered during the planning and coordination phase of the mission. Ranger units try to complete their mission quickly and then be extracted or exfiltrate from the objective area. Ranger units carry into the objective area the equipment and munitions they need. They do not normally remain long enough to need extensive resupply. Certain operations may require the use of bulky or heavy items of equipment that cannot be carried by personnel. If the ranger force is airlanded on or near the objective, one or more of the aircraft may contain palletized supplies. These can be quickly off-loaded for future use.

b. If the ranger force is inserted into the objective area by parachute assault, aircraft can also drop resupply loads either just before or just after the personnel drop. These loads might include extra water, ammunition, demolitions, or barrier material. During peacetime, preloaded airdrop pallets are stored at CONUS depots for such contingency use. These mixed loads can be quickly moved to a departure airfield and inserted along with, or following, the ranger force. By dropping a resupply of water, food, equipment, and munitions with the ranger force insertion, the heavy-drop aircraft do not have to conduct a separate penetration of the enemy air defense zone. The ECM and SEAD programs used in the insertion of the personnel aircraft mask the penetration of the heavy-drop aircraft.

c. The aircraft dropping ranger personnel can also carry door bundles to deliver bulky or heavy items to the objective area. These door bundles, normally made up of A-7 and A-21 containers, may be color-coded to identify different loads or units. The most common door bundle loads are such items as 90-mm recoilless rifles and ammunition, 60-mm mortar ammunition, Dragon missiles, SHORAD missiles, or more demolition material. These bundles are pushed from the personnel aircraft just before the lead jumper's exit. They may be marked with lights or infrared sources for locating them at night.

d. If the tactical situation requires that a ranger force receive a resupply during the course of an operation, the most common method would be by airdrop. Airdrop could be needed if the operation takes longer than planned, initial supplies are lost, or changes occur in the operational plan. The ranger unit logistics officer and operations officers normally plan for resupply needs and schedule follow up loads of Classes I, II, IV, V, and VIII. There are two different airdrop requests channels for the ranger force to use:

(1) Nonemergency airdrop requests are sent through logistics channels. The ranger LOC sends requests for the airdrop of supplies and equipment to its supporting supply unit, and the supporting unit forwards the requests to the COSCOM or TAACOM MMC. The MMC coordinates with the movement control center (MCC). Jointly they direct the actions of the supply and transportation units. The MCC sends the request to the joint transportation board (JTB). The board validates it and assigns it a priority. The JTB then sends the request to the airlift control center for action. If the items requested are not on hand in the airdrop company, the MMC directs the proper supply activity to prepare the items for shipment to the airdrop unit. The MMC coordinates with the MCC, and the MCC ships the items. The airdrop unit prepares the supplies for the mode of delivery to be used. It also makes sure the supplies are delivered to the air terminal at the right time. (See Figure 8-6.)

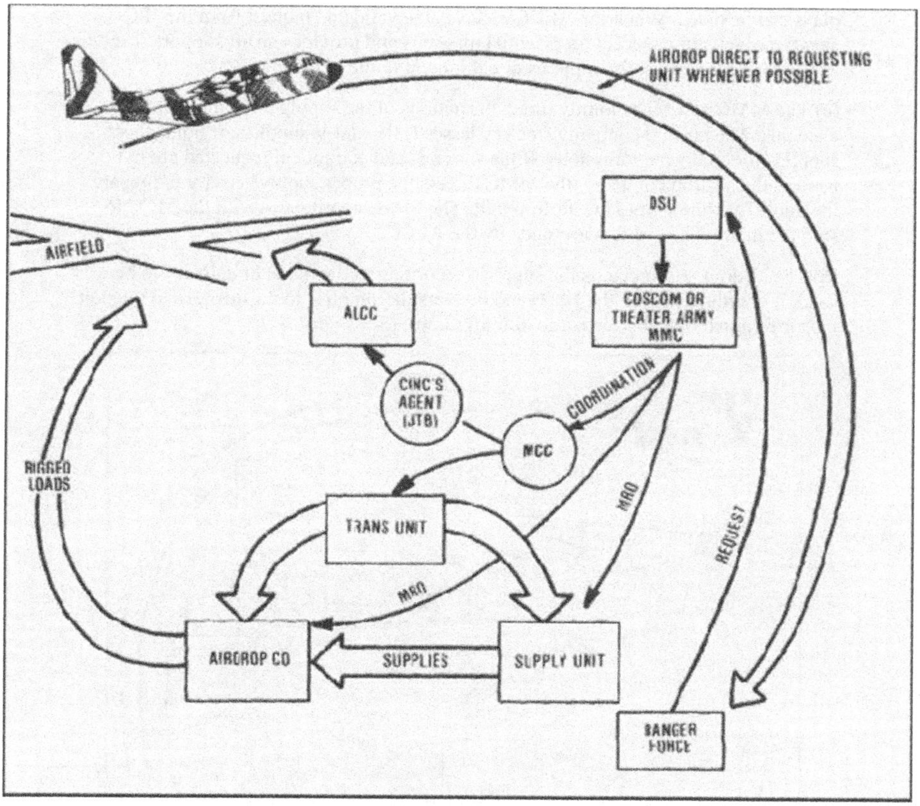

Figure 8-6. Ranger requests for nonemergency airdrop.

(2) Emergency requests for airdrop are passed through command channels in the quickest, most secure way.

(a) The request is usually sent to the command level that has authority to approve it, usually the tactical operations center at corps or above. The Air Force airlift officer at the direct air support center (DASC) coordinates with the Assistant Chief of Staff, Operations (G3) at the tactical operations center to determine whether airlift resources can be used for the mission.

(b) The TOC sends a request for an immediate airlift to the joint force commander's (JFC) designated agent for validation. At the same time, the airlift liaison officer advises the airlift control center (ALCC) that a request is coming through channels so plans can be made. When the ALCC receives the validated request from the JFC agent, it diverts or cancels less essential missions and provides airlift support. The TOC sends the request for supplies or equipment to the MMC.

(c) The MMC and MCC jointly direct the actions of the supply and transportation elements. The airdrop company stocks Classes I, III, and V supplies. It holds these supplies for emergency missions. If the supplies and equipment requested are not on hand in the airdrop company, the MMC directs the proper supply activity to prepare the items for shipment to the airdrop unit. The MMC coordinates with the MCC to ship the items. The airlift is arranged by the ALCC.

(d) The airdrop unit prepares the supplies according to the mode of delivery to be used. It coordinates with the MCC and delivers the supplies to the designated point at the air terminal (usually the cargo aircraft ramp). (See Figure 8-7.)

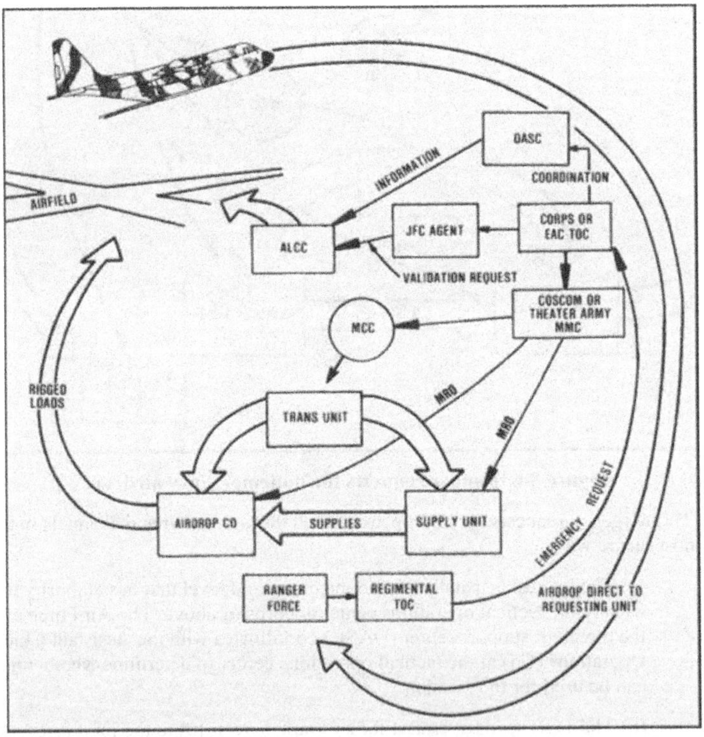

Figure 8-7. Request channels for emergency airdrop.

e. Aircraft delivering airdrop resupply deep behind enemy lines need to take precautions to avoid enemy detection and antiaircraft fire. The USAF is responsible for most of these precautions, such as ECM, suppression of enemy air defense (SEAD), and WILD WEASAL support. These do not concern the ranger commander. The choice of the type of parachute delivery system to use, however, may affect the ground commander's plan. The ranger force commander must specify any tactical restrictions that would require a specific type of delivery system. The ranger LOC and regimental logistics officer must advise the ranger force commander on the type of parachute system to be used. Conflicts must be quickly resolved so as not to interfere with the tactical operations.

(1) The safest way for the airdrop aircraft to penetrate enemy air defense and remain undetected is often by use of very low-level flight. There are four low-level parachute delivery systems that can be used without the aircraft being required to climb to a higher drop altitude and risk detection by enemy ADA.

(a) The low-altitude parachute extraction system (see Figure 8-8) needs a length of level ground with enough clearance for the delivery aircraft to make a very low pass over the extraction zone. The load is extracted by parachute from the aircraft and slides to a stop on a rugged, skid-type pallet. Heavy loads can be delivered using this system.

Figure 8-8. Low-altitude parachute extraction system.

(a) The containerized delivery system (CDS) (see Figure 8-9) consists of multiple, individually rigged A-22 containers, each with its own parachute, weighing up to 2,000 pounds each. The system provides single-pass delivery of up to 16 containers by C-130 aircraft and up to 28 containers by C-141 aircraft. The CDS loads may be delivered into drop zones using multiple points of impact to allow for tactical separation. The ranger force commander can direct the use of multiple points of impact, noting the advantages to be gained against the requirement for multiple passes.

Figure 8-9. Containerized delivery system.

(c) The high-speed, low-level, airdrop system (HSLLADS) (see Figure 8-10) is a single A-21 container specially rigged to withstand the parachute opening shock when airdropped from C-130 aircraft at high speed. This system can be used to deliver up to 600 pounds for each container, with a maximum of four containers for each pass, at speeds up to 250 knots.

Figure 8-10. High-speed, low-level, airdrop system.

(d) The CTU-2/A is a high-speed, aerial-delivery container (see Figure 8-11) that can be used to deliver supplies from high-performance aircraft flying at a minimum altitude of 300 feet above ground level and a maximum airspeed of 425 knots. The CTU-2/A is carried on the bomb racks of fighter or bomber-type aircraft. Upon release, a pilot parachute deploys the main chute and the container descends slowly.

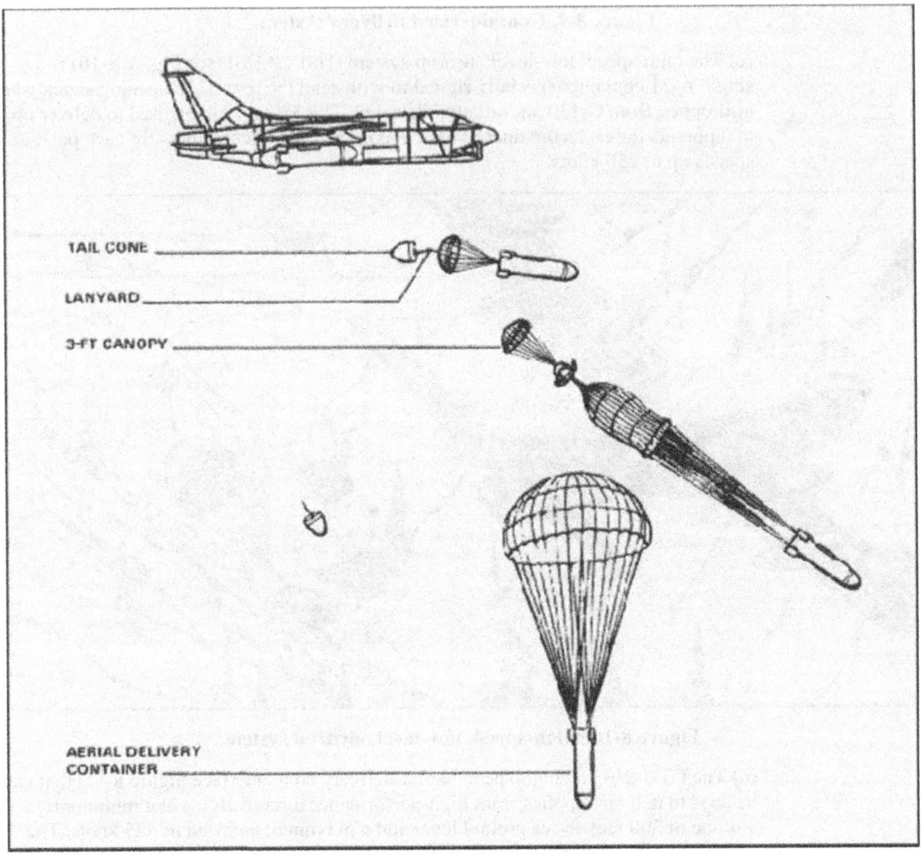

Figure 8-11. CTU-2/A high-speed, aerial-delivery container.

The CTU-2/A can be used to deliver up to 500 pounds of supplies such as weapons, water, food, or munitions. The container can be destroyed by burning. The main advantage of this system is that it can be delivered by high-performance aircraft deep behind enemy lines and in a dense air defense environment. The accuracy of this system is equal to that of a conventional bomb strike.

(2) Sometimes it may be better to drop resupply loads to the ranger force from a high altitude. The HAARS (see Figure 8-12) permits containerized unit loads weighing from 200 to 2,000 pounds to be delivered from aircraft at speeds up to 150 knots from up to 25,000 feet AGL. The HAARS consists of a cargo parachute, an airdrop container, an altitude

sensor, and a pilot chute. The pilot chute gives the descending bundle a speed slightly greater than an accompanying HALO parachutist. The system provides for the steady free-fall descent of loads from altitudes between 2,000 and 25,000 feet to an altitude at which a barometric sensor actuates deployment of the main parachute. This allows safe and accurate delivery of loads onto unprepared drop zones. The HAARS can deliver a payload to within 260 meters from a target impact point from a 10,000-foot altitude, with a proportional degree of accuracy from 25,000 feet. Ranger units can use the system for resupply of battalions and smaller units with rations, ammunitions, and medical supplies, breaking down the containerized material into man-packed loads.

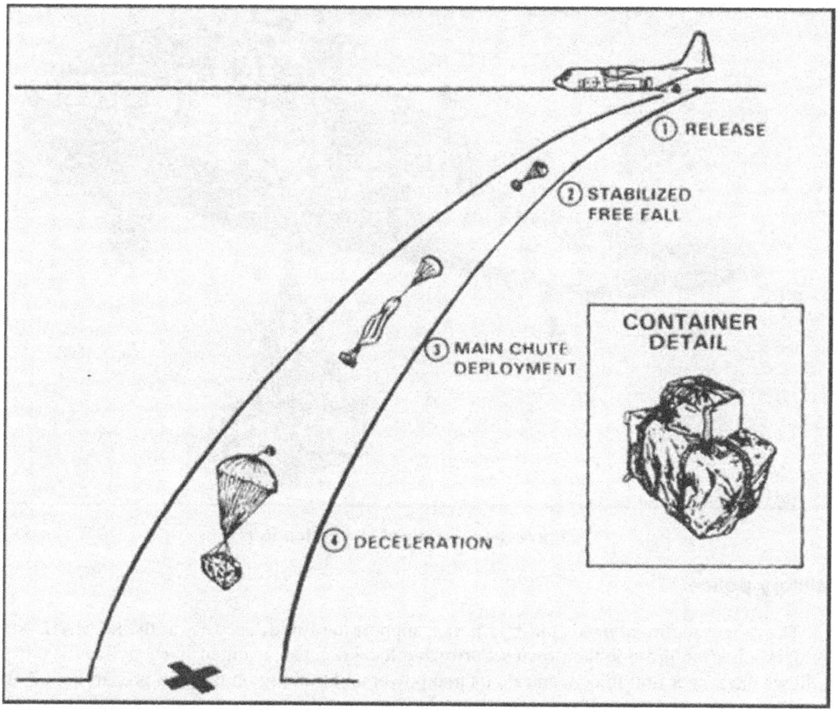

Figure 8-12. High-altitude, airdrop resupply system.

(3) The ranger force may need aircraft loads delivered during periods of poor visibility. The adverse weather aerial delivery system allows aircraft to drop resupply bundles accurately in adverse weather or darkness. These AWADS operations can be conducted safely and effectively with a minimum 300-foot ceiling AGL and a minimum visibility of less than half a mile. The minimum-drop altitude is 500 feet above the highest obstruction. These minimum ceiling or visibility restrictions do not apply to the Combat Talon aircraft.

f. Resupply to the ranger force can be provided by Army helicopters (see Figure 8-13). Normally this is done by SOA units. The danger to helicopters during long-range, low-level penetration of enemy rear areas requires that resupply be conducted when there is limited visibility.

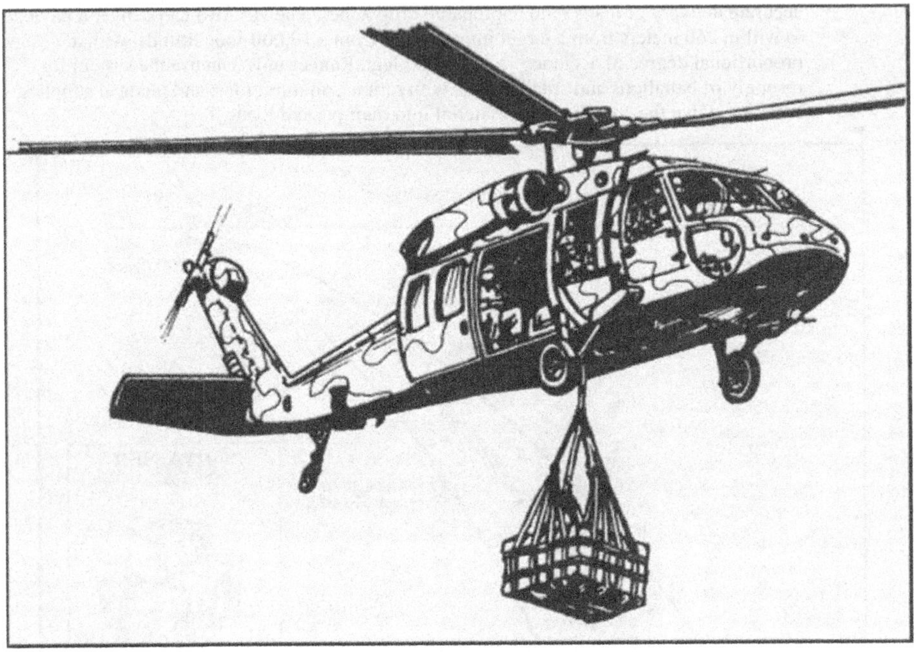

Figure 8-13. Resupply bt helicopter.

8-4. Military police.

a. The ranger regiment needs military police support to provide security at the REMAB. Although ranger units can provide their own security and local defense, using military police for this purpose allows the ranger unit to concentrate its manpower and training on mission preparation. A military police security company is normally designated to provide support to 1st SOCOM elements located in the communications zone. Military police provide security patrols of the REMAB, convoy security, and vehicular search. They provide entry and exit control of the sensitive compartmented information facility (SCIF).

b. A preemptive strike by enemy special operations forces is a threat to the ranger regiment while located in the REMAB; either by conventional attack or by terrorist-type bombing. The preparation and location of the headquarters, the type of buildings, the location of landing zones, the traffic flow, and the size of the local population all influence the size of the military police unit. A military police platoon, augmented with bomb and guard dog teams, should be enough security at the REMAB. These security forces would not normally deploy from the REMAB to an ISB with the ranger unit.

8-5. Counterintelligence.

Counterintelligence support to the ranger regiment is normally provided by a CI team from the military intelligence group at TA level. The CI team provides a liaison and informational flow from the CI agencies operating within the theater. It also acts as advisor to the ranger unit commander on OPSEC while performing CI surveillance operations in the vicinity of the REMAB. The CI support may also be provided through 1st SOCOM channels. It is through 1st SOCOM that liaison with national intelligence and counterintelligence agencies is achieved.

8-6. Personnel service support.

The personnel service support needs of a ranger unit are much like that of any other unit of the same size. To receive the support his unit needs, the ranger personnel officer must meet with the proper agencies within the theater of operations.

a. **Strength accounting.** Within the ranger regiment, strength accounting is conducted through normal personnel channels. Regimental strength figures are passed quickly to the TAACOM or COSCOM. These figures are crucial to the theater commander for planning future operations. When matched with casualty reports, they provide the information needed for replacements.

b. **Replacement operations.** Ranger units need exceptionally well-trained and physically conditioned personnel as replacements. Ranger operations require that replacements train with the unit before they can become an effective member of the team. There are five sources of replacements for ranger units:

(1) **In-theater volunteers.** Use of in-theater volunteers, preferably with experience in ranger units, is the quickest way to replace personnel. These volunteers have a knowledge of the basic skills required of rangers and have some operational experience. The ranger regimental personnel officer coordinates with the theater personnel officer to screen and select those volunteers with the required MOSs and grades. Volunteers from other special operations force components, or from other theaters of war, are identified by DA and screened for assignment to the ranger regiment. Even though these volunteers have individual training and experience, it takes several weeks to become fully integrated into the ranger unit's operations and to achieve the needed skill level.

(2) **Reorganization of units.** This method of reconstitution is used when two or more ranger units have high personnel losses. The remaining members of these units are reorganized into a single unit. This method provides the ranger regimental commander with enough ready-trained, physically fit personnel to form a full-strength unit. Extensive unit training must be conducted before such a reorganization regains the unity and esprit of its component parts. The high standards found among the subelements of the ranger regiment makes this a good method of reconstituting a ranger unit. The units that have lost all their personnel may be reconstituted later, using some other method, or left unreconstituted,

depending on the situation.

(3) **Hospital returnees.** These personnel, former members of the ranger regiment who were injured, evacuated, treated, and have now recovered, are valuable assets. They provide the unit with a major reconstitution ability. These personnel should be returned quickly to their original battalion. This enhances unity and increases morale and esprit. Returnees may not be physically able to return to full duty right away. While recuperating, they may be assigned to limited duty as instructors. The ranger unit uses the returnee's combat experience and individual skills to screen and train newly assigned volunteers. Lessons learned in combat can help prepare the unit for future operations.

(4) **Replacement volunteers from CONUS.** These personnel have training in the basic skills and may have completed the ranger course, but they do not have any ranger unit experience. They need several weeks of unit training to become prepared for combat operations.

(5) **Ranger indoctrination program.** During wartime, this program provides screening and training of newly assigned personnel. Volunteers who have neither served in a ranger unit before nor have graduated from the ranger course require extensive individual training before they are ready to begin unit training with a ranger battalion. This method of reconstitution of ranger units requires the longest training period. During this reconstitution phase, a ranger battalion needs extensive use of ranges, maneuver areas, and support by aviation and other special operations-type units. The ranger regimental commander may choose to move the program from CONUS to a location near the regiment's REMAB.

c. **Casualty reporting.** Casualty reports are handled through regimental personnel channels and must be timely and accurate. The reports are used to check strength accounting data, to assist graves registration personnel in the recovery and identification of remains, and to notify next of kin.

d. **Health service.** The health service support needed by the ranger units deployed OCONUS is usually small. Area medical support provided by host country or theater medical command assets usually meets their needs. Evacuation of wounded members of the ranger force begins upon extraction from the objective area. Personnel wounded in action are evacuated through USAF or Army channels to the nearest medical facility that can provide treatment. The ranger force medical officer and personnel officer coordinate with the area medical support facility providing preventive and routine medical care to the unit while at the REMAB. Vaccine needed for innoculation of the ranger force members and any special Class VIII supplies needed for the mission are drawn through normal medical channels.

e. **Financial service.** Financial support varies with the situation. In mid- and high-intensity conflicts, regular paydays are suspended in the theater of operations. As a nondivisional unit, the ranger regiment coordinates with the theater finance center or corps finance group for financial support. In the REMAB, ranger units are provided financial support from the closest finance support unit. The ranger unit personnel officer appoints Class A agents to help with the disbursement of funds. The ranger force may be given a specific amount of local currency or military scrip depending on the length of time in the REMAB and the operational situation. The ranger unit personnel officer also coordinates with the theater finance center to get any third-country currency that may be included as part of an escape-and-evasion kit provided to ranger elements.

f. **Postal service.** Postal service to the ranger unit is through normal channels. Mail for members of the deployed force is forwarded by the rear detachment commander to an established theater Army general support postal activity. This activity forwards the mail to the postal unit designated to support the REMAB site. The ranger unit commander appoints subordinate unit officers as censors. The exact location of the REMAB, and all aspects of the mission and its location, remains classified until released by an official Army spokesman. The ranger unit commander must ensure that the forwarding of mail, or the writing home of personal letters, does not endanger the ranger force mission. All outgoing mail is censored during wartime. For security reasons, outgoing mail may be held for a short time before it is forwarded. Mail from the ranger unit is never sent by host country or other nation's postal services.

g. **Chaplain activities.** Chaplain activities are provided by the chaplain assigned to the ranger battalion. He provides unit coverage through services, rites, and sacraments regardless of his or the unit member's religious affiliation.

h. **Legal service.** Ranger units do not need extensive legal service support. Legal service support is obtained through coordination with the theater Army staff judge advocate.

APPENDIX A

Organization and Equipment Introduction of the Ranger Regiment

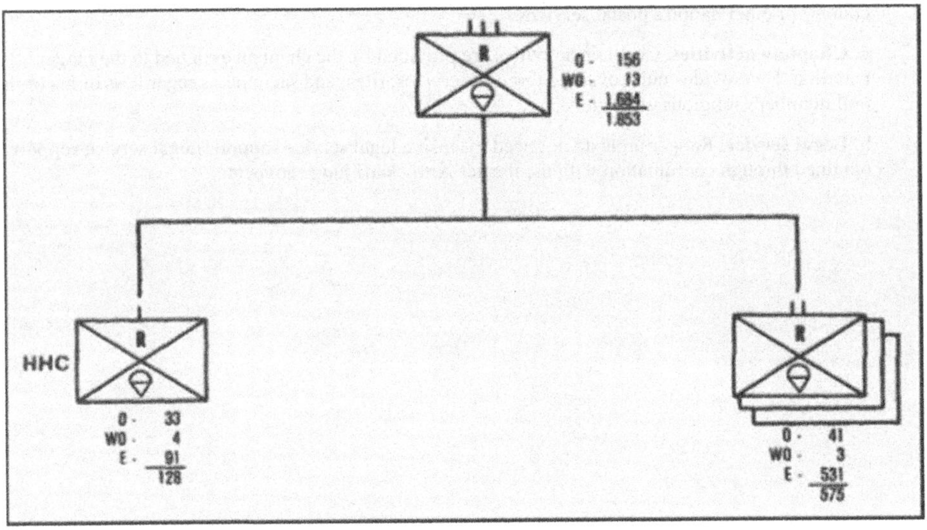

Figure A-1. Ranger regiment.

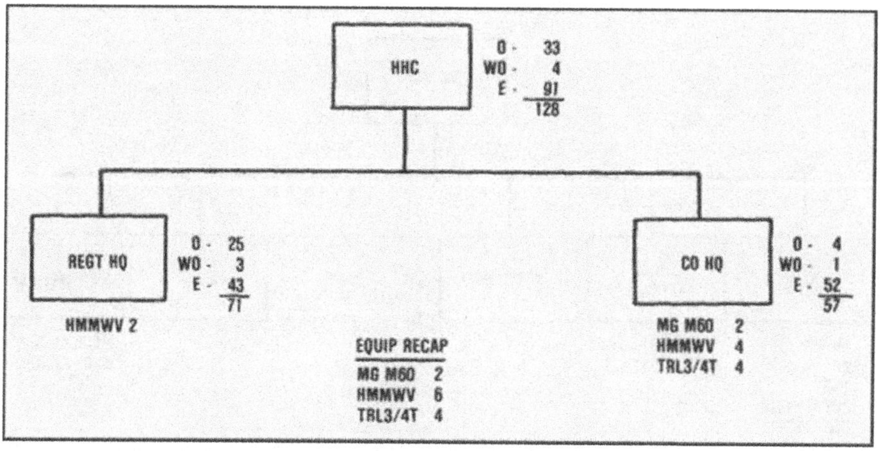

Figure A-2. Ranger regimental HHC.

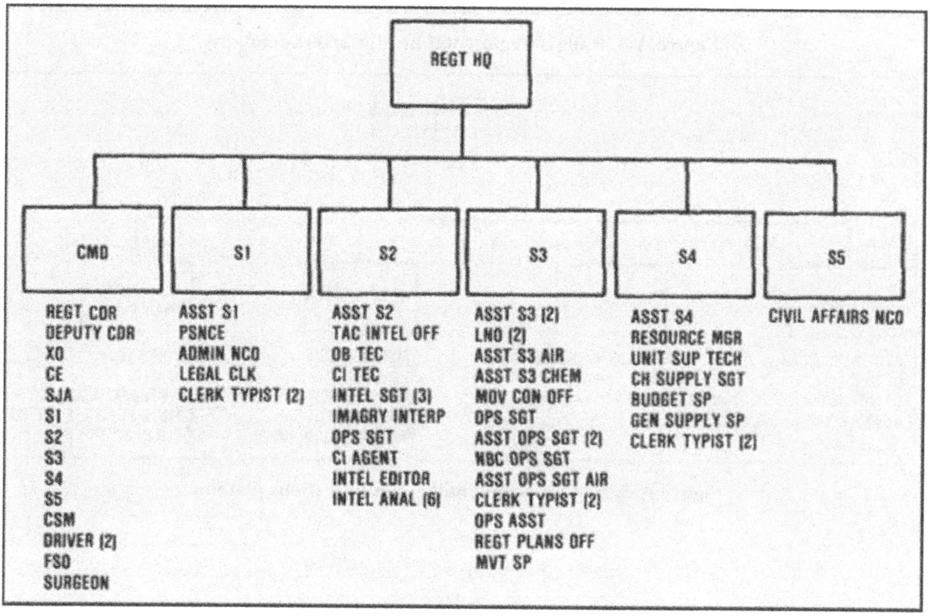

Figure A-3. Ranger regimental staff.

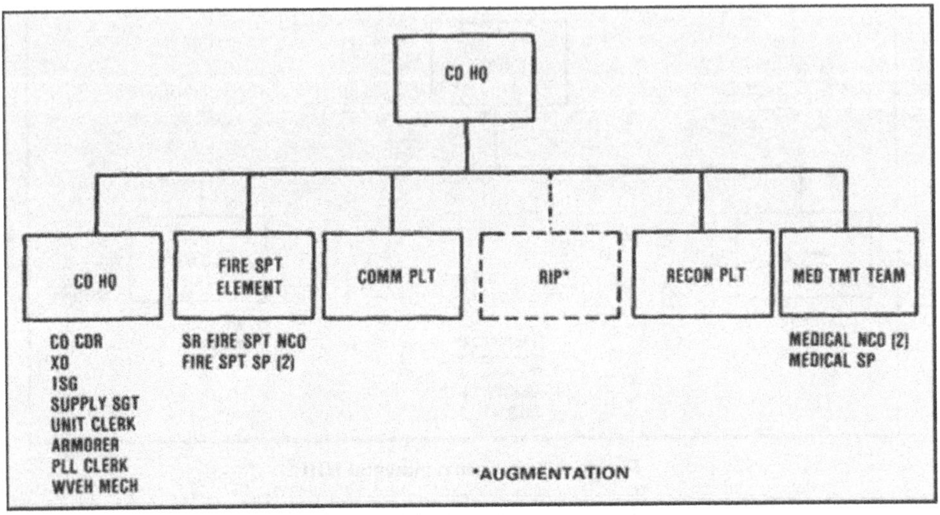

Figure A-4. Ranger regimental headquarters company.

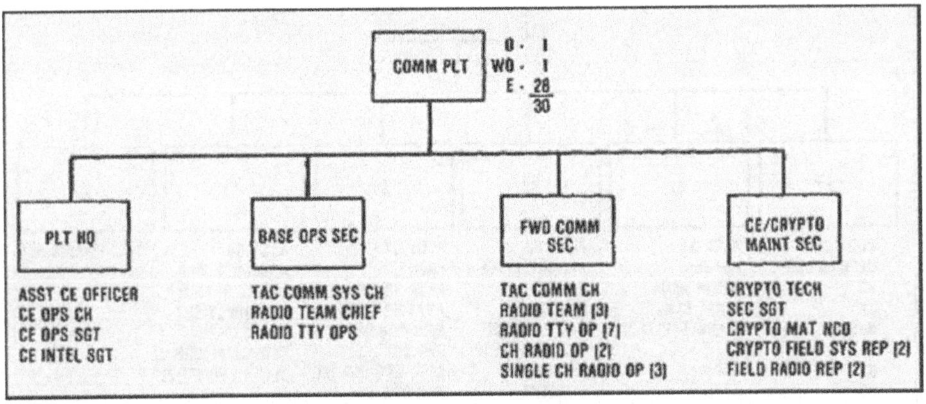

Figure A-5. Ranger regimental communications platoon.

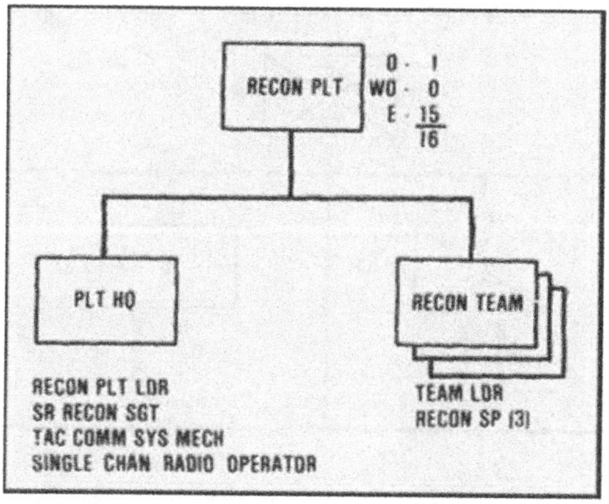

Figure A-6. Ranger regimental reconnaissance platoon.

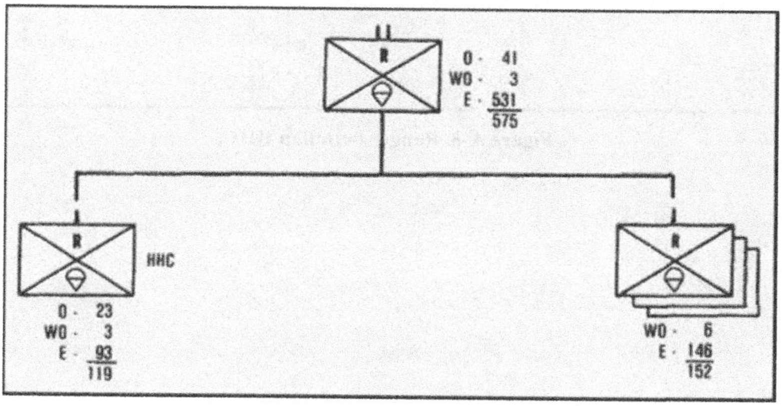

Figure A-7. The ranger battalion.

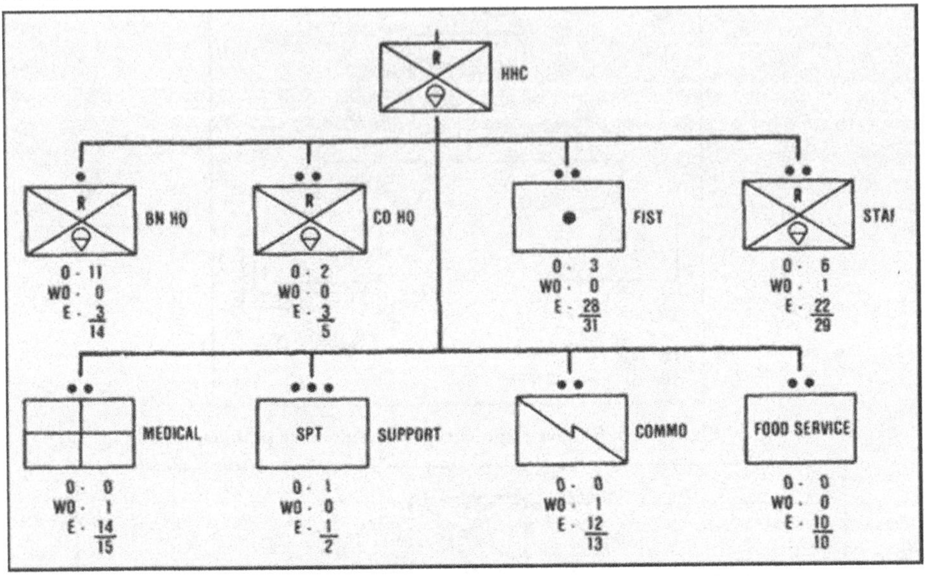

Figure A-8. Ranger battalion HHC.

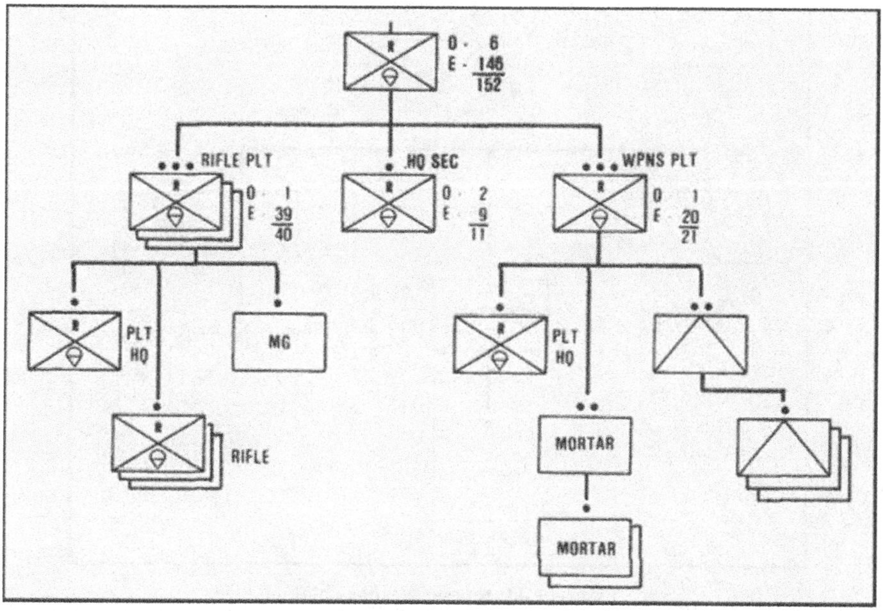

Figure A-9. Ranger rifle company.

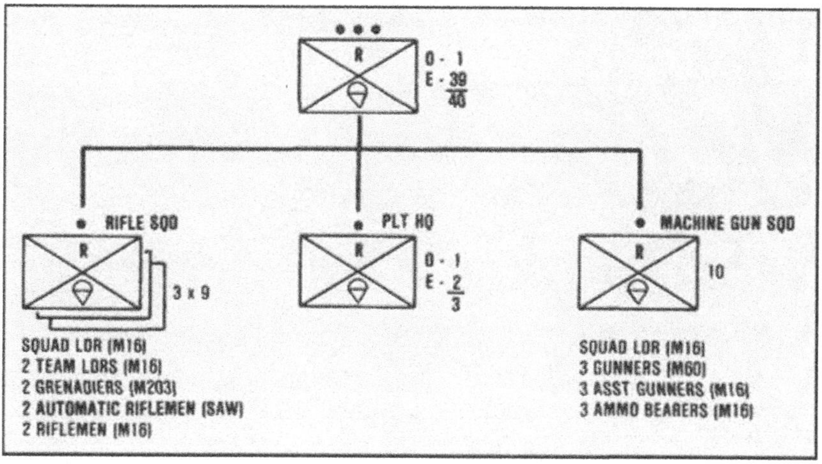

Figure A-10. Ranger rifle platoon.

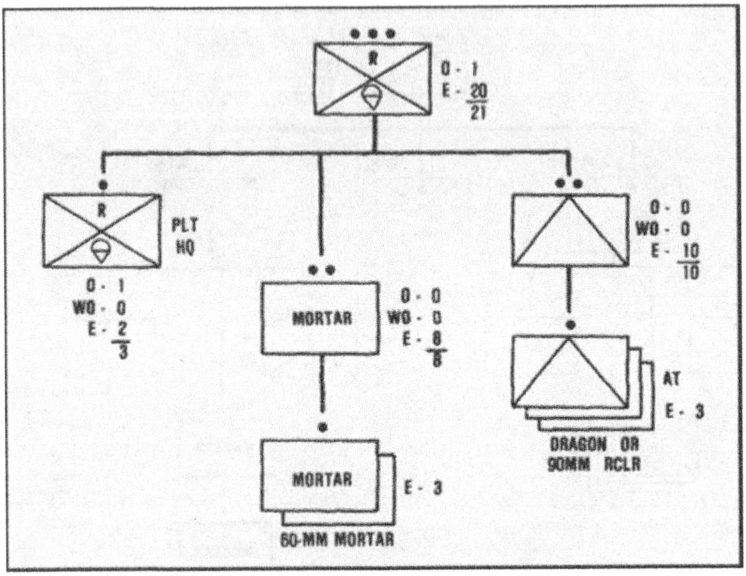

Figure A-11. Ranger weapons platoon.

APPENDIX B

Composition and Equipment of Ranger Regiment Liaison Teams

Team Red	**RANK**
Liaison Officer	MAJ
Senior Intelligence Analyst	SSG
Clerk-Typist	SP4
Tactical Communications System Operator/Mechanic	SGT
Combat Telecommunications Center Operator	SP4
Tactical Communications System Operator/Mechanic	SP4
Team White	
Liaison Officer	CPT
Senior Intelligence Analyst	SSG
Clerk-Typist	SP4
Tactical Communications System Operator/Mechanic	SGT
Combat Telecommunications Center Operator	SP4
Tactical Communications System Operator/Mechanic	SP4

Equipment (for Each Team)

HMMWV with 3/4-ton trailer	5-kw generator
Communications shelter	Tent, GP, small
AN/PCS-3 radio set (3)	AN/VSC-7 net control station (1)
MX 300 (4)	KY 57 with battery case (5)
HYX 57 with battery case (4)	HYP 57 (5)

KYK 13 (2)

KY 65 with battery case (1)

KG 84 (1)

H189 (2)

GRC 193 (1)

VGC-74 (1)

PP 1648 (1)

SB-22 switchboard (1)

KYI 18 (1)

KY 65 WLA (1)

Digital message device (1)

AN/PRC-77 radio (1)

Facsimile transmission device (1)

TA 312 with WD1 (5)

APPENDIX C
Equipment Density

C-1. Communication equipment.

	Regimental Headquarters	Battalions (x3)
AN/VSC-7 net control station	7	2
AN/PSC-3 radio set	23	13
AN/PRC-41 radio set	2	5
AN/PRC-104 radio set	9	26
AN/GRC-193A radio set	4	0
AN/PRC-70 radio set	6	0
AN/PRC-77 radio set	22	50
AN/VRC-46 radio set	2	0
AN/VRC-49 radio set	0	2
AN/PRC-68 radio set	0	66
TSEC/KY-57 speech security device	60	68
TSEC/KY-58 speech security device	6	0
TSEC/KY-65 speech security device	11	0
AO-8990/P digital message device group	15	23
AN/GXC-7A tactical FAX device	7	2
AN/UGC-74A(V)3 communications terminal	7	0
SB-22 switchboard	2	0
SB-3614(V)/TT switchboard	2	1

TSEC/KG-84	5	0
SB-993/GT switchboard	0	12

C-2. Weapons and equipment each battalion.

Infrared aiming light	96	Interrogator set, Stinger missile	3
Rifle, M16	451	Grenade launcher, M203	67
Rocket launcher, M202	13	Machine gun, M60	27
Mortar, 60-mm M224	6	Rifle, sniper, M21	24
Recoilless rifle, 90-mm	18	Viewer, infrared	9
Tracker, Dragon	9	Night vision devices (all types)	322
Truck, 1/4-ton	2	Laser target designators	11
Radio (all types)	164	Tactical satellite radios (all types)	15
Squad automatic weapon	54		
Laser range finder	14		

APPENDIX D

Standing Orders, Rogers' Rangers

1. Don't forget nothing.
2. Have your musket clean as a whistle, hatchet scoured, sixty rounds powder and ball, and be ready to march at a minute's warning.
3. When you're on the march, act the way you would if you was sneaking up on a deer. See the enemy first.
4. Tell the truth about what you see and what you do. There is an army depending on us for correct information. You can lie all you please when you tell other folks about the Rangers, but don't never lie to a Ranger or officer.
5. Don't never take a chance you don't have to.
6. When we're on the march we march single file, far enough apart so one shot can't go through two men.
7. If we strike swamps, or soft ground, we spread out abreast, so it's hard to track us.
8. When we march, we keep moving till dark, so as to give the enemy the least possible chance at us.
9. When we camp, half the party stays awake while the other half sleeps.
10. If we take prisoners, we keep 'em separate till we have had time to examine them, so they can't cook up a story between 'em.
11. Don't ever march home the same way. Take a different route so you won't be ambushed.
12. No matter whether we travel in big parties or little ones, each party has to keep a scout 20 yards ahead, 20 yards on each flank, and 20 yards in the rear so the main body can't be surprised and wiped out.
13. Every night you'll be told where to meet if surrounded by a superior force.
14. Don't sit down to eat without posting sentries.
15. Don't sleep beyond dawn. Dawn's when the French and Indians attack.
16. Don't cross a river by a regular ford.

17. If somebody's trailing you, make a circle, come back onto your own tracks, and ambush the folks that aim to ambush you.

18. Don't stand up when the enemy's coming against you. Kneel down, lie down, hide behind a tree.

19. Let the enemy come till he's almost close enough to touch. Then let him have it and jump out and finish him up with your hatchet.

MAJ Robert Rogers

APPENDIX E

The Ranger Creed

R Recognizing that I volunteered as a Ranger, fully knowing the hazards of my chosen profession, I will always endeavor to uphold the prestige, honor, and high "esprit de corps" of the Ranger Regiment.

A Acknowledging the fact that a Ranger is a more elite soldier who arrives at the cutting edge of a battle by land, sea, or air, I accept the fact that as a Ranger, my country expects me to move farther, faster, and fight harder than any other soldier.

N Never shall I fail my comrades. I will always keep myself mentally alert, physically strong, and morally straight and I will shoulder more than my share of the task, whatever it may be. One hundred percent and then some.

G Gallantly will I show the world that I am a specially selected and well-trained soldier. My courtesy to superiors, my neatness of dress, and my care of equipment shall set the example for others to follow.

E Energetically will I meet the enemies of my country. I shall defeat them on the field of battle, for I am better trained and will fight with all my might. Surrender is not a Ranger word. I will never leave a fallen comrade to fall into the hands of the enemy and under no circumstances will I ever embarrass my country.

R Readily will I display the intestinal fortitude required to fight on to the Ranger objective and complete the mission, though I be the lone survivor.

RANGERS LEAD THE WAY

APPENDIX G

Training

G-1. Application and guidance.

a. To prepare for combat, ranger units train on missions derived from their analysis of operational and contingency plans and the ranger mission training plans (MTPs). Training guidance is in doctrinal manuals, field circulars, and soldier's training publications. Training must ensure that the ranger unit is skilled in all basic light infantry skills and in ranger-unique skills. In the final analysis, good training depends upon the leadership and teamwork of ranger unit officers and NCOs.

b. Everything rangers do should prepare them to perform well in combat. Ranger units train IAW the battalion training management system (BTMS) concentrating on basic skills and on live-fire, hands-on training.

c. An effective training program breeds high morale and few discipline-related problems. Some qualities of ranger units that have good training programs are as follows:

(1) Leaders who can solve problems.

(2) A thorough and continuous analysis of the unit's mission.

(3) Validated training objectives.

(4) Attention to the basics (move, shoot, and communicate).

(5) Multiechelon sustainment training.

(6) Standardized procedures.

(7) Physical and mental strength.

(8) Live fire as an integral part of training.

(9) Safety as a basis to all operations.

(10) The use of night and adverse weather and difficult terrain as combat force multipliers,

d. When all training is planned with realism, rangers will be in training situations that require the same attention as in battle. Realism and stress in training result in faster learning and better retention of skills. Training objectives must be clear with measurable standards to determine proficiency. Rangers and ranger units must train to the standards that will be expected in combat.

e. Examples of unit training goals include:

(1) Developing each member of the chain of command to perform his duties with competence, confidence, and pride.

(2) Maintaining small unit integrity during all training and support activities. Accomplishing tasks as units.

(3) Ensuring that all tasks for training are keyed to mission needs and weaknesses. Assets--ranger's time, money, and materiel--are expended only after thorough planning.

(4) Training to standard.

(5) Developing and maintaining discipline and esprit to ensure the abilities of each ranger and all units are achieved.

(6) Emphasizing use of the ranger mission training plan (MTP) for each echelon of the organization.

(7) Emphasizing the basics.

(8) Maintaining a high level of physical fitness.

(9) Working together within the ranger regiment and other elements of the special operations forces community.

(10) Developing a unit atmosphere of quick reaction under stress. Encouraging and rewarding individual initiative. Accomplishing unit missions IAW the commander's intent.

G-2. Philosophy.

a. To be effective, training should be physically and mentally challenging. Each ranger will then have a sense of purpose, accomplishment, and satisfaction. Training is evaluated by the results rather than hours of training. Units that are well trained today need reinforcement training to sustain readiness. Training management must be continuous. Training managers must use the time available to improve training efficiency. Cyclic training will not produce the desired results. The "train-to-correct deficiencies" method is needed. The training managers must analyze weaknesses, conduct prescriptive and reinforcement training, and reevaluate their unit's training status. The process that is used to develop and maintain a continuous training and evaluation program is shown in Figure G-1.

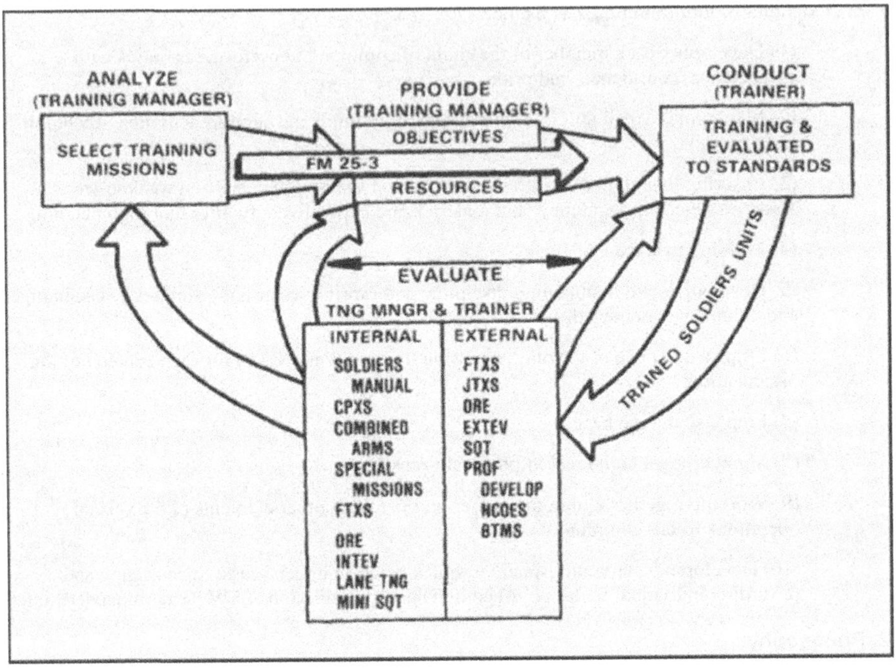

Figure G-1. The Army training and evaluation program.

b. Individual and collective training must be integrated within a unit training program. Training of the individual ranger, leaders, and units at each level must occur at the same time. The ranger commander or leader plans and conducts training for every member of his unit.

c. Ranger commanders analyze operational and contingency plans to determine unit missions and translate the associated tasks into training objectives. These objectives form the basis of the unit training program. The regimental headquarters provides training guidance in a formal regimental training plan. Commanders at lower levels program, supervise, and conduct training within their units.

d. All members of the chain of command are responsible for training. The NCO leader or first-line supervisor conducts individual and small unit collective task training. The officer plans individual training in his unit and directs and conducts collective training.

e. Centralized development of training objectives and decentralized training builds well-trained units and outstanding leaders. Authority and responsibility for the detailed planning, scheduling, conducting, evaluating, and supervising of training are delegated to the lowest element capable of managing this training. Higher headquarters is responsible for establishing objectives, monitoring and evaluating training, and issuing general guidelines.

f. Multiechelon training and evaluation means using all assets and time available to improve combat readiness. The multiechelon structure of the ranger MTP provides a basis for units to be continually trained and evaluated by internal evaluation. Each leader must know his mission tasks and train to accomplish those tasks to the set standards.

G-3. Mission analysis.

a. Mission analysis is continuous for a good training program. The selection of training missions or tasks results from analysis of the unit's overall mission. (See Figure G-1 above.) Tasks and subtasks are then placed in order of priority.

b. The trainer must identify countertasks for the opposing force (OPFOR). The countertasks should be such that an OPFOR that meets the standards for its countertask will be successful. Both sides have a chance to win.

c. The trainer should then develop combat drills and situational training exercises (STXs), if not furnished by the proponent school.

G-4. Unit training.

a. **Responsibility.** Ranger commanders at all levels are responsible for all training and management described herein and current Army training doctrine. This includes individual and collective training at all levels. Units must train to the standards specified in training objectives.

b. **Management.** The training management model for commanders is:

- Set objectives.
- Provide resources.
- Coach subordinates.
- Measure results.
- Provide feedback.

G-5. Individual training.

There are three elements basic to a well-trained individual ranger. He must be physically fit, be proficient in the use of his weapon, and have the psychological commitment to fight the enemy.

a. Individual training is the responsibility of the first-line supervisor under the supervision of the chain of command. It is conducted using performance-oriented techniques under battlefield conditions. New skills must be developed and practiced, and acquired skills must be reinforced throughout the training year. A high level of proficiency must be maintained through continuous training and practice.

b. Trainers should use peer instruction. Commanders ensure needed resources are available to conduct challenging individual training. Ranger units schedule and conduct this training and evaluate individual proficiency.

c. Flexibility in planning and scheduling is necessary. The goal of all individual training must provide for reinforcement training to offset the effects of the forgetting curve. (See Figure G-2).

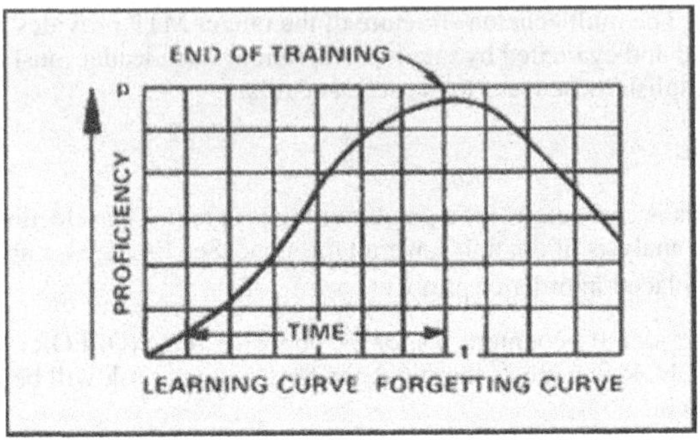

Figure G-2. The learning curve and forgetting curve.

d. The command sergeants major, unit first sergeants, and staff section sergeants are the keys to an effective training program. These senior NCOs must counsel, guide, and critique junior NCOs in training preparation and execution. Fire team leaders, section chiefs, and squad leaders coordinate and present individual training.

e. The NCO supervisors must ensure that the ranger is trained in the skills required for the specific military occupational specialty (MOS) and in those skills used daily. Not all training can be acquired in an office. For example, all rangers must maintain a high skill level with their weapons, be trained in NBC subjects, and be capable of performing contingency missions.

G-6. Collective training.

a. **Guidance.** Collective training trains crews, teams, squads, platoons, companies, and battalions to do their missions. The ranger MTP is used to plan training and evaluate the unit's weaknesses and strengths. If a ranger commander receives new training requirements or missions, they should become part of the collective training program. The ranger MTP describes the combat tasks to be performed by every echelon of the ranger battalion. It is a plan for organizing, conducting, and evaluating the year-round unit training program.

b. **Responsibility.** The ranger unit commander is responsible for the conduct of collective training and for all other training in his unit. Practical exercises emphasize the techniques of terrain use, movement, and effective employment of weapon systems. Training under stress will be conducted to move the unit to a higher plane of combat readiness.

c. **Drills.** Drills are written for small units--for example, fire team and squad, crew and section--and describe an action that is shorter than an entire STX. Drill training objectives cover both individual and collective skills (critical collective tasks) and list them together for effective group action. Drills must be mastered in order to react to rapidly changing situations in combat. The trainer can interrupt the execution of a drill to provide necessary training on individual or collective skills.

d. **Situational training exercise.**

(1) The STXs are short, mission-oriented tactical exercises in which a group of closely related collective tasks are trained. They are like drills but are more complex and flexible. They usually involve a larger unit than does a drill. They integrate leader training, leader tasks, drills, and separate individual tasks into a realistic setting. Trainers may execute STXs as written or tailor them to meet specific unit requirements. An STX is not standardized, but it must have a firm doctrinal base.

(2) Due to their design and purpose, STXs apply mainly to platoon- and company-size units. This does not preclude the development of STXs or similar exercises for unit staffs and headquarters. They combine "how-to-train" specifics for selected collective tasks with the "what-to-train" specifics found in doctrinal manuals and other documents.

G-7. Performance-oriented training.

a. Performance-oriented training emphasizes hands-on performance. Training is conducted to ensure a level of proficiency, rather than established number of hours of required training.

b. Performance-oriented training has precise training objectives and makes good use of resources. Training objectives serve as the basis for preparing, conducting, and evaluating all training.

G-8. Evaluation.

a. A training program must include an evaluation of training (planned and conducted). An evaluation plan must:

(1) Provide immediate feedback to the trainer.

(2) Recognize effective trainers.

(3) Evaluate all units in the conduct of training to provide for a comparison of training programs.

(4) Be flexible to unit needs; provide general and specific evaluations based on the level of training within the unit.

(5) Be in effect at all times.

b. Major considerations for training evaluations include:

(1) Ranger unit commanders establish and maintain an effective internal and external evaluation program.

(2) The director of each staff element is responsible for training evaluations within his area of staff responsibility with emphasis on low-density MOS groups.

(3) The operations officer of the ranger organization develops and validates training and evaluation plans.

c. Evaluation of training must be done vertically from the major task (mission) down through subtasks (collective tasks) to the leader and individual tasks. In selection of individual tasks for evaluation, priority should be given to critical individual tasks that impact on the success of the collective task or mission.

d. The critique and after-action review, after each major training event, are the most important aspects of training. The after-action review takes place immediately after completion of a training task (individual or collective). The individual(s) self-analyzes and critiques the training event. Then after a formal critique of the training, the individuals) trains to standard (not to time).

G-9. Field training.

a. **Considerations.** Ranger commanders schedule training under field conditions to permit maximum training opportunities. Field training emphasis is on mastery of a series of individual (soldier and leader) tasks and collective training tasks. These tasks are repeated until established standards are met. The goal of ranger field training is to ensure realistic, demanding, and meaningful training.

(1) When training is conducted under conditions of moderate stress, most combat critical skills are learned quicker and retained longer. (See Figure G-3 on the following page.) Stress can be incorporated in a training situation by physical deprivation (sleep, food), by live-fire exercises, by operation against an OPFOR under simulated battlefield conditions for an extended period, by the intensity of activity including demands for decision making, or by physically demanding activities. Ranger units train under stress so that they will react quickly in combat.

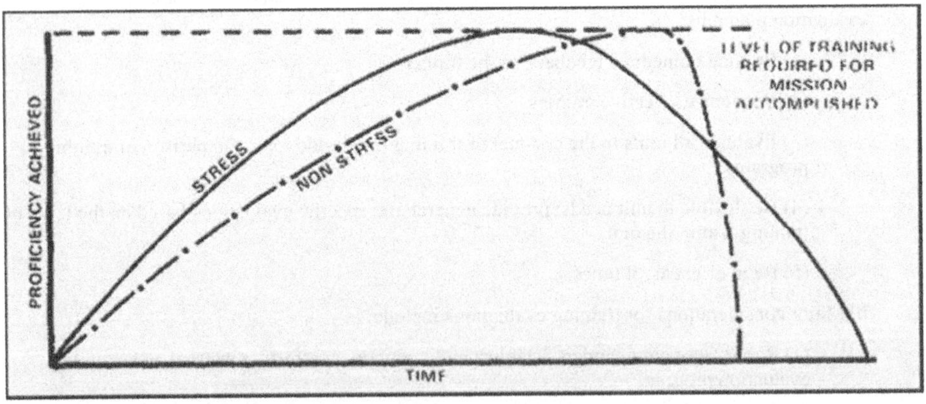

Figure G-3. The Army training and evaluation program.

(2) Antagonistic training situations are two-sided exercises conducted by opposing elements. Each element has a training objective (task and standards). The opposing force has a corresponding countertask. This training may be progressive--for example, starting with a

crawl-through, then a walk-through, and progressing to both elements operating at a fast pace. The use of multiple integrated laser equipment system (MILES) greatly enhances task and countertask training. The units should also be considered for a live-fire exercise against a realistic OPFOR target array when the basics have been mastered.

b. **Methods of training.**

(1) **Considerations.** The chance for success in combat can be increased by becoming proficient in training situations that depict realistic combat conditions. (See Figure G-4.) Meeting the standards of a realistic live-fire training exercise assures the commander that his unit could perform well in combat. Figure G-4 shows the relationship between realistic training and the challenge to the trainer.

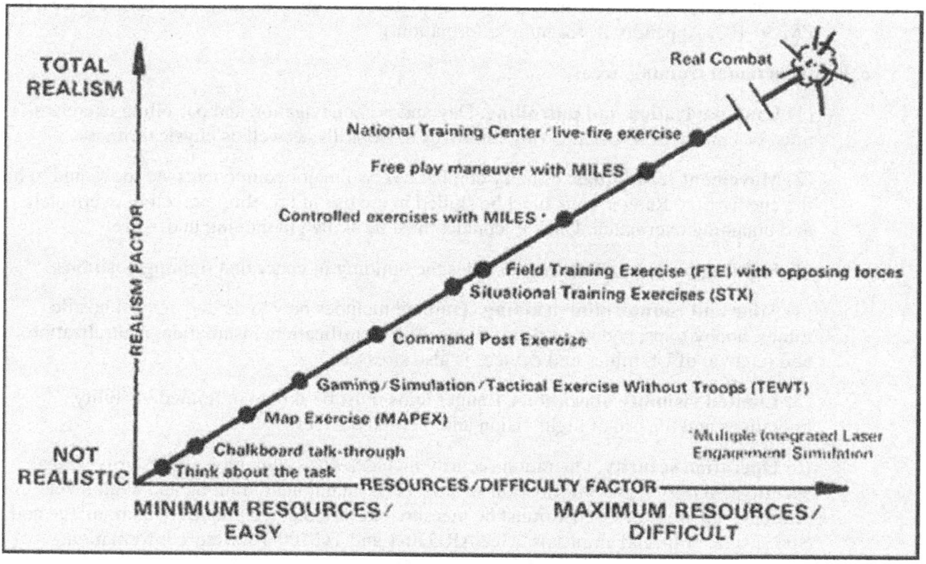

Figure G-4. Training proficiency scale.

(2) **Ranger cohesiveness training.** Ranger units promote teamwork by undergoing tough, stressful, often hazardous training. This challenges the unit, which gains in cohesiveness when it meets these challenges. This includes serving together in live-fire exercises, athletic events, precision parades, high-standard inspections, tough physical training, and combat.

(3) **Field training exercise.** An FTX is training normally conducted under simulated, realistic combat conditions. This training normally exercises command and control systems and the administrative and logistical needs of sustained combat. Stress training is part of this exercise, focusing on live-fire and MILES drills and STXS.

(4) **Combined arms training.** Combined arms training should include all means of support such as tactical air, armed helicopters, and indirect fire with emphasis on fire support coordination and airspace control.

(5) **Live-fire exercises.** Including live-fire exercises in drills and STXs offers a chance to evaluate units in their response to simulated combat missions. Ranger units train under pressure and always anticipate being selected to conduct a live-fire exercise. Live-fire exercises develop confidence and esprit, and reinforce unit discipline.

(6) **Opposing force.** Committing one unit against another adds a competitive spirit to a field training exercise (FTX). The use of MILES and squad combat operations exercises (simulation) (SCOPES) provides realism to two-sided field exercises. Trainers should use OPFOR equipment and training devices available, as well as making their own devices. (See FM 30-103, Appendix R, for more information.)

c. **Key functional training areas.**

(1) **Land navigation and patrolling.** Day and night navigation and patrolling exercises must be emphasized. Orienteering improves these skills, as well as physical fitness.

(2) **Movement techniques.** Infantry combat has two major components--to move and to hit the enemy first. Ranger units must be skilled in the use of traveling, traveling overmatch, and bounding overmatch. Units in contact must be skilled in moving under fire.

(3) **Fighting** positions. Training includes the building of concealed fighting positions.

(4) **Mine and countermine training.** Training includes how to defend against hostile mines, booby traps, and other demolitions. The identification, installation, neutralization, and removal of US mines and devices is also stressed.

(5) **Limited visibility operations.** Ranger units must be skilled in limited visibility operations and the use of night vision and thermal devices.

(6) **Operation security.** Operation security includes deception, physical security, signal security (SIGSEC), and information security. It is taught in all training and at all levels. Although all areas of OPSEC must be measured in training, techniques of camouflage and SIGSEC need special emphasis. (See AR 530-1 and TC 100-1 for more information.)

G-10. Professional development.

Ranger units at battalion and regimental levels conduct many officer and NCO professional development programs. This ensures that all officers and NCOs are skilled in their field, confident, and wellrounded. Professional readings and book reports, tactics seminars, and tactical exercises without troops are good development methods.

G-11. Physical training.

a. This program ensures that each ranger maintains the physical ability and stamina needed to do his mission. Other objectives are to promote unit cohesiveness and discipline. These are met by combining individual testing, formation, running, athletic competition, weight control, proper diet, teamwork drills, and freedom from drug and alcohol abuse. Mental fitness and toughness come from experience and success. Good physical conditioning increases a unit's chances of success.

b. Physical training (PT) programs are designed to include the elements of overload, progression, balance, variety, and regularity.

c. Ranger unit physical fitness standards are:

(1) Score 80 points for each Army physical readiness test (APRT) event and do six chinups.

(2) Pass the ranger swimming test.

(3) Complete an 8-kilometer run in 40 minutes.

(4) Complete a 12-kilometer road march in three hours (with rucksack, helmet, and weapon).

(5) Meet the Army height and weight standards.

G-12. Environmental and specialized training.

a. Rangers conduct training in environments as close to real combat missions as possible. This conditions rangers to survive and fight in a specific environment. Such training includes combat-critical tasks and technique training for that environment.

b. Each ranger battalion's environmental and specialized training includes at least the following:

(1) Jungle (once every 12 months).

(2) Desert (once every 12 months).

(3) Urban (once every 6 months).

(4) Extreme cold (once every 18 months).

(5) Mountain (once every 12 months).

(6) Amphibious (once every 18 months).

G-13. Ranger indoctrination program.

a. The RIP is a training and selection program. It is run by specially selected NCOs from the regimental headquarters. It has as its mission the training and indoctrination of newly assigned soldiers. It physically toughens and mentally sharpens the newly assigned soldier. It teaches basic skills and techniques. The RIP identifies and eliminates in three weeks any applicant who does not show dedication, motivation, physical fitness, and emotional stability.

b. An indoctrination program for newly assigned ranger-qualified officers and NCOs prepares them to assume duty as leaders within the ranger regiment. These personnel monitor and participate in RIP training, as well as undergoing individual ranger leadership training. The RIP provides them an overview of the regiment's history, its mission, and its goals. (See Appendixes C,

D, and E for further information about the history and goals of the ranger regiment.) The RIP also provides a review of the skills required in a ranger unit. Refresher training in airborne and pathfinder operations, use of indirect fires, and light infantry doctrine is conducted. During RIP all newly assigned officers and NCOs must meet the ranger physical fitness standard.

c. The RIP cadre also conducts a preranger program to prepare soldiers to attend the US Army Ranger School. It is a physically demanding course that teaches troop-leading procedures, operation orders, and small unit patrol actions.

Glossary

The following acronyms, abbreviations, and definitions are provided to aid the reader in understanding this publication.

Section I

ACRONYMS AND ABBREVIATIONS

AAT	Army assault team
ABCCC	airborne battlefield command and control center
ACL	allowable cargo load
ADA	air defense artillery
AFFOR	the USAF element of a joint task force
AGL	above ground level
AKZ	armor kill zone
ALCC	airlift control center
ALCE	airlift control element
AM	amplitude modulation
AMC	air mission commander
APERS	antipersonnel
APRT	Army physical readiness test
ARFOR	the Army element of a joint task force
ARSOC	Army special operations command
ARSOF	Army Special Operations Forces
ASG	area support group
ASPC	all-source production center

ATGM	antitank guided missile
AWADS	adverse weather aerial delivery system
BCE	battle coordination element
BTMS	battalion training management system
CA	civil affairs
CAS	close air support
CCT	combat control team
CDS	container delivery system
CE	communications-electronics
CI	counterintelligence
CINC	commander in chief
COMINT	communications intelligence
COMSEC	communications security
CONUS	continental United States
COSCOM	corps support command
CP	command post
CS	combat support; a chemical agent
CSS	combat service support
DACG	departure airfield control group
DACO	departure airfield control officer
DASC	direct air support center
DMDG	digital message device group
DOD	Department of Defense
DS	direct support
DSU	direct support unit
DZ	drop zone
EAC	echelons above corps

EACIC	Echelons Above Corps Intelligence Center
ECCM	electronic counter-countermeasure
ECM	electronic countermeasures
E&E	escape and evasion
EEFI	essential elements of friendly information
EOC	emergency operations center
EW	electronic warfare
FARP	forward area rearm and refuel point
FASCAM	family of scatterable mines
FAX	facsimile transmission
FDC	fire direction center
FEBA	forward edge of the battle area
FFL	friendly front lines
FIST	fire support team
FLOT	forward line of own troops
FM	field manual; frequency modulation
FO	forward observer
FORSCOM	US Army Forces Command
FPF	final protective fire
FPL	final protective line
FRAGO	fragmentary order
FSA	fire support area
FSCOORD	fire support coordination or coordinator
FSE	fire support element
FSO	fire support officer
FSS	fire support station
FTX	field training exercise

G3	Assistant Chief of Staff, Operations
GS	general support
GSR	ground surveillance radar
HAARS	high-altitude, airdrop resupply system
HAHO	high-altitude, high-opening
HALO	high-altitude, low-opening
HE	high explosive
HF	high frequency
HHC	headquarters and headquarters company
HSLLADS	high-speed, low-level airdrop system
HUMINT	human intelligence
IAW	in accordance with
IFF	identification-friend-or-foe
IP	initial point
IRP	initial rally point
ISB	intermediate staging base
JACC/CP	joint airborne communications center/command post
JCS	Joint Chiefs of Staff
JCSE	joint communications support element
JFC	joint force commander
JSOC	joint special operations command
JTB	joint transportation board
JTF	joint task force
kw	kilowatt
LAPES	low altitude parachute extraction system
LAW	light antitank weapon
LNO	liaison officer

LOC	logistical operations center
LRRP	long-range reconnaissance patrol
LRSU	long-range surveillance unit
LTD	laser target designator
LZ	landing zone
MAC	Military Airlift Command
MANPADS	man-portable air defense system
MCC	movement control center
METT-T	mission, enemy, terrain, troops and time available
MI	military intelligence
MILES	multiple integrated laser engagement system
mm	millimeter
MMC	Materiel Management Center
MMEE	minimum mission-essential equipment
MOS	military occupational specialty
MOUT	military operations on urbanized terrain
MRLS	multiple rocket launch system
MTP	mission training plan
NBC	nuclear, biological, chemical
NCA	National Command Authority
NCO	noncommissioned officer
NOD	night observation device
OCOKA	observation, concealment, obstacles, key terrain, and avenues of approach
OCONUS	outside continental United States
OPCOM	operational command
OPCON	operational control
OPFOR	opposing force

OPORD	operation order
OPSEC	operations security
ORP	objective rally point
PAC	personnel administration center
PDF	primary direction of fire
PIR	priority intelligence requirements
PLL	prescribed load list
POM	preparation for oversea movement
PSYOP	psychological operations
PT	physical training
PZ	pickup zone
QRE	quick-reaction element
RATELO	radiotelephone operator
REMAB	remote marshalling base
RFL	restrictive fire line
RIP	ranger indoctrination program
RP	rally point; release point
RPG	rocket-propelled grenade
RRF	ranger ready force
RRP	reentry rally point
RSE	ranger support element
RSTA	reconnaissance, surveillance, and target acquisition
S1	adjutant
S2	intelligence officer
S3	operations and training officer
S4	supply officer
S5	civil affairs officer

SAR	search and rescue
SATCOM	satellite communications)
SCIF	sensitive compartmented information facility
SCOPES	squad combat operations' exercises (simulation)
SCUBA	self-contained underwater breathing apparatus
SEAD	suppression of enemy air defense
SEMA	special electronic mission aircraft
SERE	survival, evasion, resistance, and escape
SF	special forces
SHORAD	short-range air defense
SIGSEC	signal security
SKE	station-keeping equipment
SLAR	side-looking airborne radar
SOA	special operations aviation
SOCB	special operations communications battalion
SOCOM	special operations command (usually, 1st SOCOM)
SOP	standing operating procedures
SOSB	special operations support battalion
SOTF	special operations task force
SPL	support platoon leader
STX	situational training exercise
TA	theater Army
TAACOM	Theater Army Area Command
TAC	Tactical Air Command
TACP	tactical air control party
TACSAT	tactical satellite terminal
TAMMC	Theater Army Materiel Management Center

TASG	theater Army support group
TDA	table of distribution and allowances
TOA	transfer of authority
TOC	tactical operations center
TOE	table(s) of organization and equipment
TRP	target reference point
TTY	teletype
UHF	ultrahigh frequency
US	United States
USAF	United States Air Force
USMC	United States Marine Corps
USN	United States Navy
VHF	very high frequency
WP	white phosphorus
XO	executive officer

Section II

DEFINITIONS

ADVANCE FORCE. An operation requiring the immediate deployment of US forces with the intent to negate a threat before the deployment of follow-on forces is required.

AIRLAND BATTLE DOCTRINE. An approach to military operations that realizes the full potential of US forces. Two notions (extending the battlefield and the ability to integrate conventional, nuclear, chemical, and electronics means) are combined to describe a battlefield where the enemy is attacked to the full depth of his formation.

AREA OF INFLUENCE. A geographical area wherein a commander is directly capable of influencing operations by maneuver or fire support systems normally under his command or control.

AREA OF INTEREST. That area of concern to the commander, including the area of influence, areas adjacent thereto, and extending into enemy territory to the objectives of current or planned operations. This area also includes areas occupied by enemy forces who would jeopardize the accomplishment of the mission.

AREA OF OPERATIONS. That portion of an area of conflict necessary for military operations. Areas of operations are geographical areas assigned to commanders for which they have responsibility and in which they have authority to conduct military operations.

CHALK. One specific aircraft load.

CLANDESTINE OPERATIONS. Activities to accomplish intelligence, counterintelligence, and other similar activities sponsored or conducted by governmental departments or agencies, in such a way as to ensure secrecy or concealment. (They differ from covert operations in that emphasis is placed on concealment of the operation rather than on concealment of the identity of the sponsor.)

COMMUNICATIONS INTELLIGENCE (COMINT). The interception and processing of foreign communications passed by radio, wire, or other electromagnetic means, and by the processing of foreign encrypted communications, however transmitted. Interception comprises search, intercept, operator identification, signal analysis, traffic analysis, crypto-analysis, study of plain text, the fusion of these processes, and reporting the results.

COMMUNICATIONS SECURITY. The protection resulting from all measures designed to deny to unauthorized persons information of value that might be derived from the possession and study of telecommunications, or to mislead unauthorized persons in their interpretations of the results of such study. COMSEC includes crypto-security, physical security, and transmission security.

COUNTERINTELLIGENCE. That aspect of intelligence activities, both offensive and defensive, designed to detect and neutralize or destroy the effectiveness of hostile foreign intelligence activities and to protect information against espionage, personnel against subversion, and installations and materiel against sabotage.

COVERT OPERATIONS. Operations that are so planned and executed as to conceal the identity of or permit plausible denial by the sponsor. They differ from clandestine operations in that emphasis is placed on concealment of identity of the sponsor rather than on concealment of the operation.

DECEPTION. Those measures designed to mislead the enemy by manipulation, distortion, or falsification of evidence to induce him to react in a manner prejudicial to his interests.

DEMONSTRATION. A show of force in an area where a decision is not sought. It is similar to a feint but does not make contact with the enemy.

DRILL. A drill focuses on a "chunk" or "slice" of battle and is targeted at small units. Drills are critical collective tasks that require a high degree of proficiency from a small unit. Generally, small units "drill" collective tasks that require rapid responses by unit members in the absence of detailed orders from unit leaders.

ELECTRONIC COUNTERMEASURES. That division of electronic warfare involving actions taken to prevent or reduce an enemy's effective use of the electromagnetic spectrum.

ELECTRONIC INTELLIGENCE (ELINT). The technical and intelligence information derived from foreign noncommunications electromagnetic radiations emanating from other than atomic detonations or radioactive sources.

ELECTRONIC WARFARE. Military action involving the use of electromagnetic energy to determine, exploit, reduce, or prevent hostile use of the electromagnetic spectrum and action that retains friendly use of the electromagnetic spectrum.

ELECTRONIC WARFARE SUPPORT MEASURES (ESM). That division of EW involving actions taken to search for, intercept, identify, and locate sources of radiated electromagnetic energy for the purpose of immediate threat recognition. ESM provides a source of information needed for immediate action involving electronic countermeasures, electronic counter-countermeasures, avoidance, targeting, and other tactical employment of forces.

ESPIONAGE. Actions directed toward the acquisition of information through clandestine operations.

EVACUATION/SAFEGUARD. The use of US forces to evacuate US citizens or allies from a hostile or potentially hostile area, or to safeguard US personnel or property.

EXFILTRATION. The removal of personnel or units from areas under enemy control by stealth, deception, surprise, or clandestine means. Exfiltration is a method of extracting ranger units.

EXTRACTION. The removal of personnel or units from areas under enemy control by any means, to include exfiltration, airland, airmobile, or amphibious.

FEINT. A diversionary supporting attack conducted to draw the enemy's attention from the main effort. It is normally executed by brigades and small units. Feints are usually shallow, limited objective attacks that go in before or during the main attack.

GUERRILLA WARFARE. Military and paramilitary operations conducted in enemy-held or hostile territory by irregular and predominantly indigenous forces.

HIGH-INTENSITY CONFLICT. A war between the US and any other nation or nations in which both belligerents use the most modern technology and all resources in intelligence, mobility, firepower (including NBC), command, communications and control, and service support.

HUMAN INTELLIGENCE. Intelligence derived from information provided by human resources.

INFILTRATION. The movement through or into an area of territory occupied by either friendly or enemy troops or organizations. The movement is made either by small groups or by individuals, at extended or irregular intervals. When used in conjunction with the enemy, it implies that contact is avoided. Infiltration is a subset of insertion.

INSERTION. Placement of troops and equipment into an operational area by any means.

INTERDICT. To isolate or seal off an area by military means; to prevent, hinder, or delay the use of an area or route by enemy forces.

INTERDICT LINE OF COMMUNICATIONS. An attack to seal off an area or to deny use of a route or approach.

JOINT TASK FORCE. A JTF may be constituted and designated by the Secretary of Defense or by the commander of a unified command, specified command, or an existing JTF. Normally, it performs missions having specific, limited objectives or missions of short duration. It dissolves when it has achieved its purpose. The joint task force commander is responsible to the commander, known as the establishing authority, who created the JTF. The JTF is composed of elements of two or more services operating under a single JTF commander. The JTF commander has operational control over the entire force and may have direct command of his own staff with representatives from the other services. He exercises logistics coordination or control only as necessary to meet his subordinate commanders' logistics needs.

LIAISON. That personal contact or communication maintained between elements of military forces to ensure mutual understanding and unity of purpose and effort.

LINE OF COMMUNICATIONS. All routes (land, water, and air) that connect an operating military force with one or more bases of operations and along which supplies and military forces move.

LOW-INTENSITY CONFLICT. Limited politico-military operations conducted to achieve political, social, economic, or psychological objectives. It is generally confined to a specific geographic area and is often characterized by operational constraints imposed on ranger units by political and economic considerations. It may involve intermittent combat against regular threat military forces, but it more often involves conflict with irregular, security, or paramilitary forces of a country or group involved in an operation that is counter to the best interests and national security of the United States. It is characterized by relatively low levels of combat and constraints placed on the weaponry, tactics, and movements of the ranger force. It may involve counterterrorist or antiterrorist operations and generally occurs before a pro forma declaration of war.

MID-INTENSITY CONFLICT. War, declared or undeclared, between the US and the regular forces of one or more nations and their respective allies, if any, in which the belligerents use the most modern technology and all resources in intelligence, mobility, firepower (excluding nuclear, chemical, and biological weapons), command, communications and control, and support. They are used for limited objectives under definitive policy limitations as to the extent of destructive power that can be used or the extent of geographical area that might be involved.

MILITARY STRATEGY. Art and science of using the armed forces of a nation to secure the objectives of national policy by application of force or the threat of force. Military strategy sets the fundamental conditions for operations.

NATIONAL COMMAND AUTHORITY. This includes the President and Secretary of Defense or the duly deputized alternates or successors. Commonly referred to as NCA.

OPERATIONAL COMMAND/CONTROL. Operational command (OPCOM) and operational control in joint force terminology both refer to the authority exercised by joint commanders over subordinate service components. Those terms are not interchangeable in joint operations. OPCOM applies to the authority exercised by commanders of unified commands. OPCON is the authority that subordinate joint task force commanders exercise in the conduct of specific operations. The authority that military departments exercise over their respective components is commonly referred to as command less operational command.

OPERATIONAL LEVEL OF WAR. The operational level of war uses available military resources to attain strategic goals within a theater of war. It is the theory of larger unit operations. It also involves planning and conducting campaigns. Campaigns are sustained operations designed to defeat an enemy force in a specified space and time with simultaneous and sequential battles. The disposition of forces, selection of objectives, and actions taken to weaken or to outmaneuver the enemy all set the terms of the next battle and exploit tactical gains. They are all part of the operational level of war. In AirLand Battle doctrine, this level includes the marshalling of forces and logistical support, providing direction to ground and air maneuver, applying conventional and nuclear fires in depth, and employing unconventional and psychological warfare.

OPERATIONS SECURITY. Those measures designed to protect information concerning planned, ongoing, and completed operations from unauthorized disclosure. It includes all actions a command takes to deny the enemy information about friendly units and their operations.

ORDER OF BATTLE (OB). The identification, strength, command structure, and disposition of the personnel, units, and equipment of any military force.

RAID. Usually a small-scale operation, involving a swift penetration of hostile territory to secure information, confuse the enemy, or destroy his installations. It ends with a planned withdrawal upon completion of the assigned mission.

RANGER MISSION. To conduct strike, raid, or special light infantry operations. Execution of ranger missions is normally directed by the National Command Authority. Rangers also support operational-level commanders during operations of limited duration when allocated or assigned by the theater commander.

RECONNAISSANCE. A mission undertaken to obtain, by visual observation or other detection methods, information about the activities and resources of an enemy or potential enemy; or to secure data concerning the meteorological, hydrographical, or geographical characteristics of a particular area.

SEARCH AND RESCUE. The use of aircraft, surface craft, submarines, specialized rescue teams, and equipment to search for and rescue personnel in distress on land or at sea.

SEIZURE. As an operational purpose, the capture of a voluntarily restricted portion of an enemy-controlled territory. Capture of an isolated land mass such as an island is usually categorized as a seizure.

SEIZE KEY FACILITY. An attack on an installation with the intent of holding it for a short period. Normally, rangers would make the assault to seize and secure an installation for follow-on forces.

SERVICE COMPONENT. Each service component commander is responsible for recommending the proper use of his forces and for accomplishing operational tasks assigned by the joint commander. He is also responsible for his service in matters of internal administration and discipline; training in own service doctrine, techniques, and tactics; designation of specific units to meet joint requirements; logistics functions normal to the component; tactical employment of service component forces; and service intelligence matters.

SHOW OF FORCE. A mission carried out to demonstrate US resolve, whereby US forces are deployed to defuse a situation that may be harmful to US interests or national objectives.

SIGNAL INTELLIGENCE (SIGINT). A generic term that includes COMINT and ELINT. Activities concerned with the collection and processing of foreign electromagnetic emissions for the purpose of producing intelligence information.

SITUATIONAL TRAINING EXERCISE. An STX is a short, mission-oriented exercise in which a group of closely-related collective tasks are trained, using a doctrinally preferred method of executing these tasks to established standards. An STX is drill-like in nature, but it is more complex and flexible than a drill. It usually involves a larger unit than a drill.

SPECIAL OPERATIONS. Military operations conducted by specially trained, equipped, and organized DOD forces against strategic or tactical targets in pursuit of national, military, political, economic, or psychological objectives. Special operations may support conventional military operations, or they may occur independently when the use of conventional forces is either inappropriate or infeasible. Sensitive peacetime operations, except for training, are normally authorized by the NCA and conducted under the direction of the NCA or designated commander. Special operations may include unconventional warfare, counter-terrorist operations, collective security, psychological operations, and civil affairs measures.

SPECIAL OPERATIONS FORCES. All civil affairs, psychological operations, ranger (active component), and special forces units within the US Army, US Army Reserve, and National Guard.

SPECIAL OPERATIONS COMMAND. The major subordinate command to FORSCOM exercising command and control of all active component Army special operations forces not placed under the operational command of unified or specified commands.

STRATEGIC LEVEL OF WAR. Military strategy uses the armed forces of a nation to secure the objectives of national policy by applying force or the threat of force. Military strategy sets the fundamental conditions for operations.

STRIKE. An attack that is intended to inflict damage on, seize, or destroy an objective. Sometimes referred to as direct action missions.

SURVEILLANCE. The systematic observation of aerospace, surface or subsurface areas, places, persons, or things by visual, aural, electronic, photographic, or other means.

TACTICAL. Pertaining to the employment of units in combat.

TACTICAL LEVEL OF WAR. Tactics are the specific techniques smaller units use to win battles and engagements that support operational objectives. Tactics employ all available combat, combat support, and combat service support. Tactics involve the movement and positioning of forces on the battlefield in relation to the enemy, the provisioning of fire support, and the logistical support of force before, during, and following engagements with the enemy. At corps and division, operational and tactical levels are not clearly separable, but they are guided by the same principles. An operation designed to defeat any enemy force in an extended area does so through operational maneuver and a series of tactical actions.

TACTICAL COMBAT INTELLIGENCE. That knowledge of the enemy, weather, and geographical features needed by a commander in the planning and conduct of combat operations. It is derived from the interpretation of information on the enemy and the environment.

TARGET ANALYSIS. A detailed and systematic examination of processed intelligence to identify and locate targets.

UNIFIED COMMAND. A unified command is established and designated by direction of the President. The command is organized under a single unified CINC and has a broad, continuing mission. The CINC is responsible to the Secretary of Defense for accomplishing his broad and continuing mission and operates under the strategic and operational direction of the JCS. A unified command is composed of assigned forces of two or more services. These forces, as service components, are under the operational command of the unified commander. The CINC exercises operational command authority through service component commanders. When tasked by higher authority, he may establish a subordinate JTF to conduct specific missions. Under emergency conditions, the CINC has the authority to use all facilities and supplies of assigned forces to accomplish his mission.

References

REQUIRED PUBLICATIONS

Required publications are sources that users must read in order to understand or to comply with this publication.

Field Manuals (FM)

FM 7-8	The Infantry Platoon and Squad (Infantry, Airborne, Air Assault, Ranger)
FM 7-10	The Infantry Rifle Company (Infantry, Airborne, Air Assault, Ranger)
FM 7-20	The Infantry Battalion (Infantry, Airborne, and Air Assault)

RELATED PUBLICATIONS

Related publications are sources of additional information. They are not required in order to understand this publication.

Army Regulations (AR)

AR 350-1	Army Training

Field Manuals (FM)

FM 10-52-1	Commander's Handbook for Water Usage in Desert Operations
FM 10-501	Airdrop of Supplies and Equipment: Rigging Containers
FM 10-512	Airdrop of Supplies and Equipment: Rigging Typical Supply Loads
FM 10-542	Airdrop of Supplies and Equipment: Rigging Loads for Special Operations
FM 10-547	Airdrop of Supplies and Equipment: Rigging the High Speed Aerial Delivery Container CTU-2/A
FM 10-553	Airdrop of Supplies and Equipment: Rigging Ammunition
FM 12-3-4	Echelons Above Corps (EAC) Personnel and Administrative Doctrine

FM 14-6	Comptroller/Finance Services in Theaters of Operations
FM 19-15	Civil Disturbances
FM 20-11-1	Military Diving (Volume 1)
FM 20-11-2	Military Diving (Volume 2)
FM 21-20	Physical Training Program
FM 21-26	Map Reading
FM 21-75	Combat Skills of the Soldier
FM 21-76	Survival, Evasion, and Escape
FM 21-77A(S)	Joint Worldwide Evasion and Escape Manual
FM 22-9	Soldier Performance in Continuous Operations
FM 23-67	Machinegun, 7.62-mm, M60
FM 25-1	Training
FM 25-2	Unit Training Management
FM 25-3	Training in Units
FM 25-4	How to Conduct Training Exercises
FM 25-5	Training for Mobilization and War
FM 25-7	Training Ranges
FM 26-2	Management of Stress in Army Operations
FM 29-47	Main Supply and Service Company, Forward Supply and Service Detachments, Supply and Service Battalion, Airborne Division
FM 29-51	Division Supply and Field Service Operations
FM 30-5	Combat Intelligence
FM 30-17	Counterintelligence Operations
FM 31-18	Long-Range Reconnaissance Ranger Company
FM 31-19	Special Forces Military Free-Fall Parachuting
FM 1-24	Special Forces Air Operations

FM 31-25	Special Forces Waterborne Operations
FM 31-70	Basic Cold Weather Manual
FM 34-1	Intelligence and Electronic Warfare Operations
FM 34-60A	Counterintelligence Operations (U)
FM 55-9	Unit Air Movement Planning
FM 57-38	Pathfinder Operations
FM 57-220	Basic Parachuting Techniques and Training
FM 63-3J	Combat Service Support Operations--Corps
FM 63-4	Combat Service Support Operations--Theater Army Area Command
FM 63-5	Combat Service Support Operations--Theater Army
FM 90-3	Desert Operations
FM 90-4	Airmobile Operations
FM 90-5	Jungle Operations
FM 90-6	Mountain Operations
FM 90-8	Counterguerilla Operations
FM 90-10	Military Operations on Urbanized Terrain (MOUT)
FM 90-10-1	An Infantryman's Guide to Urban Combat
FM 100-5	Operations
FM 100-20	Low Intensity Conflict
FM 100-26	The Air-Ground Operations System
FM 101-5	Staff Organization and Operations
FM 101-10-1	Staff Officers' Field Manual: Organization, Technical and Logistical Data (U)

Firing Table (FT)

FT 60-P-1	Firing Tables for Mortar, 60-mm: M224, Firing Cartridge, HE, M720, Cartridge, HE, M49A4, Cartridge, TP, M5OA3, Cartridge, WP, M302AI, Cartridge, Illuminating, M83A3

Soldier's Training Publication (STP)

STP 7-11B24-SM	Soldier's Manual MOS 11B, Infantryman, Skill Level 2/3/4

Technical Manuals (TM)

TM 10-1670-267-12&P	Operator's and Organizational Maintenance Manual Including Repair Parts and Special Tools List for High Altitude Airdrop Resupply System: 500 Pound Capacity

Training Circulars (TC)

TC 23-14	Sniper Training and Employment
TC 23-90	Mortar Training
TC 24-1	Communications-Electronics Operation Instructions, the CEOI
TC 31-20-5(C)	Special Forces Communications (U)
TC 90-6-1	Military Mountaineering

Other Related Publications

King, Michael J. Dr., *Leavenworth Papers, No. 11*, "Rangers Selected Combat Operations in World War II," Combat Studies Institute, USACGSC, Fort Leavenworth, KS

PROJECTED PUBLICATIONS

Projected publications are sources of additional information that are scheduled for printing but are not yet available. Upon print, they will be distributed automatically via pinpoint distribution. They may not be obtained from the USA AG Publications Center until indexed in DA Pamphlet 310-1.

Field Manuals (FM)

FM 7-70	Light Infantry Squad/Platoon
FM 7-71	Light Infantry Company
FM 7-72	Light Infantry Battalion
FM 7-73	Light Infantry Brigade

FM 23-36 60-mm Mortar, M224